MULTI-CRITERIA DECISION-MAKING METHODS IN MANUFACTURING ENVIRONMENTS

Models and Applications

Frontiers of Mechanical and Industrial Engineering

MULTI-CRITERIA DECISION-MAKING METHODS IN MANUFACTURING ENVIRONMENTS

Models and Applications

Shankar Chakraborty, PhD
Prasenjit Chatterjee, PhD
Partha Protim Das, PhD

AAP | APPLE ACADEMIC PRESS

First edition published 2024

Apple Academic Press Inc.
1265 Goldenrod Circle, NE,
Palm Bay, FL 32905 USA

760 Laurentian Drive, Unit 19,
Burlington, ON L7N 0A4, CANADA

CRC Press
2385 NW Executive Center Drive,
Suite 320, Boca Raton FL 33431

4 Park Square, Milton Park,
Abingdon, Oxon, OX14 4RN UK

Library and Archives Canada Cataloguing in Publication

Title: Multi-criteria decision-making methods in manufacturing environments : models and applications / Shankar Chakraborty, PhD, Prasenjit Chatterjee, PhD, Partha Protim Das, Ph.D.
Names: Chakraborty, Shankar, editor. | Chatterjee, Prasenjit, 1982- editor. | Das, Partha Protim, editor.
Description: First edition. | Series statement: Frontiers of mechanical and industrial engineering book series | Includes bibliographical references and index.
Identifiers: Canadiana (print) 20230225438 | Canadiana (ebook) 20230225497 | ISBN 9781774912614 (hardcover) | ISBN 9781774912621 (softcover) | ISBN 9781003377030 (ebook)
Subjects: LCSH: Manufacturing processes—Decision making. | LCSH: Multiple criteria decision making.
Classification: LCC TS183 .M85 2023 | DDC 670—dc23

Library of Congress Cataloging-in-Publication Data

CIP data on file with US Library of Congress

ISBN: 978-1-77491-261-4 (hbk)
ISBN: 978-1-77491-262-1 (pbk)
ISBN: 978-1-00337-703-0 (ebk)

DEDICATIONS

Prof. (Dr.) Shankar Chakraborty would like to dedicate this book to his late father Phanindranath Chakraborty, his late mother Gouri Chakraborty, his beloved wife Rupa, son Santonab, and daughter Sirin. He makes a special dedication to his Gurudev.

Dr. Prasenjit Chatterjee would like to dedicate this book to his grandparents, late father Dipak Kumar Chatterjee, mother Mrs. Kalyani Chatterjee, beloved wife Amrita, and his little angel Aheli.

Dr. Partha Protim Das would like to dedicate this book to his mother Anima Das, his father Pintu Ranjan Das, and his beloved wife Dolly.

ABOUT THE FRONTIERS OF MECHANICAL AND INDUSTRIAL ENGINEERING BOOK SERIES

EDITORS-IN-CHIEF

Dr. Prasenjit Chatterjee
Department of Mechanical Engineering, MCKV Institute of Engineering, West Bengal, India

Dr. Dilbagh Panchal
Department of Industrial and Production Engineering,
Dr. B. R. Ambedkar National Institute of Technology (NIT) Jalandhar, Punjab, India

Description of the Series

Mechanical engineering applies the principles of mechanics and materials science for analysis, design, manufacturing, and maintenance of mechanical systems, whereas industrial engineering involves figuring out how to make or do things better with minimum resource consumption. Mechanical and industrial engineering utilize a deep foundational understanding of mathematics, physics, and analysis to develop machines and systems. The former specializes in thermodynamics, combustion, and electricity to make complex machines. The latter focuses on designing workflow and making production efficient.

This series reports on the cutting-edge developments to present novel innovations in the field of mechanical and industrial engineering. Volumes published in this series aim to embrace all aspects, subfields, and new challenges of mechanical and industrial engineering and to be of interest and use to all those concerned with research in various related domains of mechanical and industrial engineering disciplines. The series aims to provide comprehensive value-added resources to the body of the literature to serve as essential references for years ahead. Features of the volumes under this series explore recent trends, model extensions, developments, solutions to real-time problems, and case studies.

The coverage of the series is committed to provide a diverse yet rigorous research experience that builds on a solid knowledge of science and engineering fundamentals and application-oriented knowledge of mechanical and industrial engineering practices for efficient functioning in today's global multidisciplinary rapidly changing work environment. It includes all new theoretical and experimental findings in mechanical, industrial engineering, and allied engineering fields.

Topics and Themes:

The primary endeavor of this series is to introduce and explore contemporary research developments in a variety of rapidly growing research areas. The volumes will deal with the following topics but not limited to:

- Automotive engineering
- Aerospace technology
- Big data in mechanical / manufacturing / industrial engineering
- Computational methods for optimizing manufacturing technology, robotics, mechatronics
- Decision science applications in mechanical, manufacturing, and industrial engineering
- Design and optimization of mechanical components
- Dynamical systems, control
- Engineering design
- Engineering thermodynamics, heat, and mass transfer
- Evolutionary computation
- Finite element method (FEM) modeling/simulation
- Fluid mechanics
- Fuzzy logic and neuro-fuzzy systems for relevant engineering applications
- Green and sustainable engineering techniques for modern manufacturing
- Innovative approaches for modeling relevant applications and real-life problems
- Life cycle engineering
- Machinery and machine elements
- Manufacturing
- Mathematical concepts and applications in mechanical and industrial engineering
- Materials engineering
- Mechanical structures and stress analysis
- MEMS
- Modeling in engineering applications
- Nanotechnology and microengineering
- Non-conventional machining in modern manufacturing systems

- Precision engineering, instrumentation and measurement
- Numerical simulations
- Soft computing techniques
- Sustainability in mechanical, manufacturing, and industrial engineering
- Theoretical and applied mechanics
- Tribology and surface technology

The series aims to serve as a valuable resource for undergraduate, postgraduate, doctoral students, researchers, academicians, and industry professionals in mechanical, manufacturing, industrial engineering, product design, management, and more.

BOOKS IN THE SERIES

Optimization Methods for Engineering Problems
Editors: Dilbagh Panchal, PhD, Prasenjit Chatterjee, PhD, Mohit Tyagi, PhD, and Ravi Pratap Singh, PhD

Modern Manufacturing Systems: Trends and Developments
Editors: Rajiv Kumar Garg, PhD, Ravi Pratap Singh, PhD, Rajeev Trehan, PhD, and Ramesh Singh, PhD

Multi-Criteria Decision-Making Methods in Manufacturing Environments: Models and Applications
Shankar Chakraborty, PhD, Prasenjit Chatterjee, PhD, and Partha Protim Das, PhD

Supply Chain Performance Measurement in Textile Enterprises
Editors: Pranav G. Charkha, PhD, Santosh B. Jaju, PhD, and Prasenjit Chatterjee, PhD

Electro-Micromachining and Microfabrication: Principles and Research Advances
Editors: Sandip Kunar PhD, Golam Kibria, PhD, and Prasenjit Chatterjee PhD

Robotics and Automation in Healthcare: Advanced Applications
Editors: R. Thanigaivelan, PhD, Sanjay Singh, PhD, and Clement Christy Deepak C., PhD

Optimization of Advanced Manufacturing Processes
Editors: Sandip Kunar, PhD, Prasenjit Chatterjee, PhD, and M. Sreenivasa Reddy, PhD

ABOUT THE AUTHORS

Shankar Chakraborty, PhD
Professor, Department of Production Engineering, Jadavpur University, India

Shankar Chakraborty, PhD, is Professor in the Department of Production Engineering at Jadavpur University, India. He has published more than 200 research papers in international journals and is an Associate Editor of the journal *OPSEARCH* and is an editorial board member of several journals. He is also a regular reviewer for numerous international and national journals. He has guided 15 PhD and 30 postgraduate theses. His research interest includes multi-criteria decision-making, operations research, and soft computing techniques. He graduated from the University of Calcutta and obtained his postgraduate degree and PhD in Engineering from Jadavpur University.

Prasenjit Chatterjee, PhD
Dean (Research and Consultancy), MCKV Institute of Engineering, West Bengal, India

Prasenjit Chatterjee, PhD, is currently the Dean (Research and Consultancy) at MCKV Institute of Engineering, West Bengal, India. He has over 100 research papers in various international journals and peer-reviewed conferences. He has authored and edited more than 15 books on intelligent decision-making, supply chain management, optimization techniques, risk, and sustainability modeling. He has received numerous awards, including Best Track Paper Award, Outstanding Reviewer Award, Best Paper Award, Outstanding Researcher Award, and University Gold Medal. Dr. Chatterjee is the Editor-in-Chief of the *Journal of Decision Analytics and Intelligent Computing.* He has also been the Guest Editor of several special issues in different SCIE/Scopus/ESCI (Clarivate Analytics) indexed journals. He is also the Lead Series Editor of the book series Smart and Intelligent Computing in Engineering; Founder and Lead Series Editor of several book series: Concise Introductions to AI and Data Science; AAP Research Notes on Optimization and Decision-Making Theories; Frontiers of Mechanical and Industrial

Engineering; and River Publishers Series in Industrial Manufacturing and Systems Engineering. Dr. Chatterjee is one of the developers of two multiple-criteria decision-making methods called Measurement of Alternatives and Ranking according to COmpromise Solution (MARCOS) and Ranking of Alternatives through Functional mapping of criterion sub-intervals into a Single Interval (RAFSI).

Partha Protim Das, PhD
Assistant Professor, Department of Mechanical Engineering, Sikkim Manipal Institute of Technology, Sikkim Manipal University, India

Partha Protim Das, PhD, is working as an Assistant Professor in the Department of Mechanical Engineering of Sikkim Manipal Institute of Technology, Sikkim Manipal University. He received his Bachelor's degree in Mechanical Engineering from Annamalai University in 2013, and Master's degree in Design and Manufacturing from National Institute of Technology, Silchar in 2015. He was awarded his PhD degree in Engineering from Jadavpur University in 2021. He has published numerous research articles in various international journals. He is also serving as a regular reviewer of various national and international journals. His research interest includes decision-making, soft computing techniques, optimization, artificial intelligence techniques, machining processes, etc.

CONTENTS

ABBREVIATIONS

AAG	available for each assembly group
AAI	anti-ideal solution
AE	aesthetic
AFF	accessibility for firefighting
AGV	automated guided vehicle
AHP	analytic hierarchy process
AJM	abrasive jet machining
ANP	analytic network process
ARAS	Additive Ratio ASsessment
AWJM	abrasive water jet machining
BH	Brinell hardness
BW	belt weight
BWM	best worst method
CAD	computer-aided design
CBN	cubic boron nitride
CC	capital cost
CFA	confirmatory factor analysis
CHM	chemical machining
CI	consistency index
CMC	capital and maintenance cost
CNC	computer numerical control
COC	comfort of crew
CoCoSo	COmbined COmpromise SOlution
CODAS	COmbinative Distance-based ASsessment
COPRAS	COmplex PRoportional ASsessment
CR	consistency ratio
CR	corner radii
CRADIS	compromise ranking of alternatives from distance to ideal solution
CRITIC	CRiteria Importance Through Intercriteria Correlation
DEMATEL	DEcision-MAking Trial and Evaluation Laboratory
DF	demand fulfillment
DFC	deviation from full consistency
DoE	design of experiments

EAC	environmental activity control
EATWOS	efficiency analysis technique with output satisfying
EBM	electron beam machining
ECG	electrochemical grinding
ECH	electrochemical honing
ECM	electrochemical machining
ECMM	electrochemical micro-machining
EDAS	evaluation based on distance from average solution
EDM	electrical discharge machining
EF	expansion flexibility
ELECTRE	ELimination and Et Choice Translating REality
EPA	extreme pressure additives
ERP	enterprise resource planning
ET	environmental training
EVAMIX	EVAluation of MIXed data
F/FLEX	flexibility
FARE	FActor RElationship
FDM	fused deposition modeling
FMS	flexible manufacturing system
FSU	floor space used
FT	fracture toughness
FUCOM	FUll COnsistency Method
GAIA	geometrical analysis for interactive aid
GIS	geographic information system
GMAW	gas metal arc welding
GRA	gray relational analysis
GTAW	gas tungsten arc welding
GTC	green technology capability
GTMA	graph theory and matrix approach
HAZ	heat affected zone
HLF	high level of flexibility
HLR	higher level of reputation
IEFD	interaction with existing facility distance
IMR	increase in market response
IoT	internet of things
IQ	improvement in quality
IS	impact strength
ITWEI	item weight
ITWI/IW	item width

JKN	Jüttler's-Körth's normalization
LBM	laser beam machining
LBWA	level-based weight assessment
LC	laser cutting
LC	load capacity
LMAW	logarithm methodology of additive weights
LP	linear programming
LTR	lead time reduction
MABAC	Multi-Attributive Border Approximation area Comparison
MACBETH	Measuring Attractiveness by a Categorical-Based Evaluation TecHnique
MADM	multi-attribute decision-making
MARCOS	Measurement Alternatives and Ranking according to COmpromise Solution
MAIRCA	multi-attributive ideal-real comparative analysis
MAUT	multi-attribute utility theory
MC	memory capacity
MCDM	multi-criteria decision-making
MCO	multiple criteria optimization
MD	maximum machining diameter
ME	modulus of elasticity
MEREC	method based on the removal effects of criteria
MG	medium good
MHPD	maximum horizontal pumping distance
ML	maximum machining length
MODM	multi-objective decision-making
MOORA	multi-objective optimization on the basis of ratio analysis
MP	machine power
MQF	material quantity flow
MR	manipulator reach
MRR	material removal rate
MSAD	maximum suitable aggregate diameter
MTS	maximum tip speed
MVPD	maximum vertical pumping distance
MWT	maximum working tension
ND	negative distance
NIS	negative-ideal solution
NSGA-II	non-dominated sorting genetic algorithm
NTM	non-traditional machining

OCRA	operational competitiveness rating analysis
OHS	occupational health and safety
ORESTE	Organization, Rangement Et Synthese de Donnes RElationnelles
PAC	plasma arc cutting
PCD	polycrystalline diamond
PD	positive distance
PF	product flexibility
PIPRECIA	PIvot Pairwise RElative Criteria Importance Assessment
PIS	positive-ideal solution
PROMETHEE	Preference Ranking Organization METHod for Enrichment Evaluation
PSI	preference selection index
QFD	quality function deployment
QI	quality improvement
RA	range
RAFSI	ranking of alternatives through functional mapping of criterion sub-intervals into a single interval
RC	resistance to corrosion
RDC	R&D capability
RFS	required floor space
RI	random index
RL	reliability
RLC	reduction in labor cost
ROV	range of value
RSC	reduction in setup cost
RWP	reduction in work-in-process
RY	repairability
SAW	simple additive weighting
SD	surface damage
SECA	simultaneous evaluation of criteria and alternatives
SF	surface finish
SiAHP	simulation-based AHP
SIR	superiority and inferiority ranking
SMAW	shielded metal arc welding
SME	small and medium-sized enterprise
SPC	speed of conveyor
SS	speed range
SWARA	step-wise weight assessment ratio analysis

SWOT	strengths-weaknesses-opportunities-threats
TC	thermal conductivity
TC	tool capacity
TEC	thermal expansion coefficient
TL	tolerance
TODIM	TOmada de Decisao Interativa Multicriterio
TOPSIS	technique for order of preference by similarity to ideal solution
TPM	total productive maintenance
TS	tensile strength
TSS	transverse shear strength
USM	ultrasonic machining
VF	volume flexibility
VG	very good
VIKOR	VIekriterijumsko KOmpromisno Rangiranje
VP	very poor
WASPAS	weighted aggregated sum product assessment
WEDM	wire electrical discharge machining
WIP	work-in-progress
WJM	water jet machining
WM	work material
WPM	weighted product model
WS	weight savings
WSM	weighted sum model
YM	Young's modulus
YS	yield strength

FOREWORD 1

Manufacturing technologies have continually gone through gradual but revolutionary changes. The advancements in manufacturing technologies have brought about metamorphism in the world's industrial scene. New engineering challenges require systematic and integrated planning and optimization approaches in the manufacturing environment. In this context, the manufacturing industries aim to achieve overall performance, utilizing resources in product design, development, production, delivery, and support of products.

To achieve this aim, the manufacturing industries must select appropriate manufacturing strategies, product designs, manufacturing processes, work and tool materials, machinery, equipment, etc. The selection decisions are complex, as decision-making is more challenging today. Decision-makers in the manufacturing industries frequently face the problem of assessing a wide range of alternative options and selecting one based on a set of conflicting criteria.

The decision-making exercise can be implemented in the manufacturing environment at different stages if appropriate procedures are available to the designers, manufacturing engineers, production planners, and managers. These aspects are considered in the present book. The purpose of this book is to demonstrate how different multi-criteria decision-making (MCDM) methods can be used for decision-making in various situations of the manufacturing environment. The authors have presented different MCDM methods, such as

- Analytic Hierarchy Process (AHP)
- Weighted Aggregated Sum Product ASsessment (WASPAS)
- Multi-Objective Optimization on the Basis of Ratio Analysis (MOORA)
- Technique for Order of Preference by Similarity to Ideal Solution (TOPSIS)
- VIekriterijumsko Kompromisno Rangiranje (VIKOR)
- ELimination Et Choice Translating REality (ELECTRE)
- Preference Ranking Organization METHod for Enrichment Evaluation (PROMETHEE)

- COmplex PRoportional ASsessment (COPRAS)
- Organization, Rangement Et Synthese De Donnes RElationnelles (ORESTE)
- Operational Competitiveness Rating Analysis (OCRA)
- Additive Ratio ASsessment (ARAS)
- Evaluation Based on Distance From Average Solution (EDAS)
- COmbinative Distance-Based ASsessment (CODAS)
- EVAluation of MIXed Data (EVAMIX)
- TOmada de Decisao Interativa Multicriterio (TODIM)
- Range of Value (ROV)
- Superiority and Inferiority Ranking (SIR)
- Measuring Attractiveness by a Categorical-Based Evaluation TecHnique (MACBETH)
- Multi-Attributive Border Approximation Area Comparison (MABAC)
- Multi-Attributive Ideal-Real Comparative Analysis (MARICA)
- Measurement Alternatives and Ranking According to COmpromise Solution (MARCOS)
- COmbined COmpromise SOlution (CoCoSO)
- Ranking of Alternatives Through Functional Mapping of Criterion
- Sub-Intervals into a Single Interval (RAFSI)
- Simultaneous Evaluation of Criteria and Alternatives (SECA)
- R
- Compromise Ranking of Alternatives From Distance to Ideal Solution (CRADIS)

The techniques used for criteria weights measurement are also presented. These include:

- entropy
- Decision-Making Trial and Evaluation Laboratory (DEMATEL)
- Semantic Web and Repurposing Applications (SWARA)
- PIvot Pairwise RElative Criteria Importance Assessment (PIPRECIA)
- CRiteria Importance Through Inter-criteria Correlation (CRITIC)
- FActor Relationship (FARE)
- best worst method (BWM)
- FUll COnsistency Method (FUCOM)
- level-based weight assessment (LBWA)
- logarithm methodology of additive weights (LMAW)
- method based on the removal effects of criteria (MEREC)

Various decision-making situations encountered in the manufacturing industries are described, along with the results of the successful application of the MCDM methods. The basket of applications treated in this book reveals a holistic understanding of the manufacturing environment, which is heterogeneous and dynamic. The methods presented in the book can be beneficial to the decision-makers in the manufacturing industries as these methods make decision-making easier, more systematic, and more effective.

The structure of the book is clear and easy to follow. It is apparent that the authors have a full grip on the subject, and therefore, the treatment of the subject matter is sound and up to mark. This book is helpful for both first-time readers in this area and those involved in high-end research in multi-criteria decision-making. This book would be an asset in any academic establishment library or an engineering manager's bookcase.

—Prof. R. Venkata Rao
Director, S. V. National Institute of Technology, Surat, Gujarat, India

FOREWORD 2

I am pleased to write this foreword for the book *Multi-Criteria Decision-Making Methods in Manufacturing Environment: Models and Applications*. The authors of this book have provided valuable insights about a topic having high relevance to everyone. In our day-to-day life, we have to solve difficult problems and make timely decisions to choose the best course of action under conflicting situations. It is nearly impossible to imagine that there is anyone who does not, at least on occasion, face challenges and struggles while making real-time decisions. Occasionally, we make impulsive decisions that may or may not lead to positive outcomes. Fortunately, in some cases, we carefully study and analyze the situation and strategically take steps toward intentional positive outcomes. Thus, it would be quite helpful if these decision-making problems could be solved using some mathematical tools and techniques.

In this book, the authors have provided insights into a comprehensive list of the available multi-criteria decision-making (MCDM) techniques to help decision-makers, mainly in the manufacturing domain, in solving complex decision-making problems. The corresponding mathematical details of each of the MCDM techniques are clearly explained. The applications of each of the MCDM techniques are exhaustively surveyed, and the mathematical steps involved are provided with the help of illustrative examples.

In addition to presenting the prevailing MCDM techniques, the authors have also provided information regarding various normalization techniques usually adopted in solving MCDM problems for data transformation. An exhaustive list of criteria weight measurement techniques, both subjective and objective methods, is also presented in this book.

This book is written in an extremely lucid language and is easily understandable for beginners in the MCDM domain. I think that this book would be an asset to the MCDM practitioners involved in manufacturing as well as service organizations. This book would also be quite helpful to MBA students and other researchers interested in solving complicated MCDM problems leading to the most compromising solutions.

I wish you great success with this book.

—Prof. M. K. Tiwari

Director, National Institute of Industrial Engineering,
Mumbai, Maharashtra, India

PREFACE

Decision-making in manufacturing often requires consideration of trade-offs between economic impacts and is complicated by the views of various stakeholders. Proceedings of the organization and effective utilization of related resources require decision-making at every step. Decision-making facilitates the entire management procedure and hence, is a crucial function of organizational management. There are a few barriers too for appropriate and cogent decisions, including insufficient and inconsistent data or information, time limitation, inadequate skill of the decision makers, interdependencies among different decisions, and subjectivity in the opinion.

Recent innovations in manufacturing processes across the globe have brought metamorphism in manufacturing and related industries. To survive in a comprehensive competitive arcade, manufacturing organizations need to function within the framework of globalization and continuous technological changes. Moreover, the dynamic requirements of customers have created an unparalleled need for quick responses from these organizations. Flawed and hurried decisions may prove to be perilous. Thus, decision-making is a major factor towards the success of manufacturing organizations.

Making a decision is a complex process irrespective of its nature and comprises professionals from different domains. It has become a mandate for decision makers of such organizations to select suitable manufacturing strategies, product designs, manufacturing processes, workpiece and tool materials, and machinery and equipment with a view to meet consumer demands and challenges to maintain quality output at minimum cost.

Decision makers of manufacturing organizations frequently come across the problem of evaluating numerous alternatives and selecting the most appropriate one based on several conflicting criteria. Thus, a scientific approach towards decision-making in the manufacturing environment has become essential to make the process easier, systematic, logical, and effective. Manufacturing engineers, strategists, and policymakers are often required to employ a range of criteria encompassing different aspects, like productivity, longevity, suitability, cost-effectiveness, and flexibility, along with many technical criteria while making decisions related to the introduction of new equipment or technology or selecting the most suitable material for any product. In this context, multi-criteria decision-making (MCDM) has emerged as a formal methodology to tackle such situations to support decisions in many fields and can be especially valuable in a manufacturing environment.

MCDM methods evaluate the performance of alternatives across several contradicting, qualitative and quantitative criteria, and result in complete ranking orders of the alternatives. Since a few decades, there has been a considerable increase in interest among researchers to apply MCDM methods in selecting proper manufacturing policies. Effective decision-making is the basis of successfully solving any problem, including selecting the most applicable strategies.

There have been consistent advances in MCDM methods in recent years. Often decision-makers adopt ad-hoc methods that may ignore either stakeholders' opinions or decision criteria. Given the novelty and inherent applicability of MCDM methods in manufacturing applications, there is a need for a case study-based book that can help the readers traverse through the recently developed MCDM methods and resolve how to apply them to specific case studies.

Through a collection of a diverse range of manufacturing applications and case studies, this book aims to aid readers in applying cutting-edge MCDM methods to their own manufacturing projects. It offers an overview of different MCDM methods developed to date and conceptual frameworks of how these methods are applied, with a special focus on their applicability and usefulness. Taking an in-depth approach for the readers, this book introduces steps of MCDM methods starting from problem formulation, model development, criteria weighting, alternative scoring, and finally, ranking of the alternatives.

The authors then explain the step-by-step processes of different MCDM methods for different case studies, which would allow readers to perceive the presented methodologies in a very articulated way. Intended as both a research and teaching tool, this book aims to create critical and logical thinking when applying MCDM methods for solving complex manufacturing decision-making problems.

The book is organized into 32 chapters. A series of case studies from the manufacturing domain collectively demonstrate the applications of different MCDM methods in situations of high variability that require integration of technical information and expert judgment – an area where MCDM clearly outperforms any other method. A brief description of each of the chapters is presented below:

- Chapter 1 discusses the fundamentals of the MCDM process, background, classification, advantages, and role of MCDM in a manufacturing environment.
- Chapter 2 introduces a very popular and basic MCDM method called "Analytic Hierarchy Process (AHP)," along with a state-of-the-art literature review and its step-by-step application for a flexible manufacturing system (FMS) selection case study.
- Chapter 3 presents a description of the "Analytic Network Process (ANP)" along with a contemporary literature review and step-by-step application of ANP for a green supplier selection problem.
- Chapter 4 deliberates on the application of the "Weighted Aggregated Sum Product ASsessment (WASPAS) Method" for an electroplating system selection case study. A literature review on recent applications of the WASPAS method is also presented in this chapter.
- Chapter 5 discusses the "Multi-Objective Optimization on the Basis of Ratio Analysis (MOORA)" method along with an up-to-date literature review and its bit-by-bit application for a rapid prototyping process selection problem.
- Chapter 6 talks about the "Technique for Order of Preference by Similarity to Ideal Solution (TOPSIS)," which is a well-applied distance-based MCDM method. The procedural application of the TOPSIS method for a conveyor belt material selection case study is also presented. Detailed methodological steps and a literature review are also given.
- Chapter 7 addresses another distance-based MCDM method called "VIekriterijumsko KOmpromisno Rangiranje (VIKOR)," along with a state-of-the-art literature review and its step-by-step application for flywheel material selection.
- Chapter 8 presents a vastly applied outranking-based MCDM method called "ELimination Et Choice Translating REality (ELECTRE)," accompanied by a review of its applications. Procedural implementation of the ELECTRE method is shown for a material handling equipment selection case study.
- Chapter 9 furnishes the applications of "Preference Ranking Organization METHod for Enrichment Evaluation (PROMETHEE)"

for a facility layout selection problem, and then a non-traditional machining (NTM) process selection case study is demonstrated using Geometrical Analysis for Interactive Aid (GAIA). This chapter also presents a literature review on the applications of the PROMETHEE method.

- Chapter 10 deliberates on a proportional ratio-based MCDM method called the "COmplex PRoportional ASsessment (COPRAS)" method, along with its contemporary literature review and step-by-step application for a naval craft material selection problem.
- Chapter 11 discusses another preference-based MCDM method called the "Organization, Rangement Et Synthese De Donnes RElationnelles (ORESTE)" method. This chapter also presents a literature review and step-by-step application of the ORESTE method for an industrial robot selection problem.
- Chapter 12 puts forward the "Operational Competitiveness Rating Analysis (OCRA)" method along with a state-of-the-art literature review and its step-by-step implementation for a flexible manufacturing system (FMS) selection case study.
- Chapter 13 reports on another ratio-based MCDM method called the "Additive Ratio ASsessment (ARAS)" method, along with a state-of-the-art literature review and its detailed application while solving a CNC machine tool selection problem.
- Chapter 14 brings up another recently developed distance-based MCDM tool called the "Evaluation Based on Distance from Average Solution (EDAS)" method, along with a literature review and its application in solving a cutting tool material selection problem.
- Chapter 15 refers to the "COmbinative Distance-Based ASsessment (CODAS)" method and its literature review. An application of renewable energy technology selection is also presented in this chapter.
- Chapter 16 presents another preference-based MCDM method called the "EVAluation of MIXed Data (EVAMIX)" method. Application of this method is demonstrated for a non-traditional machining process selection problem.
- Chapter 17 talks about the "Preference Selection Index (PSI)" method for a milling machine selection problem.

- Chapter 18 puts forward details about "TOmada de Decisao Interativa Multicriterio (TODIM)" method. A comprehensive literature review and application of the TODIM method for an automated guided vehicle (AGV) selection problem are also presented in this chapter.
- Chapter 19 discusses the "Range of Value (ROV)" method along with a state-of-the-art literature review and its step-by-step application for a sailing boat mast material selection problem.
- Chapter 20 articulates the "Superiority and Inferiority Ranking (SIR)" method for a plant layout design selection problem.
- Chapter 21 demonstrates the application of the "Measuring Attractiveness by a Categorical-Based Evaluation TecHnique (MACBETH)" method for a supplier selection problem in a discrete manufacturing environment.
- Chapter 22 discusses a less explored "Modified Similarity Index Method" method along with a state-of-the-art literature review and its application for a cutting fluid selection problem.
- Chapter 23 presents the "Multi-Attributive Border Approximation Area Comparison (MABAC)" method, and its application for a welding process consumable selection problem is subsequently exhibited.
- Chapter 24 shows the application of the "Multi-Attributive Real-Ideal Comparative Analysis (MARICA)" method for solving a battery selection problem for electric hybrid vehicles.
- Chapter 25 exhibits the application of a very recent development in the MCDM domain. This chapter reveals the implementation of the "Measurement Alternatives and Ranking According to COmpromise Solution (MARCOS)" method for a manufacturing process selection problem.
- Chapter 26 demonstrates the application of another recent development in the MCDM domain. This chapter presents the application of the "COmbined COmpromise SOlution (CoCoSo)" method for a concrete pump selection problem.
- Chapter 27 presents the "Ranking of Alternatives Through Functional Mapping of Criterion Sub-Intervals into a Single Interval (RAFSI)" method for a welding process selection problem.

- Chapter 28 puts forward the "Simultaneous Evaluation of Criteria and Alternatives (SECA)" method for a 3D printer nozzle material selection problem.
- Chapter 29 presents the use of the "R Method" for solving a conveyor selection problem.
- Chapter 30 depicts the mathematical details of an almost recent MCDM tool in the form of the "Compromise Ranking of Alternatives From Distance to Ideal Solution (CRADIS)" method. Its application is demonstrated while solving a conveyor belt material selection problem.
- Chapter 31 discusses the roles of "Normalization Techniques" in MCDM methods. It also presents applications of different normalization techniques for manufacturing applications.
- Chapter 32 introduces different objective and subjective criteria weighting techniques as frequently employed in conjunction with various ranking-based MCDM methods.

PART I

An Introduction to Multi-Criteria Decision-Making

CHAPTER 1

INTRODUCTION

1.1 MULTI-CRITERIA DECISION-MAKING (MCDM) PROCESS

We continuously strive to achieve the best in all dimensions of life. It means that there may be several options available to us, and we have to identify the most appropriate course of action in order to satisfy our requirements and desires. But, each course of action (alternative) may have its own merits and demerits, which are often evaluated based on an alternative's relative performance score pertaining to a given criterion/attribute. Again, among these attributes, some are required with their higher values (beneficial criteria), whereas lower values are required for the non-beneficial or cost criteria. Thus, the basic objective is concentrated on singling out the best option (alternative), which has the maximum number of beneficial attributes with their highest values, and non-beneficial attributes having their lowest values. Selection of the most apposite alternative from a set of available options amid a pool of conflicting attributes is basically termed as decision-making or decision analysis. Decision analysis is used for logically analyzing multifaceted problems involving quantified alternatives and some pre-defined criteria in an ambiguous and precarious situation using mathematical models. It is undoubtedly based on the principles of Operations Research and can be extensively applied in diverse fields and areas, like computer science, engineering and technology, management, and econometrics containing consistent cognitive, rational, and coherent factors. Decision theories study these factors in contemplation to interpret and elucidate a decision-making situation to increase the reliability of the decision-making process. A decision-making situation is characterized by decision alternatives, states of nature, and resulting payoffs. The decision alternatives are the different possible strategies that the decision-maker can

Multi-Criteria Decision-Making Methods in Manufacturing Environments: Models and Applications.
Shankar Chakraborty, Prasenjit Chatterjee, & Partha Protim Das (Authors)

make use of. The states of nature refer to those future events which are not under the control of the decision maker, but still eventually influence the outcomes of decisions.

In decision-making situations, alternative evaluation frequently comprises a probabilistic analysis and preference elicitation using some mathematical approaches. Differences between the desired and actual states of real-world system can be systematically approached by means of decision analysis techniques. Multi-criteria decision-making (MCDM) is a comprehensive and generic tool of Operations Research and a sub-discipline of decision analysis that unambiguously considers multiple criteria in conflicting decision-making environment. The MCDM process thus deals with situations involving selection, sorting, and ranking of the best alternative from several suitable options, subject to two or more conflicting tangible or intangible criteria with the help of some strong mathematical modeling approaches. Some mathematicians initiated mathematical treatments to decision-making problems at the beginning of the 1960s. The revolutionary work on goal programming by Simon [1] is considered to be the genesis of MCDM and since then, it has been a vigorous interdisciplinary application field in Operations Research and management science. It is basically an analytical approach to evaluate the merits and demerits of each and every alternative under consideration based on multiple criteria. In the context of present era, MCDM paradigms are gaining more and more popularity in the domains of engineering and management science as potential tools for analyzing complex real-time decision-making problems due to their innate ability to judge different alternatives based on various evaluation criteria with different units of measurement while identifying the best alternative. The key attribute of these paradigms is that the decision maker does not require optimizing any single objective, but rather targets for the accomplishment of satisfying all levels in the goals and seeks the most favorable compromise between several, often conflicting objectives. The basic theme of any MCDM method is primarily based on the concept of establishing preferences among different alternatives with respect to some pre-defined objectives and measurable attributes or criteria. In the manufacturing environment, effective decisions are intended for achieving lower manufacturing costs, higher productivity, proper production planning and scheduling systems, higher product quality, enhanced flexibility, etc. Evaluation, assortment, and assessment of manufacturing policies with confusing mutually conflicting criteria and performance measures involving tradeoffs between chosen properties, process, working condition and cost often make the decision-making process a complicated task. Due to these reasons, decision analysis techniques have been well acknowledged as a fundamental and essential module in the success of a manufacturing organization. A choice, as involved in any decision, is selected from a number of feasible alternatives. Decisions that are based on a foundation of understanding and sound reckoning of information can escort the organization towards a long-standing prosperity, while decisions made on the basis of inconsistent or uncertain judgments and partial information compel the organization out of competition.

In the most general form of MCDM methods, the best solution for a particular problem is chosen from a finite set of alternatives with precisely known qualitative or quantitative objective function values or in the presence of both. This is accomplished first by problem identification and development of feasible alternate solutions depending on decision rules and constraints of the problem under consideration. The second phase, popularly known as the development of the decision matrix, splits the problem into some appropriate smaller parts to allow data and judgments to be brought in. A decision matrix designates the influencing alternative options and respective attribute values. Finally, reassembling the component parts by the application of an analytical model is performed [2]. While doing these, a logical and consistent recommendation picture can be presented to the concerned decision maker, as shown in Figure 1.1.

Over the years, different behavioral scientists, operational researchers, and decision theorists have proposed a variety of ways to describe how a decision maker might select the best alternative in a multi-criteria environment [3]. A 'mathematically well-formulated' decision problem does not necessarily mean a 'well-structured problem' with respect to the given reality. A 'well-structured' problem can be easily understood, and an appropriate solution procedure may be formulated [4]. From a mathematical standpoint, a 'well structured' MCDM process instigates with recognizing and gathering pertinent information about the problem statement. The possible options (alternatives) are defined while eliminating those alternatives that should not be worthy for further considerations. This procedure is often called screening. It helps the decision maker to concentrate on a smaller set that contains the most favorable alternatives. Many approaches for screening the available alternatives, like sequential approach, decision information-based screening, Pareto optimality-based screening, tradeoff weight-based screening, aspiration level-based screening, distance-based screening and non-tradeoff weight-based screening have been adopted by the past researchers [5]. Next, the pre-dominant criteria affecting the performance of the shortlisted or screened alternatives are identified. A criterion can be defined as a customary statute depending on which a decision can be made, and which relates the process capabilities to the intended goals.

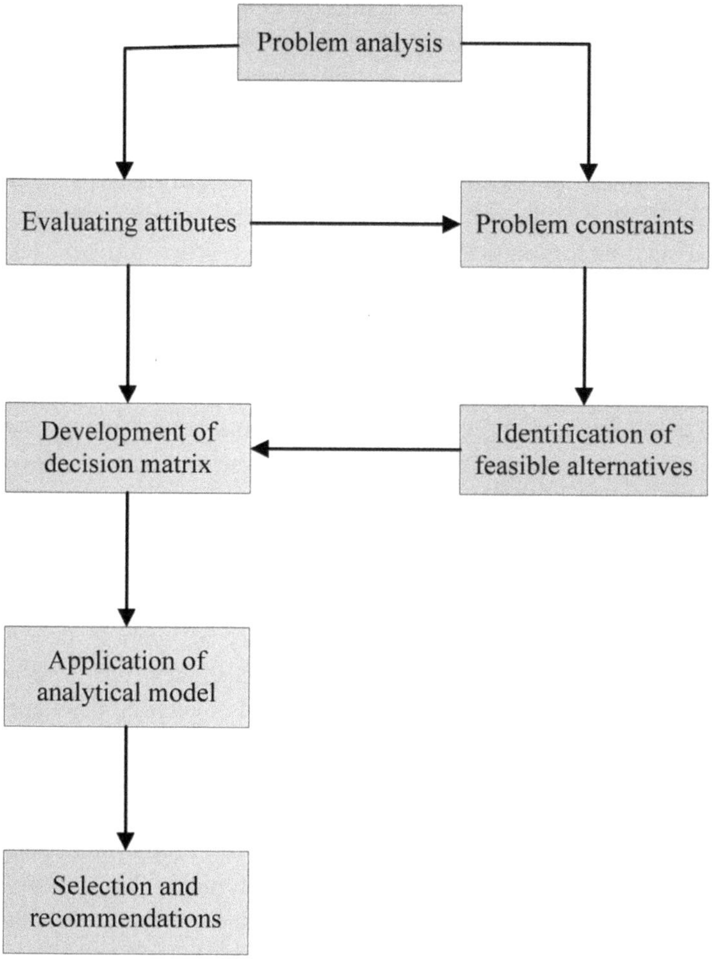

FIGURE 1.1 A typical decision-making process.

In real-time decision-making conditions, it actually becomes too intricate to define the set of applicable criteria. Inclusiveness, consistency, accuracy, independency, and non-redundancy are the most important requirements to define the pertinent criteria. Next, the alternatives' performance scores with respect to the considered criteria are represented as a decision/evaluation matrix. A typical decision matrix for MCDM problems consists of m number of alternatives ($i = 1, 2, \ldots, m$) and n number of evaluation attributes/criteria ($j = 1, 2, \ldots, n$), based on which the alternatives would be evaluated and selected. Thus, a decision problem can be represented by the following matrix:

Criteria	**1**	**2**	**…**	**j**	**…**	**n**
Alternative						
1	x_{11}	x_{12}	…	x_{1j}	…	x_{1n}
2	x_{21}	x_{22}	…	x_{2j}	…	x_{2n}
…	…	…	…	…	…	…
…	…	…	…	…	…	…
i	x_{i1}	x_{i2}	…	x_{ij}	…	x_{in}
…	…	…	…	…	…	…
m	x_{m1}	x_{m2}	…	x_{mj}	…	x_{mn}

where; x_{ij} is the performance score of ith alternative on j^{th} criterion. These ratings can take different forms depending on the nature/type of the problem and available information. Cardinal mode, where x_{ij} is expressed by a real number, is the mostly adopted and popular rating format. But, if the decision maker feels that it is difficult to derive cardinal data or direct measurable value, it can then be measured in ordinal form. Linguistic grades can be used by the decision maker to assess the alternatives' ratings in ordinal form. Sometimes, uncertainty may have to be considered in the rating process. This type of information can be expressed as interval data, fuzzy data, rough data or in some other suitable mode reflecting uncertainty in the decision-making process. Such judgments are examined by different methods in different ways to aggregate them into an overall performance score for each alternative. One very interesting fact about the developed decision matrix is that it may consist of criteria having different scales and dimensions. Hence, in order to avoid the difficulties caused by different units or dimensions of the criteria values, transformation (called normalization) of the criteria values is adopted in the modeling process for unification by making the criteria values dimensionless and comparable. Several vector and linear normalization techniques have been proposed by the past researchers to convert different units of the considered criteria into dimensionless values. In vector normalization procedure, each element of the quantified decision matrix is divided by its own Euclidian norm. The norm represents the square root of the addition of squared element values, according to each criterion. The Euclidean normalization is a non-linear transformation leading to scales in which the minimum and maximum values for each criterion are different. For linear normalization, the scale transformation is performed by dividing a criterion by its corresponding range between the maximum and minimum values. This method considers the upper and lower bounds of each criterion value during normalization and due to the presence of linear character function, ratios and intervals between the transformed values remain unaffected [6]. In most of the normalization techniques, the normalized values either range from 0 to 1 or the values are less than 1. After the scale transformation process, the overall evaluation of each alternative is done, leading to a final ranking of the alternatives. The main objective of any MCDM approach is to convert the performance of each alternative into a single value (aggregated value) to facilitate the ranking process which is the heart of the MCDM approach. Since, there is no single method of aggregation which is universally acceptable for any kind of decision problem, the decision maker has to choose the method which best corresponds to his/her purpose. Finally, a sensitivity analysis examines how robust the recommendation is even under a small change in the preference value expressed by the decision maker. If no further analysis is required, the final recommendation can be made based on the observed results. Figure 1.2 illustrates these basic steps as involved in an MCDM process. In a situation where there is a number of decision makers involved in a decision-making process, the individual decisions are taken into account, and an optimal option is chosen representing a group negotiation decision.

1.2 BACKGROUND AND CLASSIFICATION OF MCDM METHODS

'International Society on Multiple Criteria Decision-making' defined MCDM as 'the study of methods and procedures by which concern about multiple conflicting criteria can be formally incorporated into the management planning process.' Due to the ability of representing problems through conjointly conflicting criteria, MCDM has always been a well-researched area of Operations Research. The result of a multi-criteria analysis process is a function of criteria values. These criteria are simultaneously optimized within a feasible set of alternatives. In many cases, a single alternative may not actually exist which can optimize all the criteria and thus a compromise solution is aimed to be identified by most of the MCDM methods. There is a set of alternatives where an improvement in a particular criterion value may decrease the value of another criterion. Such set of alternatives is called a set of non-dominating or Pareto optimal alternatives (solutions). Each alternative in this set can be a solution of the multi-criteria problem.

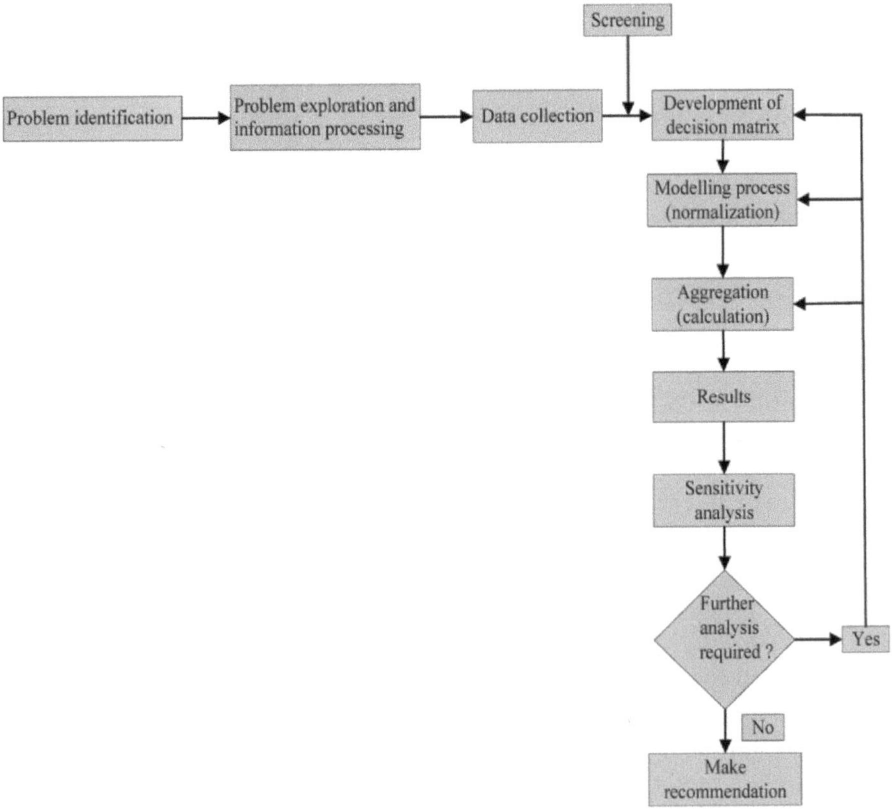

FIGURE 1.2 Step-by-step illustration of an MCDM process.

In order to choose only a particular alternative, some additional information is required to be provided by the decision maker. The information that the decision-maker provides reflects the global preference with respect to the quality of the alternative sought. As already mentioned, an MCDM problem necessarily has some feasible list of alternatives along with a number of evaluation criteria, so it becomes very important to understand whether a particular decision-making problem is single or multiple criteria based. While formulating a multi-criteria problem, the decision maker must consider the relevant criteria which generally depend on the nature of the problem to be analyzed [7]. In the presence of a single criterion or single assessment in a decision-making problem, the decision can then be made while searching the alternative with the best value for that criterion. This situation can be considered as a traditional optimization problem where the objective function is the single criterion, and the constraints are the criteria values of the alternatives. The case, when the decision problem has a finite number of criteria, but the number of feasible alternatives is infinite, belongs to the field of multiple criteria optimization (MCO). Depending on the nature and structure of the problem, different optimization techniques, like linear programming (LP), non-linear programming, discrete optimization, etc., can be adopted for solution. The MCO techniques can also be used when the number of feasible alternatives is finite, but they are expressed only in implicit form [8]. In this way, it can be said that there are three major influencing factors to be observed in an emblematic decision-making problem, i.e., objectives, alternatives, and attributes. More generally, there exist two distinctive types of MCDM problems due to different problem settings, involving a finite number of alternatives and an infinite number of solutions. Normally in problems associated with selection and assessment of alternatives, the number of alternative solutions is limited. In design related problems, there can be an infinite number of alternative solutions as criteria values may be expressed within a specific range. Based on this, MCDM problems can be classified into two preliminary categories, i.e., multi-objective decision-making (MODM) and multi-attribute decision-making (MADM) or MCDM. MODM focuses on a continuous decision space subject to associated problem constraints, while MADM helps for alternative evaluation in a discrete or discontinuous decision space of pre-determined set of alternatives. In real-time decision-making situations, MODM often demands for formulation of a decision model based on target objectives and interrelationship between the evaluating criteria, whereas, MADM usually consists of identification of the alternatives, specification of each alternative with respect to each evaluation criterion indicating the relative preference of each alternative over another and finally, estimating overall performance scores of the alternatives. In a nutshell, MADM methods select the best alternative among a finite number of choices, unlike MODM where the best alternative is projected with several objectives based on continuous decision variables subject to some constraints. Table 1.1 summarizes the main features of these two approaches [9].

There are several methods in each of the above-mentioned categories. Each of these methods has its own characteristics, and can be classified as deterministic, probabilistic, and fuzzy or with their different combinations. Depending on the number of decision-makers, these methods can further be classified as single (number of decision makers is only one) or group (number of decision makers is multiple) decision-making techniques. Decision-making under uncertainty and decision support systems are also prominent decision-making tools [10]. Based on the mathematical modeling and operational

TABLE 1.1 Characteristics of MADM and MODM Techniques

Feature	MADM	MODM
Criteria Defined by	Attribute/Criterion	Objective
Model basis	Alternative identification and specification	Target objectives and inter-relationship between evaluating criteria
Objectives	Implicit	Explicit
Alternatives	Finite number	Infinite number
Attributes	Finite field	Infinite field
Decision space	Discrete	Continuous
Decision maker's control	Limited	Significant
Decision-making paradigm	Outcome-oriented	Process-oriented
Relevant application	Selection/evaluation	Design/research

procedures, MADM tools can be broadly categorized as elemental approaches, pair-wise comparison-based approaches, unique synthesizing approaches, outranking approaches, distance-based approaches, and dominance-based approaches. Elementary methods reduce a complicated problem to the singular basis for selection of a preferred alternative. They are simple requiring minimum of the computer software support and are best suited for single decision-making problems with few alternatives and criteria [11]. Simple additive method, weighted sum method and weighted product method are the unique examples of elementary methods. In pair-wise comparison approaches, all the criteria under consideration are first pairwise compared to determine their relative importance (weights) in the decision-making process. Thereafter, the alternatives are pairwise compared with respect to each of the criteria to evaluate their relative performance. The relative performance of the alternatives and the corresponding criteria weights are finally aggregated together to rank the candidate alternatives. Unique synthesizing approaches adopt some special mathematical and analytical techniques in the modeling and execution phase. Distance-based approaches employ positive-ideal, negative-ideal, and reference point concepts to arrive at a preference order of the considered alternatives [12]. The best value in the decision matrix is assumed to be the imaginary positive-ideal or reference point solution, while the negative- or anti-ideal solution is an imaginary solution for which all the criteria values correspond to the minimum values in the decision matrix. On the other hand, in dominance-based approaches, the degree by which one alternative dominates another alternative in terms of a specific criterion or is being dominated by other alternative is estimated. These dominance degrees are later aggregated together to aid in the final ranking of the candidate alternatives. In outranking approaches, a series of pair-wise alternative comparisons for each criterion is employed to develop an outranking relation indicating the degree of dominance of one alternative over the other [13, 14]. The prevalent outranking approaches are ELimination and Et Choice Translating REality (ELECTRE), Preference Ranking Organization METHod for Enrichment Evaluation (PROMETHEE), Organization Rangement Et Synthese De Donnes Relationnelles (ORESTE), etc., and the most commonly used MODM tools are linear programming (LP), goal programming and lexicographic vector-max approaches. Based on the above discussions, MADM/MCDM methods can be classified into different categories, as shown in Figure 1.3. As this book deals with the applications of some particular types of MADM tools in diverse manufacturing decision-making situations, so further discussions are restricted only to those tools. In MADM methods, the best alternative among the available ones is obtained while providing some mathematical treatments to the decision matrix. The MADM covers a wide range of techniques for assessing the decision problems characterized by a large number of qualitative and quantitative data, and preferences regarding the relative importance of the evaluation criteria. The relative importance of the evaluation criteria with respect to criteria weights in MADM though does not have any clear economic significance, but it plays an important role by providing the opportunity to represent the actual aspects of decision-making (the preference structure). The criterion weight indicates a criterion's relative importance over the others. It often becomes difficult for the decision maker to provide exact values for criteria weights due to existence of inaccuracy, inconsistency, and lack of agreement [15].

Several approaches have been developed and proposed for determining the corresponding criteria weights. Analytic hierarchy process (AHP) [16] is one of the mostly adopted MADM tools for calculating subjective weights of the criteria, whereas, for objective weighting, entropy method [17] is frequently used. There are also some newly developed criteria weight measurement techniques having their own advantageous features. An MADM problem can comprise both objective (cardinal or quantitative) and subjective (qualitative or ordinal or linguistic) criteria which can further be classified as beneficial (higher values indicate better performance) and non-beneficial (lower values denote better performance). It is expected that an ideal solution would maximize all the beneficial criteria and at the same time, it would minimize all the non-beneficial criteria. Normally, this type of ideal solution is often unobtainable. In case, no ideal solution is obtained, then some non-dominated solutions may be considered. If there are other alternatives that are better than the solution on at least one criterion and as excellent as it on other criteria, the alternative (solution) is said to be dominated. An alternative is called non-dominated if it is not dominated by any other alternative on any criterion and best fulfills the decision maker's expectations and objectives. Attributing to such advantages and properties, MADM tools are vastly applied in manufacturing, materials selection, advanced manufacturing process selection, environment management, plant layout design selection, supplier evaluation, energy management, maintenance strategy selection, transportation, logistics, project investment decision, financial and strategic decision-making situations, enterprise resource planning (ERP) system selection, water management, agricultural, and forestry management, etc. [18].

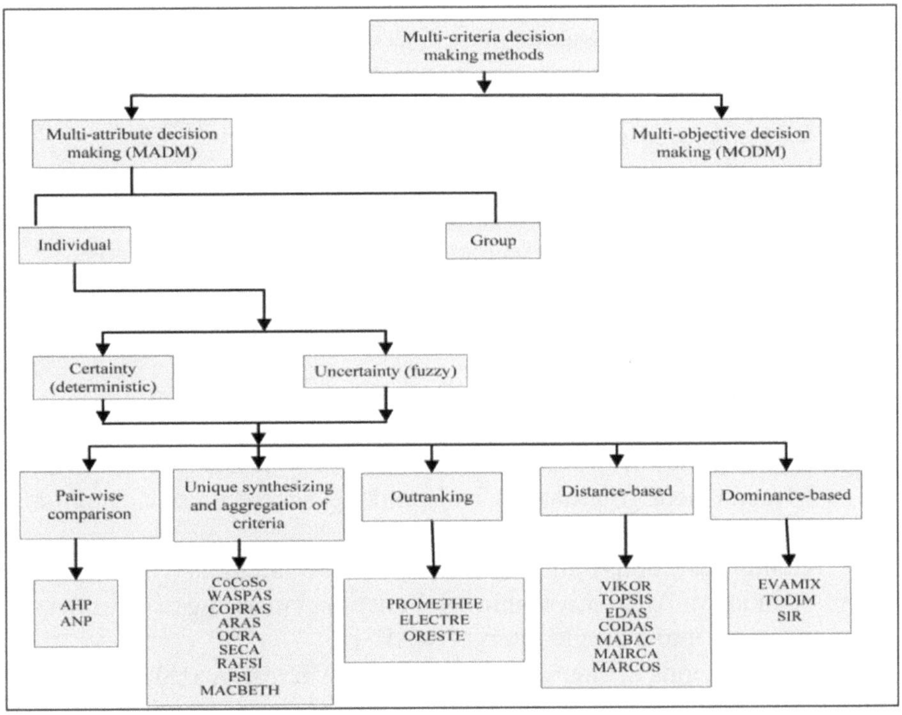

FIGURE 1.3 Classification of MCDM methods.

In this book, the following exhaustive list of the existent MCDM techniques is considered along with their mathematical formulations, applications in diverse fields of manufacturing and illustrative examples:

- Analytic Hierarchy Process (AHP);
- Analytic Network Process (ANP);
- Weighted Aggregated Sum Product ASsessment (WASPAS);

- Multi-Objective Optimization on the basis of Ratio Analysis (MOORA);
- Technique for Order of Preference by Similarity to Ideal Solution (TOPSIS);
- VIekriterijumsko KOmpromisno Rangiranje (VIKOR);
- ELimination and Et Choice Translating REality (ELECTRE);
- Preference Ranking Organization METHod for Enrichment Evaluation (PROMETHEE);
- COmplex PRoportional ASsessment (COPRAS);
- Organization, Rangement Et Synthese De Donnes Relationnelles (ORESTRE);
- Operational Competitiveness Rating Analysis (OCRA);
- Additive Ratio ASsessment (ARAS);
- Evaluation based on Distance from Average Solution (EDAS);
- COmbinative Distance-based ASsessment (CODAS);
- EVAluation of MIXed data (EVAMIX);
- Preference Selection Index (PSI);
- TOmada de Decisao Interativa Multicriterio (TODIM);
- Range of Value (ROV) method;
- Superiority and Inferiority Ranking (SIR) method;
- Measuring Attractiveness by a Categorical-Based Evaluation TecHnique (MACBETH);
- Modified Similarity Index method;
- Multi-Attributive Border Approximation area Comparison (MABAC);
- Multi-Attributive Ideal-Real Comparative Analysis (MAIRCA);
- Measurement Alternatives and Ranking according to COmpromise Solution (MARCOS);
- Combined Compromise Solution (CoCoSo) method;
- Ranking of Alternatives through Functional mapping of criterion subintervals into a Single Interval (RAFSI);
- Simultaneous Evaluation of Criteria and Alternatives (SECA) method;
- R method;
- Compromise Ranking of Alternatives from Distance to Ideal Solution (CRADIS) method.

It has already been mentioned that the relative importance (weight) assigned to the considered evaluation criteria plays an important role in the decision-making process. A slight variation in the criteria weights may significantly influence the final ranking order of the candidate alternatives. Besides AHP method, there are also several criteria weight measurement techniques available, as presented here-in-under:

- Entropy method;
- DEcision-MAking Trial and Evaluation Laboratory (DEMATEL) method;
- Step-wise Weight Assessment Ratio Analysis (SWARA) method;
- PIvot Pairwise RElative Criteria Importance Assessment (PIPRECIA) method;
- CRiteria Importance Through Intercriteria Correlation (CRITIC) method;
- FActor RElationship (FARE) method;
- Best Worst Method (BWM);
- FUll COnsistency Method (FUCOM);
- Level-Based Weight Assessment (LBWA) method;
- Logarithm Methodology of Additive Weights (LMAW);
- MEthod based on the Removal Effects of Criteria (MEREC).

There are also different normalization techniques deployed for making the elements of the original decision matrixes dimensionless and comparable. It has also been observed that while developing the corresponding decision matrix for a specific problem, the decision maker may often feel confused to evaluate the performance of an alternative using a single cardinal number or absolute value. The concerned decision maker may be more relaxed in expressing his/her judgment by choosing any value within a given range. The values in the range may be numerical or linguistic. There are several approaches, like fuzzy set theory, rough set theory, intuitionistic fuzzy set theory, interval numbers, etc., to deal with the confusion or ambiguity of the decision maker while developing the corresponding decision matrixes. Effective integration of these tools with the existing MCDM techniques would help in providing more accurate, robust, and reliable ranking results.

1.3 ADVANTAGES OF MCDM METHODS

The MCDM methods provide a quantitative resource to aid in decision-making situations where there are multiple and conflicting goals measured in different units. The main benefits of MCDM methods include making a decision clearer to others, giving an appropriate way for structuring and information gathering, and assisting people in better recognizing a problem from different perspectives. Some comprehensive advantages that MCDM approaches have are enlisted as below:

- These methods are used in an uncluttered and unambiguous manner.

- Objectives and criteria selection for any decision-making problem are open to analysis and change, if it is felt to be inappropriate.
- It allows a variety of objectives to be examined at the same time.
- Being a multi-disciplinary approach, it can be applied to a wide field of real-life decision-making situations.
- It provides a systemic approach that demonstrates tradeoffs (compromises) between the conflicting issues.
- It serves as a key means of communication inside the decision-making group and on occasion, amongst members of the general public.
- When scores and weights are employed, they are also explicit, and developed using well-formulated and mathematically sound methods.
- It incorporates the subjective experience of the decision makers, resulting in a practical decision-making process.
- One can judge what would happen if different objectives are given different importance or assessments are made about the effectiveness of available options in relation to different objectives.
- It is easier to adopt appraisal-led design, the appraisal being refined as the design process develops.
- An effective management problem is properly structured through the help of MCDM methods.
- It provides a framework for discussion.
- It is based on a process that leads to sensible and comprehensible judgments.
- Any organization that adopts it, is able to solve problems faster.
- It deals with mixed sets of data involving quantitative and qualitative measurements.
- Plans and decision-making can be structured in a participatory manner.
- The arrived decision is transparent and constructive, as it enables the decision maker to better understand the situation and helps to arrive at a better and negotiated solution.
- It effectively eliminates personal favoritism in the choice of any particular alternative.
- It facilitates a better grasp of objectives and challenges.
- It accurately assesses the criteria that have been selected for consideration.
- An option whose consensus is reached is deemed to be the best option.
- It also measures a decision maker's preference towards each decision criterion and makes a balance between the conflicting criteria while providing the most cost-effective decision.

- It helps to build consensus among the decision makers, facilitating compromise among them which favors in competing alternatives.

Decision-making methodologies can assist the decision makers in making multifaceted decisions while providing adequate scopes to different aspects of the problem that can be additionally harnessed for enhancing impact on practical decision-making situations.

1.4 MCDM IN MANUFACTURING ENVIRONMENT

Decision-making is the primary job of administration of any organization. Proceedings of the organization and effective utilization of the scarce resources require decision-making at every step. It facilitates the entire management procedure and hence, is a crucial function of management. However, there are few barriers to appropriate and cogent decisions, like insufficient and inconsistent data or information, time limitation, fuzzy environment, inadequate skill of the decision maker, interdependencies among different decisions, subjectivity in the opinion, etc. Recent innovations in the manufacturing processes across the globe have brought about metamorphism in the manufacturing and allied industries. In order to survive in the market, the manufacturing organizations need to function within the framework of globalization and continuous technological changes. Moreover, the dynamic and changing requirements of the customers have created a need for quick response from the manufacturing organizations. On the other hand, flawed and hurried decisions may prove to be perilous. Thus, effective decision-making is a major contributor towards the success of any manufacturing organization. Taking decision is a complex process irrespective of its nature and comprises professionals from different domains. Therefore, it has become a mandate for the managers of such organizations to select suitable manufacturing strategies, product designs, manufacturing processes, workpiece and tool materials, and machinery and equipment with a view to meet diverse upcoming demands and challenges to maintain competitive output. The decision makers in the manufacturing organizations repeatedly come across the problem of evaluating numerous alternatives and selecting the most appropriate option based on several conflicting criteria. Thus, a scientific approach towards decision-making in the manufacturing environment would be quite helpful as it makes the process easier, systematic, more logical, and effective.

Only a few of the decisions in a manufacturing organization are straightforward. The majority of the decisions taken would affect different stakeholders of the organization, like customers, suppliers, employees, investors, etc. Moreover, all of them have their discrete sets of evaluation criteria for choosing from an assortment of alternatives. Managers are often required to employ a wide range of criteria encompassing different aspects, like productivity, longevity, timeliness, cost-effectiveness, flexibility, etc., while taking decision related to introduction of a new equipment or technology in the organization. The MCDM is an important tool that can be applied to rejoinder these types of complicated decisions. It is a technique for evaluation of the alternatives using several, contradictory criteria resulting in consensus-based solutions [19]. It is a proficient technique to face these challenges while selecting suitable manufacturing strategies, product designs, manufacturing processes, materials for workpiece and tools, machinery, and equipment, etc. Formulating a complex decision-making problem correctly while explicitly considering multiple criteria for evaluation is the first step to have more informed and better decisions. There have been consistent advances in MCDM since MCDM tools have been first used in real-life decision-making problems. Numerous decision-making software using different MCDM techniques are being continuously developed for their application in a wide array of disciplines, ranging from engineering [20], management [21], social sciences [22], and even policy decisions by government [23]. There have been a variety of MCDM methods used to date in the manufacturing domain for solving problems of computer-aided manufacturing software selection [24], material selection [25], machine tool selection [26], non-traditional machining (NTM) process selection [27], cutting fluid selection [28], cutting tool selection [29], robot selection [30], inspection device selection [31], supplier selection [32], etc.

KEYWORDS

- **analytic hierarchy process**
- **enterprise resource planning**
- **multi-attribute decision making**
- **multi-criteria decision making**
- **multi-objective decision making**
- **multiple criteria optimization**

REFERENCES

1. Simon, H. A., (1955). *Models of Man*. Wiley, New York.
2. Mourits, M. C. M., & Oude, L. A. G. J. M., (2006). Multi-criteria decision making to evaluate quarantine disease control strategies. In: Oude, L. A. G. J. M., (ed.), *New Approaches to the Economics of Plant Health*. Springer, Netherlands.
3. Tretheway, K. R., Wood, R. J. K., & Roberge, P. R., (1998). Development of a knowledge-based system for materials management. *Materials & Design, 19*(1, 2), 39–56.
4. Scheubrein, R., & Zionts, S., (2006). A problem structuring front end for a multiple criteria decision support system. *Computers & Operations Research, 33*(1), 18–31.
5. Chen, Y., Kilgour, D. M., & Hipel, K. W., (2008). Screening in multiple criteria decision analysis. *Decision Support Systems, 45*(2), 278–290.
6. Schneider, F., (2010). *Multiple Criteria Decision Making in Application Layer Networks*. Grin Verlag, Germany.
7. Yoon, K. P., & Hwang, C. L., (1995). *Multiple Attribute Decision Making - An Introduction*. SAGE Publications Inc., California.
8. Steuer, R. E., (1986). *Multiple Criteria Optimization: Theory, Computation and Application*. Wiley, New York.
9. Bernroider, E., & Stix, V., (2007). A method using weight restrictions in data envelopment analysis for ranking and validity issues in decision making. *Computers & Operations Research, 34*(9), 2637–2647.
10. Pohekar, S. D., & Ramachandran, M., (2004). Application of multi-criteria decision making to sustainable energy planning: A review. *Renewable and Sustainable Energy Reviews, 8*(4), 365–381.
11. Linkov, I., Varghese, A., Jamil, S., Seager, T. P., Kiker, G., & Bridges, T., (2004). Multi-criteria decision analysis: A framework for structuring remedial decisions at contaminated sites. In: Linkov, I., & Ramadan, A. B., (eds.), *Comparative Risk Assessment and Environmental Decision Making*. Kluwer Academic Publishers, Netherlands.
12. Mardani, A., Juson, A. Md. N. K., Khalifah, Z., Zakwani, N., & Valipour, A., (2015). Multiple criteria decision-making techniques and their applications: A review of the literature from 2000 to 2014. *Economic, 28*(1), 516–571.
13. Martel, J. M., & Matarazzo, B., (2005). Other outranking approaches. In: Figueira, J., Salvatore, G., & Ehrgott, M., (eds.), *Multiple Criteria Decision Analysis: State of the Art Surveys*. Springer, New York.
14. Kangas, A., Kangas, J., & Pykalainen, J., (2001). Outranking methods as tools in strategic natural resources planning. *Silva Fennica, 35*(2), 215–227.
15. Hyde, K. M., Maier, H. R., & Colby, C. B., (2005). A distance-based uncertainty analysis approach to multi-criteria decision analysis for water resource decision making. *Journal of Environmental Management, 77*(4), 278–290.
16. Saaty, T. L., (1980). *The Analytical Hierarchy Process*. McGraw-Hill, New York.
17. Zhi-hong, Z., Yi, Y., & Jing-Nan, S., (2006). Entropy method for determination of weight of evaluating indicators in fuzzy synthetic evaluation for water quality assessment. *Journal of Environmental Sciences, 18*(5), 1020–1023.
18. Eshlaghy, A. T., & Homayonfar, M., (2011). MCDM methodologies and applications: A literature review from 1999 to 2009. *Research Journal of International Studies, 21*, 86–137.

19. Kolios, A., Mytilinou, V., Lozano-Minguez, E., & Salonitis, K., (2016). A comparative study of multiple criteria decision-making methods under stochastic inputs. *Energies, 9*(7), 566.
20. Zavadskas, E., Antucheviciene, J., Vilutiene, T., & Adeli, H., (2018). Sustainable decision-making in civil engineering, construction and building technology. *Sustainability, 10*(1), 14.
21. Pangsri, P., (2015). Application of the multi-criteria decision-making methods for project selection. *Universal Journal of Management, 3*(1), 15–20.
22. Chaghooshi, A. J., Janatifar, H., & Dehghan, M., (2014). An application of AHP and similarity-based approach to personnel selection. *International Journal of Business Management and Economics, 1*(1), 24–32.
23. Rehman, T., & Romero, C., (1993). The application of the MCDM paradigm to the management of agricultural systems: Some basic considerations. *Agricultural Systems, 41*(3), 239–255.
24. Ayağ, Z., (2020). An approach to evaluate CAM software alternatives. *International Journal of Computer Integrated Manufacturing, 33*(5), 504–514.
25. Emovon, I., & Oghenenyerovwho, O. S., (2020). Application of MCDM method in material selection for optimal design: A review. *Results in Materials, 7*, 100115.
26. Funda, S., & Ayağ, Z., (2016). Fuzzy ANP-based PROMETHEE II approach for evaluation of machine tool alternatives. *Journal of Intelligent & Fuzzy Systems, 30*(4), 2223–2235.
27. Chakraborty, S., Dandge, S. S., & Agarwal, S., (2019). Non-traditional machining processes selection and evaluation: A rough multi-attributive border approximation area comparison approach. *Computers & Industrial Engineering, 139*, 106201.
28. Prasad, K., & Chakraborty, S., (2016). A quality function deployment-based model for cutting fluid selection. *Advances in Tribology* (p. 10). Article ID: 3978102 https://doi.org/10.1155/2016/3978102.
29. Prasad, K., & Chakraborty, S., (2016). A knowledge-based system for end mill selection. *Advances in Production Engineering & Management, 11*(1), 15–28.
30. Suszyński, M., & Rogalewicz, M., (2020). Selection of an industrial robot for assembly jobs using multi-criteria decision-making methods. *Management and Production Engineering Review, 11*(1), 62–72.
31. Kumru, M., & Kumru, P. Y., (2015). A fuzzy ANP model for the selection of 3D coordinate-measuring. *Journal of Intelligent Manufacturing, 26*, 999–1010.
32. Govindan, K., Rajendran, S., Sarkis, J., & Murugesan, P., (2015). Multi-criteria decision-making approaches for green supplier evaluation and selection: A literature review. *Journal of Cleaner Production, 98*, 66–83.

PART II

MCDM Methods and Their Applications

CHAPTER 2

ANALYTIC HIERARCHY PROCESS (AHP)

The analytic hierarchy process (AHP) method, formulated by Saaty [1], is one of the initially developed MCDM techniques, in which an overall goal is followed by criteria, sub-criteria, and alternatives in a hierarchical arrangement. In this method, the problem hierarchy provides a visual and structured way of modeling the decision-making problem being considered. The basic step of this method thus involves the creation of a hierarchy to model the decision problem enabling the decision makers to increase their understanding of the problem, its context, and in the case of group decision-making environment, depicts the alternative approaches to the problem across different stakeholders. The AHP hierarchy consists of a goal (the decision), a number of alternatives for reaching that goal, and a number of criteria on which the alternatives can be judged that relate to the goal. It has already found wide-ranging applications in diverse domains of engineering, management, and social science. It attempts to analyze the impact of the alternatives or decisions on the overall goal of the problem at the top of the hierarchy. The hierarchical structure of a typical AHP problem is shown in Figure 2.1.

The application procedure for AHP comprises the following four basic steps:

- Define the problem and set the overall goal;
- Structure the hierarchy from the top down to bottom consisting of the overall goal, criteria, sub-criteria, and alternatives considered in the problem;
- Develop a set of pair-wise comparisons for each level expressing the preference between two different criteria at the same level of the hierarchy. These numerical judgments are then summarized in the form of a comparison matrix for each hierarchy level. The AHP

Multi-Criteria Decision-Making Methods in Manufacturing Environments: Models and Applications.
Shankar Chakraborty, Prasenjit Chatterjee, & Partha Protim Das (Authors)

technique makes it possible to take into account some inconsistencies in judgments. If the inconsistency level exceeds a value of 0.10, some modifications in the judgments may be required; and

- By calculating the normalized eigenvectors of the pair-wise comparison matrixes, it is possible to estimate the final priorities of the decision options or alternatives.

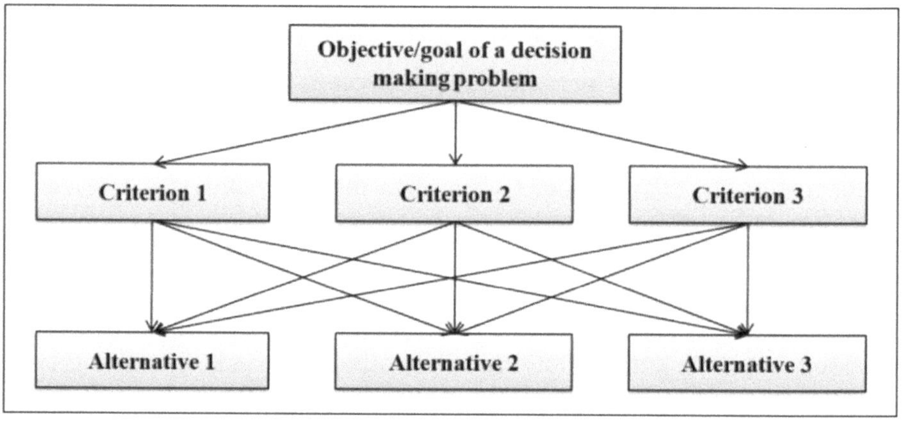

FIGURE 2.1 A typical hierarchical structure.

The essential feature of these pair-wise comparisons lies in quantifying the linguistic variables of the participating decision makers during the evaluation process. Usually, the decision makers' qualitative responses are represented by some numbers, most of the time, ratios of integers. Pair-wise comparisons are eventually expressed using a scale [2]. Such a scale actually represents a one-to-one mapping between a set of linguistic choices available and a discrete set of integers that describe the weights of those choices. The original scale proposed by Saaty is provided in Table 2.1. The next step involves extracting the relative importance implied in the pair-wise comparisons. Saaty [3] asserted that in order to determine the priority weights of different criteria, it is necessary to estimate the right principal eigenvector of the pair-wise comparison matrix. One of the most practical issues in AHP technique is that it allows slight inconsistencies in the pair-wise judgments. If all the judgments are perfectly consistent, then the relation $a_{ij} = a_{ik}.a_{kj}$ applies to any pair-wise comparison matrix. In practice, however, absolute consistency is uncommon. In AHP, the consistencies of the developed pair-wise comparison matrixes are usually checked with the help of consistency ratio (CR) values. The synthesis stage is undertaken after the alternatives have been evaluated against each other on each criterion and the individual priority values have been determined. Columns of the decision matrix represent the priority values of different criteria and finally, the preference values of the alternatives are computed using the decision matrix. In AHP, all the developed pair-wise comparisons are square, positive, and reciprocal, and in each of the matrixes, exactly $m(m-1)/2$ number of pair-wise comparisons are required (where m indicates the number of elements in the decision-making process). In this technique, there may be inconsistencies in the judgments for varying decision makers and it is also plagued by the issue of rank reversal, i.e., when a new alternative is added to the considered problem or an existing alternative is discarded from the problem, the derived ranking order of the alternatives may alter.

TABLE 2.1 Scale of Relative Importance in AHP Method

Scale Value	Interpretation
1	Equal importance of *i* and *j*
2	Between equal and weak importance of *i* over *j*
3	Weak importance of *i* over *j*
4	Between weak and strong importance of *i* over *j*
5	Strong importance of *i* over *j*
6	Between strong and demonstrated importance of *i* over *j*
7	Demonstrated importance of *i* over *j*
8	Between demonstrated and absolute importance of *i* over *j*
9	Absolute importance of *i* over *j*
Reciprocals of the above non-zero numbers	When comparing *i* and *j*, if *i* has one of the above-mentioned non-zero values assigned to it, then *j* has the reciprocal value.

The procedural steps involved in AHP are presented as below:

➤ **Step 1:** Develop the following ($m \times m$) pair-wise comparison matrix A for m objectives:

$$A = \begin{bmatrix} a_{11} & a_{12} & \cdots & a_{1m} \\ a_{21} & a_{22} & \cdots & a_{2m} \\ \cdots & \cdots & \cdots & \cdots \\ a_{m1} & a_{m2} & \cdots & a_{mm} \end{bmatrix} \quad (2.1)$$

where; a_{ij} denotes the relative importance of ith objective over j^{th} objective. For all i and j, it is essential that $a_{ii} = 1$ and $a_{ij} = 1/a_{ji}$.

- **Step 2:** Divide each entry in the jth column of A by the sum of the entries in the jth column. This results in a new matrix, Aw, in which the sum of the entries in each column should be 1.

$$Aw = \begin{bmatrix} \frac{a_{11}}{\sum a_{i1}} & \frac{a_{12}}{\sum a_{i2}} & \cdots & \frac{a_{1m}}{\sum a_{im}} \\ \frac{a_{21}}{\sum a_{i1}} & \frac{a_{22}}{\sum a_{i2}} & \cdots & \frac{a_{2m}}{\sum a_{im}} \\ \cdots & \cdots & \cdots & \cdots \\ \frac{a_{m1}}{\sum a_{i1}} & \frac{a_{m2}}{\sum a_{i2}} & \cdots & \frac{a_{mm}}{\sum a_{im}} \end{bmatrix} \quad (2.2)$$

- **Step 3:** Compute c_i as the geometric mean of the entries in row i of Aw to yield the column vector C.

$$GM_j = \left[\prod_{j=1}^{m} a_{ij} \right]^{1/m} \quad (2.3)$$

- **Step 4:** Normalize the elements of vector C to determine the relative importance of the considered objectives.

$$w_j = GM_j \Big/ \sum_{j=1}^{m} GM_j \quad (2.4)$$

- **Step 5:** Calculate matrixes A_2 and A_3 such that:

$$A_2 = A \times A_1 \text{ and } A_3 = A_2/A_1$$

where; $A_1 = [w_1, w_2, \ldots, w_m]^T$.

- **Step 6:** Check the consistency of judgments in the pair-wise comparison matrix.
  i. Compute the maximum eigenvalue of the pair-wise comparison matrix (λ_{max}) which is the average of matrix A_3. The principal eigenvalue can also be obtained from the summation of products between each element of the weight vector and the sum of columns of the reciprocal matrix.
  ii. Compute the consistency index (CI).

$$\text{CI} = \frac{\lambda_{max} - m}{m - 1}$$

  iii. To establish whether the degree of consistency is satisfactory or not, compare CI to the random index (RI) for the suitable value of m. If the CI value is found to be sufficiently small, the decision maker's judgment may be consistent enough to provide useful weighting estimates for different objectives. If CI/RI (i.e., CR) ≤ 0.10, the degree of consistency is acceptable. If CR > 0.10, there are inconsistencies. In this situation, AHP may not produce useful findings unless the decision makers re-examine their judgments and make the necessary adjustments to reduce the inconsistency below 0.10.

For matrixes involving human judgment, the condition $a_{ij} = a_{ik}.a_{kj}$ does not hold as human judgments are usually inconsistent to some extent. The difference, if any, between λ_{max} and m is an indication of inconsistency of judgments in decision-making. If $\lambda_{max} = m$, the judgments are supposed to be totally consistent. The computed CI value needs to be assessed against judgments made completely at random, and Saaty [1] calculated large samples of random matrixes of increasing order and CI values of those matrixes. A true CR is estimated while dividing CI for the set of judgments by the corresponding random index. If CR equals to 0, the judgments are perfectly consistent. Table 2.2 provides the RI values for varying values of the number of objectives in a decision-making problem.

- **Step 7:** Pairwise compare the alternatives in terms of how much better they are at meeting each of the criteria.
- **Step 8:** Calculate the composite performance scores of the alternatives.

The composite scores of the alternatives are obtained by multiplying the elements of the criteria weight matrix with the elements of the matrix derived after pairwise comparing all the considered alternatives with respect to each criterion. Thus, in a decision-making problem with m alternatives and n criteria, the criteria weight matrix is of $(1\times n)$ order and the corresponding matrix obtained after pair-wise comparing m alternatives against n criteria is of $(m\times n)$ order, the final matrix after multiplication would be of $(m\times 1)$ order showing the overall performance scores of the considered alternatives. The alternatives are ranked in descending order based on their composite performance scores, which indicate the best and the worst options for a specific problem.

TABLE 2.2 Values of RI

m	2	3	4	5	6	7	8	9	10
RI	0	0.58	0.90	1.12	1.24	1.32	1.41	1.45	1.51

2.1 REVIEW OF LITERATURE

Stephen [4] presented the application of AHP technique in four steps, i.e., (a) shortlisting of the relevant criteria and alternatives; (b) estimation of criteria weights by pair-wise comparisons; (c) rating of each of the alternatives with respect to the considered criteria; and (d) computation of the overall weighted score for each alternative. A software prototype (AutoMan) was also developed to ease out the computational complexities related to AHP calculations. Alberto [5] selected the best location for a combined manufacturing and distribution facility based on several traditional as well as logistic service attributes using AHP method while taking into account the dynamic changes in business environmental conditions. Pineda-Henson et al. [6] demonstrated the applicability of AHP technique along with a life-cycle assessment approach for appraising the environmental performance of a pulp and paper manufacturing process with respect to environmental impacts and process improvement strategies. Yurdakul [7] developed the corresponding hierarchical decision structures for application of AHP technique for selection of machine tools in a manufacturing organization while modeling the links between machine tool alternatives and manufacturing strategy. Considering three alternative maintenance strategies, i.e., failure-based maintenance, time-based maintenance, and condition-based maintenance, Goossens & Basten [8] identified the most appropriate policy for maintaining naval ships using AHP technique. For that purpose, a five-step structured approach was proposed to provide a better understanding of the maintenance policy selection procedure.

Vaidya & Kumar [9] reviewed the applications of AHP technique in diverse domains, like forecasting, benefit-cost analysis, planning and development, allocation, medicine, quality function deployment (QFD), etc. Ayağ [10] integrated AHP technique with simulation to solve a machine tool selection problem while satisfying the needs and expectations of a manufacturing unit. The best machine tool was chosen based on the unit investment cost ratio, estimated after dividing the investment cost per year of each alternative by the additional number of units produced. Ahn & Choi [11] employed a simulation-based AHP (SiAHP) method for selection of the most appropriate ERP system under group decision-making environment for a Korean home-shopping company to enhance its profitability and effectiveness. Buyurgan & Saygin [12] suggested an AHP-based framework to aid in real-time scheduling and part routing choices based on pair-wise evaluations of various manufacturing systems. To aid dynamic short-term decision-making, the system incorporated an enhanced finite state machine and a scheduler model. It was finally tested in a simulated environment to evaluate how well it would perform over a variety of time horizons.

While applying AHP technique, Lin & Juang [13] attempted to choose the best green supplier keeping in mind the environmental protection concept within supplier evaluation processes in a biotechnology industry. Aguilar-Lasserre et al. [14] optimally designed batch plants having imprecise demands in product amounts, and developed a software based on the AHP technique to analyze genetic algorithm-based Pareto optimal solutions for the optimal batch plant design. While considering Kanban, CONWIP, and Hybrid as the three alternative production policies, and binomial, exponential, lognormal, and Poisson distributions to represent the probabilistic product demand situations, Sharma & Agrawal [15] applied AHP technique to choose the most apposite pull production control policy for a manufacturing system. Gupta et al. [16] considered four alternative manufacturing simulation software, i.e., NX-IDEAS, Star-CD, Micro Saint Sharp and ProModel, and applied the AHP technique to comparatively evaluate their relative performance while identifying the best software to fulfill the specified requirements.

Sipahi & Timor [17] critically reviewed the recent applications of AHP technique in manufacturing and service sectors and observed that its use had continued to increase exponentially during the considered review period. It was propounded that the adoption of AHP technique would help the decision-making community in arriving at the best courses of action. Ishizaka & Labib [18] extensively reviewed the applications of AHP technique with respect to problem modeling, pair-wise comparisons, judgment scales and many other parameters. Borchardt et al. [19] adopted AHP method to evaluate the degree of implementation of eco-design in three furniture manufacturing companies based on some specified eco-design constructs related to environmental practices. Subramanian & Ramanathan [20] reviewed the applications of AHP technique in operations strategy, product, and process design.

Using AHP technique, Dey & Cheffi [21] proposed a methodology for assessing the environmental performance of green manufacturing supply chains. Agarwal et al. [22] presented the application of AHP technique to help the decision makers in supplier evaluation and selection while considering the specific requirements of a manufacturing organization. Taking into account different sustainable manufacturing practices, Gupta et al. [23] compared the relative performance of some Indian electrical panel industries using AHP technique. Based on 12 design parameters, Jayant & Singh [24] applied AHP method to identify the best welding process to be adopted in a high-pressure welding environment. Alhuraish et al. [25] employed AHP technique to examine the relative performance of six sigma and lean strategies in different manufacturing organizations. Applying AHP technique, Yadav & Sharma [26] solved a supplier selection problem for an automobile industry under heterogeneous supply environments while considering 22 sub-criteria under six criteria. A sensitivity analysis study was also conducted to judge the effects of changing criteria weights on the final rankings of the considered suppliers. Durão et al. [27] considered several evaluation criteria, such as reliability, security, business, mobility, and heterogeneity, and applied the AHP method for selection of the most suitable internet of things (IoT) process in two manufacturing units. Deja et al. [28] applied AHP technique as a decision support tool to select the most effective production process between computer numerical control (CNC) milling and direct metal laser sintering for producing offshore equipment components. The relative performance of subtractive and additive manufacturing technologies was also compared. Using AHP method, Pandey & Litoriya [29] proposed a framework for identification of the most attractive software project among 20 competing alternatives. Horňáková et al. [30] presented the application of AHP method for selection of the most appropriate handling equipment for transfer of materials from the manufacturing area to warehouse, while providing due stresses on increased automation, enhanced productivity, and maximum safety.

Due to its simplicity and apprehensiveness, the possibility of the AHP technique to be integrated with other MCDM models has been explored since the two decades [31, 32]. In this direction, attempts have been put forward to find out answers to the following queries, like a) Which type of integrated AHP approaches has become more popular? b) In which application areas the integrated AHP approaches were applied to? c) For which specific problems AHP has been integrated with other models?

2.2 FMS SELECTION PROBLEM

A flexible manufacturing system (FMS) is a manufacturing technique that can quickly adjust to changes in the product's variety and quantity. It can substantially improve the efficiency and flexibility of a production system while decreasing production cost. It has now become a fundamental component of make-to-order approach that allows consumers to create the items they desire in today's production environment. An FMS is a set of networked processing workstations with computer terminals that control the whole production process, from loading and unloading to machining and assembly, storage, quality inspection, and data processing. Karsak & Kuzgunkaya [33], while suggesting a fuzzy multi-objective programming approach for selection of an FMS, considered eight alternative FMSs and seven criteria affecting the FMS selection process. These criteria are reduction in labor cost (RLC), reduction in work-in-process (RWP), reduction in setup cost (RSC), increase in market response (IMR), improvement in quality (IQ), capital and maintenance cost (CMC), and floor space used (FSU). Among these, five criteria are quantitative in nature and the remaining two are qualitative. Table 2.3 represents the performance characteristics of the considered FMSs with respect to all the evaluation criteria. The qualitative information of the two criteria, i.e., IMR and IQ are converted into appropriate quantitative data using a fuzzy conversion scale [34, 35]. The RLC, RWP, RSC, IMR, and IQ are beneficial attributes requiring higher values; on the other hand, CMC and FSU are non-beneficial attributes for which lower values are always preferable. The hierarchical structure for this FMS selection problem is shown in Figure 2.2.

TABLE 2.3 Quantitative Data for the FMS Selection Problem [34]

FMS	RLC (%)	RWP (%)	RSC (%)	IMR	IQ	CMC ($000)	FSU (sqft.)
1	30	23	5	0.745	0.745	1,500	5,000
2	18	13	15	0.745	0.745	1,300	6,000
3	15	12	10	0.500	0.500	950	7,000
4	25	20	13	0.745	0.745	1,200	4,000
5	14	18	14	0.255	0.745	950	3,500
6	17	15	9	0.745	0.500	1,250	5,250
7	23	18	20	0.500	0.745	1,100	3,000
8	16	8	14	0.255	0.500	1,500	3,000

Source: Reprinted with permission from Ref. [34]. © 2021 Taylor and Francis.

Now, setting the basic objective of this decision-making problem as the selection of the most apposite FMS for an organization so as to enhance its manufacturing effectiveness, all the seven evaluation criteria are pairwise compared based on the 1–9 scale to determine the corresponding criteria weights. Table 2.4 provides this comparison matrix. For a particular decision maker, RLC appears to have equal importance with respect to RWP, it is three times more important with respect to RSC, and so on. Following the procedural steps of the AHP method, this matrix is first normalized using Eqn. (2.2), and the corresponding criteria weights are determined as w_{RLC} = 0.1129, w_{RWP} = 0.1129, w_{RSC} = 0.0445, w_{IMR} = 0.1129, w_{IQ} = 0.2861, w_{CMC} = 0.2861 and w_{FSU} = 0.0445. It signifies that both IQ and CMC are the most important evaluation criteria for this problem. The corresponding λ_{max} value is estimated as 7.0879 which provides the CR value equal to 0.0104. As this CR value is less than the critical value of 0.10, it can be concluded that the judgments of the concerned decision maker while developing this comparison matrix are entirely consistent.

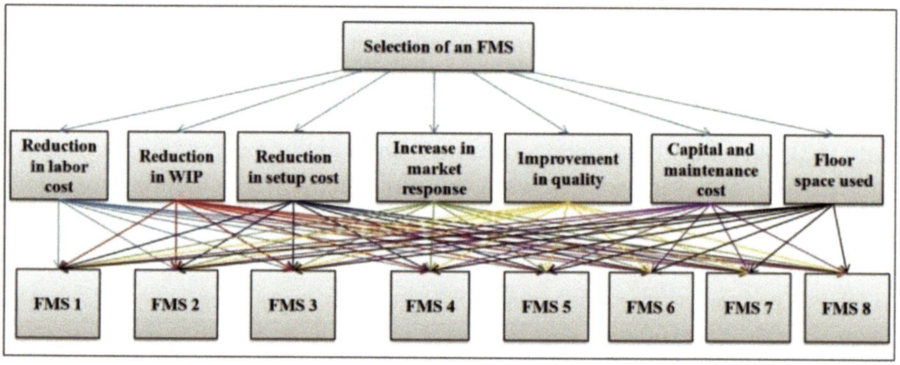

FIGURE 2.2 Hierarchical structure for the FMS selection problem.

TABLE 2.4 Pair-Wise Comparison Matrix between the FMS Evaluation Criteria

Criteria	RLC	RWP	RSC	IMR	IQ	CMC	FSU
RLC	1	1	3	1	1/3	1/3	3
RWP	1	1	3	1	1/3	1/3	3
RSC	1/3	1/3	1	1/3	1/5	1/5	1
IMR	1	1	3	1	1/3	1/3	3
IQ	3	3	5	3	1	1	5
CMC	3	3	5	3	1	1	5
FSU	1/3	1/3	1	1/3	1/5	1/5	1

In a similar direction, the performance of the FMS alternatives under consideration is now pair-wise contrasted with respect to seven FMS evaluation criteria, as shown in Tables 2.5–2.11. The priority weight vectors are computed along with the determination of the corresponding

CR values, which also signify consistency of the judgments. The elements of the composite priority weight vector for the FMS alternatives against the evaluation criteria are multiplied by the elements of the related criteria weight vector to determine the performance scores (P_i) of the candidate FMS alternatives, as provided in Table 2.12, indicating FMS 4 with the maximum performance score (P_i) of 0.2236 as the best-chosen alternative, followed by FMS 7. On the other hand, FMS 8 is the least preferred alternative for the said manufacturing organization. Conversely, using a decision-making technique based on combinatorial mathematics, Rao & Parnichkun [34] determined the ranking of the FMS alternatives as FMS 7 → FMS 4 → FMS 1 → FMS 2 → FMS 5 → FMS 6 → FMS 3 → FMS 8. It can be noticed that the top two and the last FMS alternatives match with the AHP-based ranking results.

TABLE 2.5 Pair-Wise Comparison Matrix with Respect to 'RLC' Criterion

FMS	FMS 1	FMS 2	FMS 3	FMS 4	FMS 5	FMS 6	FMS 7	FMS 8
FMS 1	1	2	1	1/5	1	2	1	2
FMS 2	1/2	1	3	1/3	1/2	1	1	1
FMS 3	1	1/3	1	1/3	1	3	1/3	2
FMS 4	5	3	3	1	2	1	1	1
FMS 5	1	2	1	1/2	1	1	1	1
FMS 6	1/2	1	1/3	1	1	1	1	2
FMS 7	1	1	3	1	1	1	1	1
FMS 8	1/2	1	1/2	1	1	1/2	1	1

TABLE 2.6 Pair-Wise Comparison Matrix with Respect to 'RWP' Criterion

FMS	FMS 1	FMS 2	FMS 3	FMS 4	FMS 5	FMS 6	FMS 7	FMS 8
FMS 1	1	2	1	1/5	1	2	1	2
FMS 2	1/2	1	3	1/3	1/2	1	1	1
FMS 3	1	1/3	1	1	1	2	1/3	2
FMS 4	5	3	1	1	2	1	1	1
FMS 5	1	2	1	1/2	1	1	1	1
FMS 6	1/2	1	1/2	1	1	1	1	2
FMS 7	1	1	3	1	1	1	1	1
FMS 8	1/2	1	1/2	1	1	1/2	1	1

TABLE 2.7 Pair-Wise Comparison Matrix with Respect to 'RSC' Criterion

FMS	FMS 1	FMS 2	FMS 3	FMS 4	FMS 5	FMS 6	FMS 7	FMS 8
FMS 1	1	2	1	1/5	1	2	1	2
FMS 2	1/2	1	3	1/3	1/2	1	1	1
FMS 3	1	1/3	1	1/3	1	2	1/3	3
FMS 4	5	3	3	1	3	1	1	1
FMS 5	1	2	1	1/3	1	1	1	1
FMS 6	1/2	1	1/2	1	1	1	1	2
FMS 7	1	1	3	1	1	1	1	1
FMS 8	1/2	1	1/3	1	1	1/2	1	1

TABLE 2.8 Pair-Wise Comparison Matrix with Respect to 'IMR' Criterion

FMS	FMS 1	FMS 2	FMS 3	FMS 4	FMS 5	FMS 6	FMS 7	FMS 8
FMS 1	1	2	1	1/5	1	2	2	2
FMS 2	1/2	1	3	1/3	1/2	1	1	1
FMS 3	1	1/3	1	1/3	1	2	1/3	2
FMS 4	5	3	3	1	2	1	1	1
FMS 5	1	2	1	1/2	1	1	1	1
FMS 6	1/2	1	1/2	1	1	1	1	2
FMS 7	1/2	1	3	1	1	1	1	1
FMS 8	1/2	1	1/2	1	1	1/2	1	1

TABLE 2.9 Pair-Wise Comparison Matrix with Respect to 'IQ' Criterion

FMS	FMS 1	FMS 2	FMS 3	FMS 4	FMS 5	FMS 6	FMS 7	FMS 8
FMS 1	1	2	1	1/5	1	2	1	2
FMS 2	1/2	1	3	1/3	1/2	1	1	1
FMS 3	1	1/3	1	1/3	1	2	1/3	3
FMS 4	5	3	3	1	3	1	1	1
FMS 5	1	2	1	1/3	1	1	1	1
FMS 6	1/2	1	1/2	1	1	1	1/3	2
FMS 7	1	1	3	1	1	3	1	1
FMS 8	1/2	1	1/3	1	1	1/2	1	1

TABLE 2.10 Pair-Wise Comparison Matrix with Respect to 'CMC' Criterion

FMS	FMS 1	FMS 2	FMS 3	FMS 4	FMS 5	FMS 6	FMS 7	FMS 8
FMS 1	1	2	1	1/5	1	2	1	2
FMS 2	1/2	1	3	1/2	1/2	1	1	1
FMS 3	1	1/3	1	1/3	1	2	1/2	3
FMS 4	5	2	3	1	2	1	1	1
FMS 5	1	2	1	1/2	1	1	1	1
FMS 6	1/2	1	1/2	1	1	1	1/3	2
FMS 7	1	1	2	1	1	3	1	1
FMS 8	1/2	1	1/3	1	1	1/2	1	1

TABLE 2.11 Pair-Wise Comparison Matrix with Respect to 'FSU' Criterion

FMS	FMS 1	FMS 2	FMS 3	FMS 4	FMS 5	FMS 6	FMS 7	FMS 8
FMS 1	1	2	1	1/5	1	2	1	2
FMS 2	1/2	1	3	1/3	1/2	1	1	1
FMS 3	1	1/3	1	1/3	1	3	1/3	3
FMS 4	5	3	3	1	2	1	1	1
FMS 5	1	2	1	1/2	1	1	1	1
FMS 6	1/2	1	1/3	1	1	1	1/3	1
FMS 7	1	1	3	1	1	3	1	1
FMS 8	1/2	1	1/3	1	1	1	1	1

$$\begin{bmatrix} 0.1278 & 0.1300 & 0.1268 & 0.1410 & 0.1258 & 0.1280 & 0.1265 \\ 0.0981 & 0.0997 & 0.0973 & 0.0980 & 0.0965 & 0.1040 & 0.0971 \\ 0.0964 & 0.1083 & 0.0957 & 0.0909 & 0.0949 & 0.1023 & 0.1011 \\ 0.2273 & 0.1977 & 0.2390 & 0.2272 & 0.2371 & 0.2148 & 0.2250 \\ 0.1195 & 0.1216 & 0.1119 & 0.1194 & 0.1110 & 0.1197 & 0.1183 \\ 0.1022 & 0.1101 & 0.1074 & 0.1082 & 0.0911 & 0.0927 & 0.0783 \\ 0.1398 & 0.1422 & 0.1387 & 0.1266 & 0.1610 & 0.1546 & 0.1620 \\ 0.0888 & 0.0903 & 0.0832 & 0.0887 & 0.0825 & 0.0839 & 0.0916 \end{bmatrix} \times \begin{bmatrix} 0.1129 \\ 0.1129 \\ 0.0445 \\ 0.1129 \\ 0.2861 \\ 0.2861 \\ 0.0445 \end{bmatrix}^{T} = \begin{bmatrix} 0.1289 \\ 0.0994 \\ 0.0986 \\ 0.2236 \\ 0.1170 \\ 0.0970 \\ 0.1498 \\ 0.0856 \end{bmatrix}$$

TABLE 2.12 Performance Scores of the FMS Alternatives

FMS	1	2	3	4	5	6	7	8
P_i	0.1289	0.0994	0.0986	0.2236	0.1170	0.0970	0.1498	0.0856
Rank	3	5	6	1	4	7	2	8

KEYWORDS

- **analytic hierarchy process**
- **consistency index**
- **consistency ratio**
- **quality function deployment**
- **random index**
- **simulation-based AHP**

REFERENCES

1. Saaty, T. L., (1980). *The Analytic Hierarchy Process.* McGraw-Hill Company, New York.
2. Saaty, T. L., (1977). A scaling method for priorities in hierarchical structures. *Journal of Mathematical Psychology, 15*(3), 234–281.
3. Saaty, T. L., (2003). Decision-making with the AHP: Why is the principal eigenvector necessary. *European Journal of Operational Research, 145,* 85–91.
4. Stephen, F., (1993). A modified analytic hierarchy process for automated manufacturing decisions. *Interfaces, 23*(4), 75–84.
5. Alberto, P., (2000). The logistics of industrial location decisions: An application of the analytic hierarchy process methodology. *International Journal of Logistics, 3*(3), 273–289.
6. Pineda-Henson, R., Culaba, A. B., & Mendoza, G. A., (2002). Evaluating environmental performance of pulp and paper manufacturing using the analytic hierarchy process and life-cycle assessment. Journal of Industrial Ecology, 6(1), 15–28.
7. Yurdakul, M., (2004). AHP as a strategic decision-making tool to justify machine tool selection. *Journal of Materials Processing Technology, 146,* 365–376.
8. Goossens, A. J. M., & Basten, R. J. I., (2005). Exploring maintenance policy selection using the analytic hierarchy process; An application for naval ships. *Reliability Engineering and System Safety, 142,* 31–41.
9. Vaidya, O. S., & Kumar, S., (2006). Analytic hierarchy process: An overview of applications. *European Journal of Operational Research,* 169, 1–29.
10. Ayağ, Z., (2007). A hybrid approach to machine-tool selection through AHP and simulation. *International Journal of Production Research, 45*(9), 2029–2050.
11. Ahn, B. S., & Choi, S. H., (2008). ERP system selection using a simulation based AHP approach: A case of Korean home shopping company. *Journal of the Operational Research Society, 59,* 322–330.
12. Buyurgan, N., & Saygin, C., (2008). Application of the analytical hierarchy process for real-time scheduling and part routing advanced manufacturing systems. *Journal of Manufacturing Systems, 27,* 101–110.
13. Lin, S. S., & Juang, Y. S., (2008). Selecting green suppliers with analytic hierarchy process for biotechnology industry. *Operations and Supply Chain Management, 1*(2), 115–129.

14. Aguilar-Lasserre, A. A., Bautista, M. A. B., Ponsich, A., & Huerta, M. A. G., (2009). An AHP-based decision-making tool for the solution of multiproduct batch plant design problem under imprecise demand. *Computers & Operations Research, 36*, 711–736.
15. Sharma, S., & Agrawal, N., (2009). Selection of a pull production control policy under different demand situations for a manufacturing system by AHP-algorithm. *Computers & Operations Research, 36*, 1622–1632.
16. Gupta, A., Singh, K., & Verma, R., (2010). A critical study and comparison of manufacturing simulation software using analytic hierarchy process. *Journal of Engineering Science and Technology, 5*(1), 108–129.
17. Sipahi, S., & Timor, M., (2010). The analytic hierarchy process and analytic network process: An overview of applications. *Management Decision, 48*(5), 775–808.
18. Ishizaka, A., & Labib, A., (2011). Review of the main developments in the analytic hierarchy process. *Expert Systems with Applications, 38*(11), 14336–14345.
19. Borchardt, M., Sellitto, M. A., Pereira, G. M., & Gomes, L. P., (2012). Ecodesign case studies for furniture companies using the analytic hierarchy process. *International Journal of Industrial Engineering, 19*(8), 330–340.
20. Subramanian, N., & Ramanathan, R., (2012). A review of applications of analytic hierarchy process in operations management. *International Journal of Production Economics, 138*, 215–241.
21. Dey, P. K., & Cheffi, W., (2013). Green supply chain performance measurement using the analytic hierarchy process: A comparative analysis of manufacturing organizations. *Production Planning & Control, 24*(8, 9), 702–720.
22. Agarwal, P., Sahai, M., Mishra, V., Bag, M., & Singh, V., (2014). Supplier selection in dynamic environment using analytic hierarchy process. *International Journal of Information Engineering and Electronic Business, 4*(1), 20–26.
23. Gupta, S., Dangayach, G. S., Singh, A. K., & Rao, P. N., (2015). Analytic hierarchy process (AHP) model for evaluating sustainable manufacturing practices in Indian electrical panel industries. *Procedia - Social and Behavioral Sciences, 189*, 208–216.
24. Jayant, A., & Singh, M., (2015). Use of analytic hierarchy process (AHP) to select welding process in high pressure vessel manufacturing environment. *International Journal of Applied Engineering Research, 10*(8), 5869–5884.
25. Alhuraish, I., Robledo, C., & Kobi, A., (2016). Assessment of lean manufacturing and six sigma operation with decision making based on the analytic hierarchy process. *IFAC-Papers OnLine, 49*(12), 59–64.
26. Yadav, V., & Sharma, M. K., (2016). Multi-criteria supplier selection model using the analytic hierarchy process approach. *Journal of Modeling in Management, 11*(1), 326–354.
27. Durão, L. F. C. S., Carvalho, M. M., Takey, S., Cauchick-Miguel, P. A., & Zancul, E., (2018). Internet of things process selection: AHP selection method. *International Journal of Advanced Manufacturing Technology, 99*, 2623–2634.
28. Deja, M., Siemiątkowski, M. S., & Zieliński, D., (2020). Multi-criteria comparative analysis of the use of subtractive and additive technologies in the manufacturing of offshore machinery components. *Polish Maritime Research, 27*(3), 71–81.
29. Pandey, P., & Litoriya, R., (2020). Software process selection system based on multi-criteria decision making. *Journal of Software Evolution and Process, e2305*, 10. https://doi.org/10.1002/smr.2305.

30. Horňáková, N., Jurík, L., Chovanová, H. H., Cagáňová, D., & Babčanová, D., (2021). AHP method application in selection of appropriate material handling equipment in selected industrial enterprise. *Wireless Networks, 27*, 1683–1691.
31. Ho, W., (2008). Integrated analytic hierarchy process and its applications: A literature review. *European Journal of Operational Research, 186*(1), 211–228.
32. Ho, W., & Ma, X., (2018). The state-of-the-art integrations and applications of the analytic hierarchy process. *European Journal of Operational Research, 267*, 399–414.
33. Karsak, E. E., & Kuzgunkaya, O., (2002). A fuzzy multiple objective programming approach for the selection of a flexible manufacturing system. *International Journal of Production Economics, 79*, 101–111.
34. Rao, R. V., & Parnichkun, M., (2009). Flexible manufacturing system selection using a combinatorial mathematics-based decision-making method. *International Journal of Production Research, 47*, 6981–6998.
35. Rao, R. V., (2007). *Decision Making in the Manufacturing Environment Using Graph Theory and Fuzzy Multi-Attribute Decision Making Methods*. Springer-Verlag, London.

CHAPTER 3

ANALYTIC NETWORK PROCESS (ANP)

The AHP and ANP are the two most popularly employed MCDM techniques to aid the decision-makers in selecting the best choice under situations characterized by multiple conflicting criteria. Although the AHP method, developed by Saaty [1], has been successfully applied to different situations with reasonably satisfactory results, but it fails to handle the intricacies of a wide range of real-time decision-making challenges because of its strictly hierarchical structure. In order to deal with those complex problems, Saaty [2, 3] proposed ANP approach to depict a decision-making issue as a network of criteria and alternatives that are clustered. In ANP, all of the elements in the network can be linked in any feasible way, signifying that a network can incorporate feedback and interdependent relationships within and between clusters. A supermatrix can be formulated to elucidate the effects of inter-network elements. It yields a two-dimensional element-by-element matrix that modifies the relative weights in individual pair-wise comparison matrixes to generate a new overall supermatrix with the adjusted weights. The ANP is a generic form of AHP to allow more complex interdependence relationships among different elements. It is also called a system with feedback approach.

The ANP provides a broad framework for dealing with decision-making problems that does not rely on assumptions about the independence of higher-level components from lower-level elements or independence of elements within a level. It works on the basis of a network, but without the necessity for levels as in a hierarchy. The essential notion of ANP technique is influence. It consists of two components: a) a control hierarchy or network of criteria and sub-criteria that regulates interactions, and b) a network of influences between elements and clusters. The network topology differs from criterion to criterion, and each control criterion has its own supermatrix of limiting impact. Finally, the priority of each control criterion is considered to weigh each of these supermatrixes.

Multi-Criteria Decision-Making Methods in Manufacturing Environments: Models and Applications.
Shankar Chakraborty, Prasenjit Chatterjee, & Partha Protim Das (Authors)

Saaty [2, 3] proposed the following four main steps of ANP method, presented as below:

- **Step 1: Construction of the Model and Problem Structuring:** The considered decision-making problem needs to be clearly stated and decomposed into a rational system in the form of a network. The framework can be structured based on the decision maker's opinion.
- **Step 2: Development of Pair-Wise Comparison Matrixes and Priority Vectors:** Like AHP, pairs of decision elements in each cluster are compared in respect of how important they are in their control criteria. The clusters themselves are compared pair by pair in terms of their contribution to the goal or objective. The decision maker responds to a series of pair-wise comparisons of two elements or two clusters that must be evaluated in regard to their contribution to their respective upper-level criteria. Furthermore, interdependencies among cluster elements must be studied pairwise, with each element's effect on other elements represented by an eigenvector.

  The relative importance values are calculated using Saaty's 1–9 scale, where 1 indicates equal significance between two elements and 9 shows extreme importance of one element compared to the other. The inverse comparison is provided by the corresponding reciprocal value, i.e., $a_{ij} = 1/a_{ji}$, where a_{ij} reflects the relative significance of ith element over j^{th} one. A local priority vector can be derived as an estimate of the relative importance associated with the elements (or clusters) being compared by solving the following equation:

$$A \times w = \lambda_{max} \times w \tag{3.1}$$

  where; A is the matrix of pair-wise comparison, w is the eigenvector; and λ_{max} is the maximum eigenvalue of A. If A denotes a consistency matrix, then the eigenvector X can be determined using the following expression:

$$(A - \lambda_{max} I)X = 0 \tag{3.2}$$

  where; I is an identity matrix. Similar to AHP method, values of CI and CR are determined in order to verify the consistency of the pair-wise comparison matrix. From the pair-wise comparison matrix, the relative priorities or relative weights of the components being compared with their upper-level criteria are estimated. Priority vectors need to be determined for all the comparison matrixes.

- **Step 3: Supermatrix Formation:** The local *priori* is entered in the relevant columns of a matrix to derive global priorities in a system with interdependent impacts or influences. Consequently, a supermatrix, also called a partitioned matrix, is formed, with each matrix segment representing a connection between two clusters in a system. Let C_k ($k = 1, 2, \ldots, n$) be the clusters of a decision system, and each cluster k has m_k elements, represented by $e_{k1}, e_{k2}, \ldots, e_{kmk}$. The local priority vectors, acquired in Step (2), are aggregated and arranged in suitable locations in a supermatrix depending on the flow of influence from one cluster to another or from itself. The following is an example of a supermatrix [2].

$$W = \begin{array}{c c} & \begin{array}{ccccccc} & C_1 & \cdots & C_k & \cdots & C_n & \\ e_{11} \cdots e_{1\,n1} & \cdots & e_{k1} \cdots e_{k\,nk} & \cdots & e_{n1} \cdots e_{n\,nn} \end{array} \\ \begin{array}{cc} C_1 & \begin{array}{c} e_{11} \\ \vdots \\ e_{1\,n1} \end{array} \\ \vdots & \vdots \\ C_k & \begin{array}{c} e_{k1} \\ \vdots \\ e_{k\,nk} \end{array} \\ \vdots & \vdots \\ C_n & \begin{array}{c} e_{n1} \\ \vdots \\ e_{n\,nn} \end{array} \end{array} & \begin{bmatrix} W_{11} & \cdots & W_{1k} & \cdots & W_{1n} \\ \vdots & & \vdots & & \vdots \\ W_{k1} & \cdots & W_{kk} & \cdots & W_{kn} \\ \vdots & & \vdots & & \vdots \\ W_{n1} & \cdots & W_{nk} & \cdots & W_{nn} \end{bmatrix} \end{array} \tag{3.3}$$

As an example, the supermatrix representation for a hierarchy with three levels, as shown in Figure 3.1(a), can be given as below:

$$W_h = \begin{bmatrix} 0 & 0 & 0 \\ w_{21} & 0 & 0 \\ 0 & W_{32} & I \end{bmatrix} \tag{3.4}$$

In this matrix, w_{21} is a vector, which represents the impact of the goal on the criteria. On the other hand, in W_{32} matrix, the entries are arranged according to the influence of each criterion on the considered alternatives and zero entries represent those elements having no influence. When the criteria are interrelated in the given example, Figure 3.2(b) shows a network instead of a hierarchy. A matrix element W_{22} of the supermatrix reveals the interdependency in the following matrix:

$$W_n = \begin{bmatrix} 0 & 0 & 0 \\ w_{21} & W_{22} & 0 \\ 0 & W_{32} & I \end{bmatrix} \tag{3.5}$$

If there is an association or interrelationship between the components inside a cluster or between two clusters, a matrix can replace any zero value in the supermatrix. As the clusters in a network are normally interdependent, the columns of a supermatrix may add to more than one. The supermatrix, on the other hand, must be changed such that each column of the matrix adds to one.

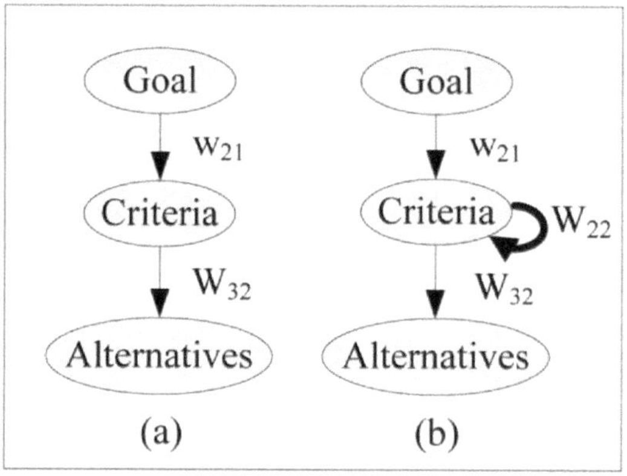

FIGURE 3.1 Hierarchy and network.

- **Step 4: Selection of the Best Candidate Alternative:** The priority weights of the alternatives may be found in the column of alternatives in the normalized supermatrix, if the supermatrix, developed in Step (3), considers the whole network. If the supermatrix contains only interconnected clusters, however, additional calculations are required to determine the overall priorities of the alternatives. The best alternative should be chosen as the one with the highest overall priority.

In any decision-making process, it is very important to consider the interdependent relationship among criteria that exists in many real-time problems The AHP method, on the other hand, is limited in its ability to resolve problems with a linear unidirectional hierarchical connection among the criteria. This strict hierarchical structure is not required for ANP, and consequently, it can deal with problems having complex interrelationships among criteria (dependencies and feedbacks) as often encountered while making societal, governmental, and corporate decisions. It may allow the inclusion of both physical and intangible variables that are relevant in making optimal decisions. The flexible network structure of ANP makes it possible to represent any decision problem without concern for what criteria comes first and what comes next as in a hierarchy. The application of ANP is unique in nature as it generates synthetic scores, measuring the relative importance of several alternatives accessible to the decision maker. Due to its complexity and considerable computing effort, ANP applications have been considerably limited in comparison to AHP.

3.1 SURVEY OF THE LITERATURE

Meade & Presley [4] first stressed on the requirements for selection of R&D projects and later applied ANP technique to solve a R&D project selection problem while taking into account the considered organization's strategic objectives and structure by integrating financial and strategic benefits of each alternative project. Yurdakul [5] proposed the application of ANP technique to provide a multicriteria performance measurement model for a manufacturing organization. The developed model would incorporate the competitive strategies and interdependence between various criteria in a hierarchical structure while achieving a more realistic representation of the organizational performance. Bayazit [6] considered a supplier selection problem having interdependent qualitative and quantitative criteria and solved the same while employing ANP method. Jharkharia & Shankar [7] first shortlisted a set of logistic service providers and later applied ANP method for final selection of the most apposite service provider. Based on the available data from a medium-sized valve manufacturer, Anand & Kodali [8] applied the ANP technique for the implementation of a lean manufacturing system while analyzing the capabilities, practices, tools, and techniques of the candidate alternatives besides investigating their influences on the final selection decision.

Using the ANP technique, Hsu & Hu [9] considered the problem of hazardous substance management in supplier selection and identified the best supplier in a situation having interdependencies among the decision criteria while fulfilling the requirements of hazardous substance minimization for environmental regulations. Kodali & Anand [10] adopted ANP method for effective design of an FMS from a pool of alternative configurations. Flexible production system emerged out as the best FMS configuration. Sipahi & Timor [11] conducted a survey of literature on the applications of ANP technique. It was concluded that ANP had become quite popular among the decision-making community for its capability to deal with interdependency and interrelationship among various quantitative as well as qualitative evaluation criteria. Using ANP technique, Anand et al. [12] identified the most suitable material handling system during FMS design while considering the existent dependencies among various criteria. Based on ANP technique, Das & Chakraborty [13] developed a software prototype in Visual Basic with a suitable graphical user interface for identification of the most appropriate non-traditional machining (NTM) process for a given shape feature and work material combination.

Anand et al. [14] considered 40 interdependent and interrelated evaluation criteria for arriving at the most appropriate plant location selection decision while employing ANP technique. Its ranking performance was also contrasted with other popular MCDM approaches. Taking into account seven interdependent criteria, Giner-Santonja et al. [15] applied the ANP method to evaluate the best available technique for improving implementation of an integrated pollution prevention and control plan. Using the ANP technique, Ordoobadi [16] developed a decision tool to help the policy makers to have more informed decisions regarding their investments in advanced manufacturing technologies. It was emphasized that the selection of a new technology was multidimensional, involving interdependent relationships among various qualitative and quantitative criteria. Govindan et al. [17] presented the multi-step application of ANP method for identification of the most appropriate third-party reverse logistic provider (3PRLP) for an automobile manufacturing unit. Al-Hawari et al. [18] adopted the ANP technique for the selection of the best facility layout plan taking into account multiple dependent and independent criteria having interactions among themselves. Kodali et al. [19] considered a hypothetical case study to demonstrate the application of ANP technique to select the best cellular manufacturing system while taking into account interdependency among different evaluation criteria. Van Horenbeek & Pintelon [20] proposed the application of ANP method based on a maintenance performance measurement framework for the development of appropriate maintenance performance indicators to assist the maintenance managers while fulfilling the corporate objectives and strategies of an organization.

Ocampo et al. [21] first aimed at identifying the content of a sustainable manufacturing strategy along with sustainability and later presented the application of ANP method in a complex manufacturing strategy formulation process. The robustness of the derived results was also validated using Monte Carlo simulation. Employing the ANP technique, Barzegar et al. [22] endeavored to identify the most important drivers for the adoption of green manufacturing concept in small- and medium-sized manufacturing organizations. Kurian et al. [23] demonstrated the application of ANP technique as an efficient MCDM tool for selection of the most appropriate maintenance strategy for a typical manufacturing organization. The influences of various evaluation criteria on the selection decision were also studied. Asadabadi et al. [24] critically reviewed the application potentialities of both the AHP and ANP techniques. It was mentioned that the ranking results provided by AHP method would not be sometimes acceptable by the rational decision maker due to its simple hierarchical structure. Kheybari et al. [25] extensively reviewed the applications of ANP technique in diverse domains considering complex and interrelated relationships between different quantitative and qualitative evaluation criteria. It was observed that this method was mainly employed in eight areas, i.e.: (a) health, safety, and environmental management; (b) water management; (c) financial management; (d) human resources management; (e) tourism management; (f) logistics and supply chain management; (g) engineering and manufacturing systems; and (h) energy management.

3.2 GREEN SUPPLIER SELECTION PROBLEM

In today's highly competitive environment, a manufacturing organization must focus on satisfying its customers with low-cost products having high quality and prompt services. Thus, the production system is to be designed so as to decrease the related manufacturing cost, increase its flexibility, and meet the quality standards. As the raw materials are generally transformed into final products in any of the manufacturing systems, any deviation in the quality of the input materials may cause the end product's quality to deteriorate.

Thus, in supply chain management, selection of the appropriate suppliers and evaluation of their performance are identified as one the crucial strategic issues for overall survival of the concerned organization. To fulfill its long-term objective and enhance the efficiency of the supply chain, the selection of the most reliable suppliers for varying input materials has been recognized to be of immense importance. However, the industrial organizations must now pay greater attention to a variety of environmental concerns mandated by the concerned governments. The continual depletion of raw materials, increased pollution and greenhouse gas emissions are the primary causes of these environmental challenges.

Those organizations must streamline their manufacturing processes to minimally affect the environment. This can only be accomplished by enhancing a green manufacturing system that includes green suppliers across the supply chain. Customers are now being more focused on obtaining better ecologically sensitive items, in addition to their basic demand of low-cost, high-quality products. With respect to environmental friendliness, green supply chain has been observed to be extremely important [26]. Besides the environmental awareness, the notion of green suppliers should also encompass management and organizational methods. Effective green supplier selection is also influenced by government legislation, social duty, consumer demand and business rewards. Selection of the most appropriate green supplier based on a variety of green characteristics often helps the concerned manufacturing organization to outperform its competitors. It not only influences the organization's profitability and competitiveness, but also improves the performance of the whole supply chain. A wrong green supplier selection decision may adversely affect the health of the organization as well as its goodwill.

Green supplier selection has now become essential in today's manufacturing environment considering increasing pollution levels worldwide, which can be attributed to increased consumption as well as revolutions in manufacturing. Manufacturers as well as suppliers are more worried about the environmental effect of the products they make and consume as public knowledge and attention on climate change grows. Consumers' demand for environmentally friendly products, as well as manufacturers' attention to carbon footprint and notion of sustainable development has prompted the manufacturers to rethink and refocus their manufacturing processes, beginning with raw material acquisition. As a result, in today's highly competitive industrial environment, selecting suppliers with a shared vision of eco-friendliness and 'green' thinking has become critical. It not only helps in winning the battle against pollution, but also aids the organizations in developing green marketing strategies while enhancing public perception and trust in their products.

For any manufacturing organization, selection of the most apposite green supplier is a complex decision-making task due to the involvement of various experts (decision makers) from different related departments, like procurement, planning, production, quality control, etc. It has been identified as an MCDM problem where the best supplier needs to be identified in the presence of a set of conflicting criteria [27]. In green supplier selection process, a group of experts having dissimilar backgrounds, experiences, expertise, and stature usually participate. They frequently express their subjective opinions on the relative performance of the possible green suppliers based on a variety of evaluation criteria. It has also been observed that in a typical green supplier selection problem, there exist interrelationships between different evaluation criteria which need to be taken into consideration before arriving at the most accurate decision.

In this illustrative example, a green supplier selection problem related to a mining industry [28] is solved while employing ANP technique. In this problem, all the considered evaluation criteria are first divided into two groups, i.e., benefit factors and risk factors. Benefit factors include high level of flexibility (HLF) (level of flexibility with less restriction to sustain the relationship with the core organization), green technology capability (GTC) (using green concept, planning to reduce energy consumption and greenhouse gas emission control), R&D capability (RDC) (engagement of skilled manpower and technologies to further develop the process and access to the innovations), higher level of reputation (HLR) (with green production capability, ISO standard, financial and management strength, reduction of liability and risk) and cost reduction (RC) (ability to reduce cost involved in the mining operations). On the other hand, the risk factors consist of demand fulfillment (DF) (inability to meet demand fluctuations), environmental training (ET) (lack of process-based ET to the employees), occupational health and safety (OHS) (lack of occupational health and safety policy implementation) and environmental activity control (EAC) (inability to control hazardous environmental activity). To have reliable results for this green supplier selection problem, it is assumed that all the evaluation criteria under benefit factors are dependent to each other, i.e., all these factors influence each other. Similarly, under the risk factors, DF is treated as an independent criterion, whereas, the remaining three criteria are dependent. The network structure for this green supplier selection problem for the mining industry is exhibited in Figure 3.2.

Now, in order to represent the inner dependency of the benefit factors with respect to each of the considered evaluation criteria, the corresponding pair-wise comparison matrixes of Tables 3.1–3.5 are developed along with the determination of the related priority weights. In a similar direction, Tables 3.6–3.8 show the pair-wise comparison matrixes highlighting the inner-dependency of the risk factors with respect to each of the selection criteria (DF is treated as an independent criterion).

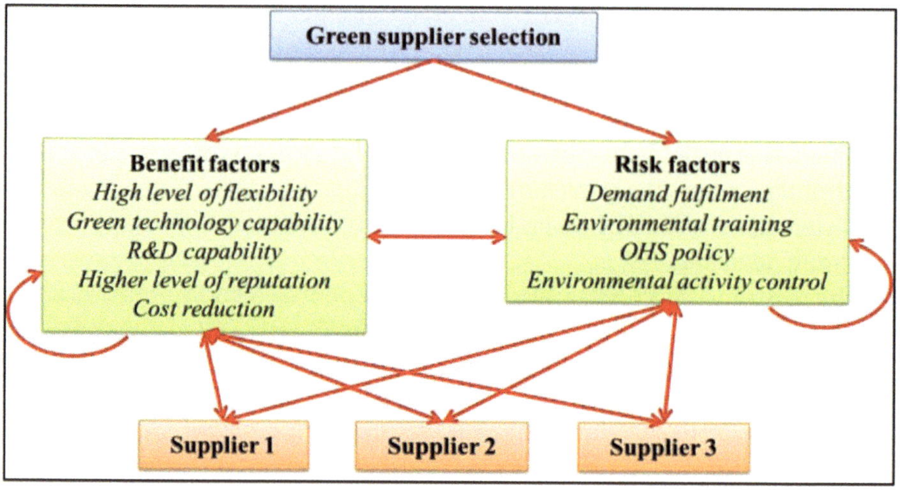

FIGURE 3.2 Network structure for the green supplier selection problem.

TABLE 3.1 Inner-Dependence Matrix of Benefit Factors with Respect to HRL

Benefit Factor	GTC	RDC	HLR	RC	Priority Weight
GTC	1	1/2	1/5	1/4	0.081
RDC	2	1	1/3	1/5	0.154
HLR	5	3	1	1/3	0.476
RC	4	5	3	1	0.288

TABLE 3.2 Inner-Dependence Matrix of Benefit Factors with Respect to GTC

Benefit Factor	HLF	RDC	HLR	RC	Priority Weight
HLF	1	5	3	2	0.483
RDC	1/5	1	1/3	1/2	0.087
HLR	1/3	3	1	2	0.252
RC	1/2	2	1/2	1	0.178

TABLE 3.3 Inner-Dependence Matrix of Benefit Factors with Respect to RDC

Benefit factor	HLF	GTC	HLR	RC	Priority Weight
HLF	1	1/3	1/3	1/3	0.098
GTC	3	1	2	4	0.462
HLR	3	1/2	1	2	0.268
RC	3	1/4	1/2	1	0.172

TABLE 3.4 Inner-Dependence Matrix of Benefit Factors with Respect to HLR

Benefit factor	HLF	GTC	RDC	RC	Priority Weight
HLF	1	1/5	1/5	1/3	0.067
GTC	5	1	1/3	3	0.291
RDC	5	3	1	3	0.491
RC	3	1/3	1/3	1	0.151

TABLE 3.5 Inner-Dependence Matrix of Benefit Factors with Respect to RC

Benefit Factor	HLF	GTC	RDC	HLR	Priority Weight
HLF	1	5	3	3	0.505
GTC	1/5	1	1/3	1/2	0.087
RDC	1/3	3	1	1/3	0.164
HLR	1/3	2	3	1	0.244

TABLE 3.6 Inner-Dependence Matrix of Risk Factors with Respect to ET

Risk Factor	OHS	EAC	Priority Weight
OHS	1	4	0.8
EAC	1/4	1	0.2

TABLE 3.7 Inner-Dependence Matrix of Risk Factors with Respect to OHS

Risk Factor	ET	EAC	Priority Weight
ET	1	1/5	0.167
EAC	5	1	0.833

TABLE 3.8 Inner-Dependence Matrix of Risk Factors with Respect to EAC

Risk Factor	ET	OHS	Priority Weight
ET	1	2	0.667
OHS	1/2	1	0.333

Tables 3.9–3.17 provide the priority weights when the relative performance of all the three green suppliers under consideration is pairwise compared with respect to each of the criteria under both the benefit and risk factors.

TABLE 3.9 Pair-Wise Comparison of Alternatives with Respect to HLF

Supplier	S_1	S_2	S_3	Priority Weight
S_1	1	5	3	0.633
S_2	1/5	1	1/3	0.106
S_3	1/3	3	1	0.26

TABLE 3.10 Pair-Wise Comparison of Alternatives with Respect to GTC

Supplier	S_1	S_2	S_3	Priority Weight
S_1	1	1/3	3	0.272
S_2	3	1	4	0.608
S_3	1/3	1/4	1	0.12

TABLE 3.11 Pair-Wise Comparison of Alternatives with Respect to RDC

Supplier	S_1	S_2	S_3	Priority Weight
S_1	1	1/5	1/2	0.118
S_2	5	1	4	0.681
S_3	2	1/4	1	0.201

TABLE 3.12 Pair-Wise Comparison of Alternatives with Respect to HLR

Supplier	S_1	S_2	S_3	Priority Weight
S_1	1	1/3	1/2	0.164
S_2	3	1	2	0.539
S_3	2	1/2	1	0.297

TABLE 3.13 Pair-Wise Comparison of Alternatives with Respect to RC

Supplier	S_1	S_2	S_3	Priority Weight
S_1	1	4	2	0.571
S_2	1/4	1	1/2	0.143
S_3	1/2	2	1	0.286

TABLE 3.14 Pair-Wise Comparison of Alternatives with Respect to DF

Supplier	S_1	S_2	S_3	Priority Weight
S_1	1	2	3	0.539
S_2	1/2	1	2	0.297
S_3	1/3	1/2	1	0.164

TABLE 3.15 Pair-Wise Comparison of Alternatives with Respect to ET

Supplier	S_1	S_2	S_3	Priority Weight
S_1	1	1/5	1/2	0.122
S_2	5	1	3	0.648
S_3	2	1/3	1	0.23

TABLE 3.16 Pair-Wise Comparison of Alternatives with Respect to OHS

Supplier	S_1	S_2	S_3	Priority Weight
S_1	1	1/4	1/2	0.137
S_2	4	1	3	0.623
S_3	2	1/3	1	0.239

TABLE 3.17 Pair-Wise Comparison of Alternatives with Respect to EAC

Supplier	S_1	S_2	S_3	Priority Weight
S_1	1	1/6	1/3	0.096
S_2	6	1	3	0.653
S_3	3	1/3	1	0.251

In a similar direction, all the benefit factors are again pairwise compared with regard to each of the considered green suppliers in Tables 3.18–3.20. On the other hand, Tables 3.21–3.23 exhibit the corresponding pair-wise comparison matrixes when all the risk factors are compared among themselves with respect to each of the green suppliers under consideration. In order to adopt green supply chain in this mining industry, amongst all the benefit criteria, the capability of the suppliers in establishing green technology (criterion GTC) plays the most important role. Hence, the performance of the risk factors (keeping DF as an independent criterion) is appraised with respect to criterion GTC in Table 3.24.

TABLE 3.18 Pair-Wise Comparison of Benefit Factors with Respect to Supplier 1

Benefit Factor	HLF	GTC	RDC	HLR	RC	Priority Weight
HLF	1	5	4	3	2	0.394
GTC	1/5	1	3	2	1/5	0.126
RDC	1/4	1/3	1	1/2	1/4	0.065
HLR	1/3	1/2	2	1	1/4	0.099
RC	1/2	5	4	4	1	0.316

TABLE 3.19 Pair-Wise Comparison of Benefit Factors with Respect to Supplier 2

Benefit Factor	HLF	GTC	RDC	HLR	RC	Priority Weight
HLF	1	1/5	1/4	1/3	1/3	0.058
GTC	5	1	3	2	4	0.412
RDC	4	1/3	1	2	2	0.227
HLR	3	1/2	1/2	1	3	0.194
RC	3	1/4	1/2	1/3	1	0.11

TABLE 3.20 Pair-Wise Comparison of Benefit Factors with Respect to Supplier 3

Benefit Factor	HLF	GTC	RDC	HLR	RC	Priority Weight
HLF	1	1/4	3	2	1/3	0.158
GTC	4	1	3	3	4	0.434
RDC	1/3	1/3	1	1/2	1/2	0.084
HLR	1/2	1/3	2	1	1/2	0.118
RC	3	1/4	2	2	1	0.206

TABLE 3.21 Pair-Wise Comparison of Risk Factors with Respect to Supplier 1

Risk Factor	DF	ET	OHS	EAC	Priority Weight
DF	1	4	3	4	0.53
ET	1/4	1	2	2	0.208
OHS	1/3	1/2	1	2	0.16
EAC	1/4	1/2	1/2	1	0.102

TABLE 3.22 Pair-Wise Comparison of Risk Factors with Respect to Supplier 2

Risk Factor	DF	ET	OHS	EAC	Priority Weight
DF	1	1/4	1/4	1/5	0.069
ET	4	1	1/2	1/2	0.209
OHS	4	2	1	1/2	0.29
EAC	5	2	2	1	0.431

TABLE 3.23 Pair-Wise Comparison of Risk Factors with Respect to Supplier 3

Risk Factor	DF	ET	OHS	EAC	Priority Weight
DF	1	1/2	1/3	1/3	0.108
ET	2	1	1/2	1/2	0.187
OHS	3	2	1	1/2	0.293
EAC	3	2	2	1	0.412

TABLE 3.24 Pair-Wise Comparison of Risk Factors with Respect to GTC

Risk Factor	ET	OHS	EAC	Priority Weight
ET	1	1/2	1/4	0.137
OHS	2	1	1/3	0.239
EAC	4	3	1	0.623

In this green supplier selection problem for the mining industry, the developed ANP formulation consists of three major clusters, i.e., green suppliers, benefit factors and risk factors. In Table 3.25, two clusters, i.e., benefit factors and risk factors are pairwise compared while treating green suppliers as an independent cluster. It can be revealed from this table that benefit factors are more important than the risk factors for this green supplier selection problem. Similarly, in Tables 3.26 and 3.27, all the three clusters are respectively pairwise compared with respect to benefit factors and risk factors. The comprehensive cluster weight matrix is provided in Table 3.28.

TABLE 3.25 Pair-Wise Comparison Between Clusters with Respect to Suppliers

Cluster	Benefit Factor	Risk Factor	Priority Weight
Benefit factor	1	3	0.667
Risk factor	1/3	1	0.333

TABLE 3.26 Pair-Wise Comparison Between Clusters with Respect to Benefit Factors

Cluster	Supplier	Benefit Factor	Risk Factor	Priority Weight
Supplier	1	2	3	0.539
Benefit factor	1/2	1	2	0.297
Risk factor	1/3	1/2	1	0.164

TABLE 3.27 Pair-Wise Comparison Between Clusters with Respect to Risk Factors

Cluster	Supplier	Benefit Factor	Risk Factor	Priority Weight
Supplier	1	3	3	0.589
Benefit factor	1/3	1	1/2	0.159
Risk factor	1/3	2	1	0.252

TABLE 3.28 Cluster Weight Matrix

Cluster	Supplier	Benefit Factor	Risk Factor
Supplier	0	0.539	0.589
Benefit factor	0.667	0.297	0.159
Risk factor	0.333	0.164	0.252

Now, based on all the above-developed matrixes, the corresponding supermatrix for this green supplier selection problem for the considered mining industry is constructed in Table 3.29. Using this supermatrix, the related weighted supermatrix and normalized weighted supermatrix are respectively developed in Tables 3.30 and 3.31. Table 3.32 provides the limiting supermatrix for this problem where all of its elements are observed to remain constant after 62 iterations. Based on this limiting supermatrix, the final priorities of the green suppliers, benefit factors and risk factors are determined, as shown in Table 3.33. Using the normalized priority values, it can be unveiled that Supplier 2 emerges outs as the best green supplier while fulfilling all the requirements of this mining industry, followed by Supplier 1. Supplier 3 is the worst preferred alternative. Among the benefit factors, GTC is the most important criterion, followed by HLF. From this table, it can also be observed that EAC is the most important risk factor, followed by OHS.

KEYWORDS

- **analytic network process**
- **green technology capability**
- **high level of flexibility**
- **higher level of reputation**
- **non-traditional machining**
- **R&D capability**

TABLE 3.29 Supermatrix for the Green Supplier Selection Problem

Cluster		Supplier			Benefit Factor					Risk Factor			
		S_1	S_2	S_3	HLF	GTC	RDC	HLR	RC	DF	ET	OHS	EAC
Supplier	S_1	0	0	0	0.633	0.272	0.118	0.164	0.571	0.539	0.122	0.137	0.096
	S_2	0	0	0	0.106	0.608	0.681	0.539	0.143	0.297	0.648	0.623	0.653
	S_3	0	0	0	0.26	0.12	0.201	0.297	0.286	0.164	0.23	0.239	0.251
Benefit factor	HLF	0.394	0.058	0.158	0	0.483	0.098	0.067	0.505	0	0	0	0
	GTC	0.126	0.412	0.434	0.081	0	0.462	0.291	0.087	0	1	1	1
	RDC	0.065	0.227	0.084	0.154	0.087	0	0.491	0.164	0	0	0	0
	HLR	0.099	0.194	0.118	0.476	0.252	0.268	0	0.244	0	0	0	0
	RC	0.316	0.11	0.206	0.288	0.178	0.172	0.151	0	0	0	0	0
Risk factor	DF	0.53	0.069	0.108	0	0	0	0	0	0	0	0	0
	ET	0.208	0.209	0.187	0	0.137	0	0	0	0	0	0.167	0.667
	OHS	0.16	0.29	0.293	0	0.239	0	0	0	0	0.8	0	0.333
	EAC	0.102	0.431	0.412	0	0.623	0	0	0	0	0.2	0.833	0

TABLE 3.30 Weighted Supermatrix

Cluster		Supplier			Benefit Factor					Risk Factor			
		S_1	S_2	S_3	HLF	GTC	RDC	HLR	RC	DF	ET	OHS	EAC
Supplier	S_1	0	0	0	0.341	0.147	0.064	0.088	0.308	0.317	0.072	0.081	0.057
	S_2	0	0	0	0.057	0.328	0.367	0.291	0.077	0.175	0.382	0.367	0.385
	S_3	0	0	0	0.14	0.065	0.108	0.16	0.154	0.097	0.135	0.141	0.148
Benefit factor	HLF	0.263	0.039	0.105	0	0.143	0.029	0.02	0.15	0	0	0	0
	GTC	0.084	0.275	0.289	0.024	0	0.137	0.086	0.026	0	0.159	0.159	0.159
	RDC	0.043	0.151	0.056	0.046	0.026	0	0.146	0.049	0	0	0	0
	HLR	0.066	0.129	0.079	0.141	0.075	0.08	0	0.072	0	0	0	0
	RC	0.211	0.073	0.137	0.086	0.053	0.051	0.045	0	0	0	0	0
Risk factor	DF	0.176	0.023	0.036	0	0	0	0	0	0	0	0	0
	ET	0.069	0.07	0.062	0	0.022	0	0	0	0	0	0.042	0.168
	OHS	0.053	0.097	0.098	0	0.039	0	0	0	0	0.202	0	0.084
	EAC	0.034	0.144	0.137	0	0.102	0	0	0	0	0.05	0.21	0

TABLE 3.31 Normalized Weighted Supermatrix

Cluster		Supplier			Benefit Factor					Risk Factor			
		S_1	S_2	S_3	HLF	GTC	RDC	HLR	RC	DF	ET	OHS	EAC
Supplier	S_1	0	0	0	0.408	0.147	0.077	0.105	0.368	0.538	0.072	0.081	0.057
	S_2	0	0	0	0.068	0.328	0.439	0.348	0.092	0.297	0.382	0.367	0.385
	S_3	0	0	0	0.168	0.065	0.129	0.191	0.184	0.165	0.135	0.141	0.148
Benefit factor	HLF	0.263	0.039	0.105	0	0.143	0.035	0.024	0.179	0	0	0	0
	GTC	0.084	0.275	0.289	0.029	0	0.164	0.103	0.031	0	0.159	0.159	0.159
	RDC	0.043	0.151	0.056	0.055	0.026	0	0.175	0.059	0	0	0	0
	HLR	0.066	0.129	0.079	0.169	0.075	0.096	0	0.086	0	0	0	0
	RC	0.211	0.073	0.137	0.103	0.053	0.061	0.054	0	0	0	0	0
Risk factor	DF	0.176	0.023	0.036	0	0	0	0	0	0	0	0	0
	ET	0.069	0.07	0.062	0	0.022	0	0	0	0	0	0.042	0.168
	OHS	0.053	0.097	0.098	0	0.039	0	0	0	0	0.202	0	0.084
	EAC	0.034	0.144	0.137	0	0.102	0	0	0	0	0.05	0.21	0

TABLE 3.32 Limiting Supermatrix (W^{62})

Cluster		Supplier			Benefit Factor				Risk Factor			
		S_1	S_2	S_3	HLF	GTC	RDC	HLR	RC	DF	ET	OHS	EAC
Supplier	S_1	0.12	0.12	0.12	0.12	0.12	0.12	0.12	0.119	0.12	0.12	0.12	0.12
	S_2	0.179	0.179	0.179	0.179	0.179	0.179	0.179	0.179	0.179	0.179	0.179	0.179
	S_3	0.086	0.086	0.086	0.086	0.086	0.086	0.086	0.086	0.086	0.086	0.086	0.086
Benefit factor	HLF	0.083	0.083	0.083	0.083	0.083	0.083	0.083	0.083	0.083	0.083	0.083	0.083
	GTC	0.133	0.133	0.133	0.133	0.133	0.133	0.133	0.133	0.133	0.133	0.133	0.133
	RDC	0.062	0.062	0.062	0.062	0.062	0.062	0.062	0.062	0.062	0.062	0.062	0.062
	HLR	0.074	0.074	0.074	0.074	0.074	0.074	0.074	0.074	0.074	0.074	0.074	0.074
	RC	0.073	0.073	0.073	0.073	0.073	0.073	0.073	0.073	0.073	0.073	0.073	0.073
Risk factor	DF	0.028	0.028	0.028	0.028	0.028	0.028	0.028	0.028	0.028	0.028	0.028	0.028
	ET	0.043	0.043	0.043	0.043	0.043	0.043	0.043	0.043	0.043	0.043	0.043	0.043
	OHS	0.052	0.052	0.052	0.052	0.052	0.052	0.052	0.052	0.052	0.052	0.052	0.052
	EAC	0.068	0.068	0.068	0.068	0.068	0.068	0.068	0.068	0.068	0.068	0.068	0.068

TABLE 3.33 Final Priorities of the Factors and Green Suppliers

Cluster	Factor	Priorities from Limiting Matrix	Normalized Priorities
Supplier	Supplier 1	0.120	0.312
	Supplier 2	0.179	0.465
	Supplier 3	0.086	0.223
Benefit factors	High level of flexibility	0.083	0.195
	Green technology capability	0.133	0.313
	R&D capability	0.062	0.146
	Higher level of reputation	0.074	0.174
	Cost reduction	0.073	0.172
Risk factors	Demand fulfillment	0.028	0.147
	Environmental training	0.043	0.225
	OHS policy	0.052	0.272
	Environmental activity control	0.068	0.356

REFERENCES

1. Saaty, T. L., (1980). *The Analytic Hierarchy Process*. McGraw-Hill, New York.
2. Saaty, T. L., (2005). *Theory and Applications of the Analytic Network Process: Decision Making with Benefits, Opportunities, Costs, and Risks*. RWS Publications, USA.
3. Saaty, T. L., (2005). Making and validating complex decisions with the AHP/ANP. *Journal of Systems Science and Systems Engineering, 14*, 1–36.
4. Meade, L. M., & Presley, A., (2002). R&D project selection using the analytic network process. *IEEE Transactions on Engineering Management, 49*(1), 59–66.
5. Yurdakul, M., (2003). Measuring long-term performance of a manufacturing firm using the analytic network process (ANP) approach. *International Journal of Production Research, 41*(11), 2501–2529.
6. Bayazit, O., (2006). Use of analytic network process in vendor selection decisions. *Benchmarking: An International Journal, 13*(5), 566–579.
7. Jharkharia, S., & Shankar, R., (2007). Selection of logistics service provider: An analytic network process (ANP) approach. *Omega, 35*, 274–289.
8. Anand, G., & Kodali, R., (2009). Selection of lean manufacturing systems using the analytic network process-a case study. *Journal of Manufacturing Technology Management, 20*(2), 258–289.
9. Hsu, C. W., & Hu, H. A., (2009). Applying hazardous substance management to supplier selection using analytic network process. *Journal of Cleaner Production, 17*, 255–264.
10. Kodali, R., & Anand, G., (2010). Application of analytic network process for the design of flexible manufacturing systems. *Global Journal of Flexible Systems Management, 11*(1, 2), 39–54.
11. Sipahi, S., & Timor, M., (2010). The analytic hierarchy process and analytic network process: An overview of applications. *Management Decision, 48*(5), 775–808.
12. Anand, G., Kodali, R., & Santosh, K. B., (2011). Development of analytic network process for the selection of material handling systems in the design of flexible manufacturing systems (FMS). *Journal of Advances in Management Research, 8*(1), 123–147.
13. Das, S., & Chakraborty, S., (2011). Selection of non-traditional machining processes using analytic network process. *Journal of Manufacturing Systems, 30*, 41–53.
14. Anand, G., Kodali, R., & Dhanekula, C. S., (2012). An application of analytic network process for selection of a plant location: A case study. *International Journal of Services and Operations Management, 12*(1), 35–66.
15. Giner-Santonja, G., Aragonés-Beltrán, P., & Niclós-Ferragut, J., (2012). The application of the analytic network process to the assessment of best available techniques. *Journal of Cleaner Production, 25*, 86–95.
16. Ordoobadi, S. M., (2012). Application of ANP methodology in evaluation of advanced technologies. *Journal of Manufacturing Technology Management, 23*(2), 229–252.
17. Govindan, K., Sarkis, J., & Palaniappan, M., (2013). An analytic network process-based multicriteria decision making model for a reverse supply chain. *International Journal of Advanced Manufacturing Technology, 68*, 863–880.
18. Al-Hawari, T., Mumani, A., & Momani, A., (2014). Application of the analytic network process to facility layout selection. *Journal of Manufacturing Systems, 33*, 488–497.
19. Kodali, R., Anand, G., & Reddy, M. U., (2014). An application of analytic network process for the selection of cellular manufacturing systems. *International Journal of Services and Operations Management, 19*(3), 287–318.

20. Van, H. A., & Pintelon, L., (2014). Development of a maintenance performance measurement framework-using the analytic network process (ANP) for maintenance performance indicators election. *Omega, 42*, 33–46.
21. Ocampo, L. A., Angelo, M., & Promentilla, B., (2016). Development of a sustainable manufacturing strategy using analytic network process. *International Journal of Business and Systems Research, 10*(2, 3, 4), 262–290.
22. Barzegar, M., Rasi, R. E., & Niknamfar, A. H., (2018). Analyzing the drivers of green manufacturing using an analytic network process method: A case study. *International Journal of Research in Industrial Engineering, 7*(1), 61–83.
23. Kurian, M. C., Shalij, P. R., & Pramod, V. R., (2019). Maintenance strategy selection in a cement industry using analytic network process. *Journal of Quality in Maintenance Engineering, 26*(4), 509–525.
24. Asadabadi, M. R., Chang, E., & Saberi, M., (2019). Are MCDM methods useful? A critical review of analytic hierarchy process (AHP) and analytic network process (ANP). *Cogent Engineering, 6*(1), 1623153.
25. Kheybari, S., Rezaie, F. M., & Farazmand, H., (2020). Analytic network process: An overview of applications. *Applied Mathematics and Computation, 367*, 124780.
26. Gurel, O., Acar, A. Z., Onden, I., & Gumus, I., (2015). Determinants of the green supplier selection. *Procedia - Social and Behavioral Sciences, 181*, 131–139.
27. Govindan, K., Rajendran, S., Sarkis, J., & Murugesan, P., (2015). Multi-criteria decision-making approaches for green supplier evaluation and selection: A literature review. *Journal of Cleaner Production, 98*, 66–83.
28. Sivakumar, R., Kannan, D., & Murugesan, P., (2015). Green vendor evaluation and selection using AHP and Taguchi loss functions in production outsourcing in mining industry. *Resources Policy, 46*, 64–75.

CHAPTER 4

WEIGHTED AGGREGATED SUM PRODUCT ASSESSMENT (WASPAS) METHOD

The application of any of the MCDM problems starts with the following decision/evaluation matrix, where the performance of every candidate alternative is appraised against each of the evaluation criteria with respect to a performance score.

$$X = \begin{bmatrix} x_{11} & x_{12} & \cdots & x_{1n} \\ x_{21} & x_{22} & \cdots & x_{2n} \\ \cdots & \cdots & \cdots & \cdots \\ x_{m1} & x_{m2} & \cdots & x_{mn} \end{bmatrix} \tag{4.1}$$

where; m is the number of the alternatives; n is the number of evaluation criteria; and x_{ij} is the performance score assigned to i^{th} alternative with respect to the jth criterion.

The WASPAS method is a unique combination of two well-known MCDM approaches, i.e., weighted sum model (WSM) and weighted product model (WPM). While employing WASPAS method for solving any MCDM method, it is first required to make all the elements of the decision matrix dimensionless and comparable with the help of a linear normalization technique based on the following two equations:

i. **For Beneficial Criterion:**

$$\bar{x}_{ij} = \frac{x_{ij}}{\max_i x_{ij}} \tag{4.2}$$

Multi-Criteria Decision-Making Methods in Manufacturing Environments: Models and Applications.
Shankar Chakraborty, Prasenjit Chatterjee, & Partha Protim Das (Authors)

ii. For Non-Beneficial (cost) Criterion:

$$\bar{x}_{ij} = \frac{\min_i x_{ij}}{x_{ij}} \tag{4.3}$$

where; $\bar{x}_{ij}$ is the normalized value of x_{ij}.

In this method, a joint criterion of optimality is sought based on two criteria of optimality. The first criterion of optimality, i.e., criterion of a mean weighted success is similar to the WSM method, which has already become a popular and well-accepted MCDM tool adopted for evaluating a number of alternatives with respect to a given set of decision criteria. Based on WSM method, the total relative importance of ith alternative is calculated as follows [1, 2]:

$$Q_i^{(1)} = \sum_{j=1}^{n} \bar{x}_{ij} w_j \tag{4.4}$$

where; w_j is the weight (relative importance) of the jth criterion.

On the other hand, based on WPM method, the total relative importance of ith alternative is computed using the following expression [2, 3]:

$$Q_i^{(2)} = \prod_{j=1}^{n} (\bar{x}_{ij})^{w_j} \tag{4.5}$$

A joint generalized criterion of weighted aggregation of additive and multiplicative models can then be formulated as below [4, 5]:

$$Q_i = 0.5Q_i^{(1)} + 0.5Q_i^{(2)} = 0.5\sum_{j=1}^{n} \bar{x}_{ij} w_j + 0.5\prod_{j=1}^{n} (\bar{x}_{ij})^{w_j} \tag{4.6}$$

For having increased ranking accuracy and effectiveness of the decision-making process, in WASPAS method, a more generalized equation for determining the total relative importance of ith alternative is developed [6, 7].

$$Q_i = \lambda Q_i^{(1)} + (1-\lambda)Q_i^{(2)} = + \ \lambda\sum_{j=1}^{n} \bar{x}_{ij} w_j + 0.5\prod_{j=1}^{n} (\bar{x}_{ij})^{w_j} \ (\lambda = 0, 0.1, \ldots, 1) \tag{4.7}$$

Now, the candidate alternatives are ranked based on the computed Q values, i.e., the alternative having the maximum priority (best option) should have the highest Q value. When the value of λ is 0, WASPAS method is transformed to WPM, and when λ is 1, it becomes similar to the WSM method. Till date, this method has been successfully adopted in various domains of manufacturing environment.

4.1 LITERATURE REVIEW

Chakraborty & Zavadskas [8] explored the applicability of WASPAS method as an effective MCDM tool in solving eight manufacturing decision-making problems, i.e., selection of cutting fluid, electroplating system, forging condition, arc welding process, industrial robot, milling condition, machinability of material, and electrical discharge micro-machining process parameters. Its robustness with respect to changing values of the corresponding tuning parameter (λ) was also proved. Madić [9] considered eight feasible NTM processes and five economical criteria, and subsequently adopted WASPAS method for identification of the most suitable process for a given machining operation. Using WASPAS method, Chakraborty et al. [10] solved five real-time decision-making problems from manufacturing domain, i.e., selection of an FMS, a machine in a flexible manufacturing cell, an automated guided vehicle (AGV), an automated inspection system and an industrial robot. Chakraborty et al. [11] applied the WASPAS method for finding out the optimal parametric combinations for five NTM processes, and claimed it as an effective tool for solving both single- and multi-response optimization problems.

While treating laser power, cutting speed, assist gas pressure and focus position as the input parameters, and burr height, dragline separation, depth of separation line, surface roughness and perpendicularity of the cut as the outputs, Madić et al. [12] conducted 27 experiments using a CO_2 laser cutting (LC) process. The considered process parameters were later optimized based on WASPAS method. Petković et al. [13] considered 10 alternative NTM processes for machining of ceramic materials and employed WASPAS method to single out the most feasible process while taking into account 10 machining criteria. Karabašević [14] solved a personnel selection problem while applying WASPAS method. Considering two real-time industrial robot selection problems, Karande et al. [15] compared the ranking performance of WASPAS method with other popular MCDM techniques. A sensitivity analysis was also performed to validate the robustness of those MCDM tools. Madic et al. [16] performed LC operation considering laser power, cutting speed, assist gas pressure and focus position as the input parameters, and applied WASPAS method for identification of the optimal parametric intermix for the said process.

Mardani et al. [17] extensively reviewed the applications of WASPAS method and explored the possibility of its deployment under fuzzy environment. It was claimed that the derived observations would assist the decision makers in dealing with interconnected or conflicting criteria in uncertain environments. Rudnik [18] adopted WASPAS method as an MCDM tool for flow control in a manufacturing system having three production lines. The derived results were later validated based on simulation outputs. Suresh et al. [19] studied the machinability characteristics of hardened AISI D3 tool steel using vegetable oils as the cutting fluid under minimum quantity lubrication and applied WASPAS method for parametric optimization of the considered turning process. Surface roughness, material removal rate (MRR), interface temperature, specific energy and flank wear were treated as the responses. Emovon et al. [20] proposed the application of WASPAS method for enhancing the reliability centered maintenance process so as to improve its efficacy in marine maintenance system applications. Stojić et al [21] applied WASPAS method for selection of the most apposite supplier in an organization manufacturing polyvinyl chloride carpentry products. Sensitivity analysis was also performed to verify its robustness and its ranking performance was finally compared against other MCDM techniques.

Considering nine human risk alternatives, Bid & Siddique [22] adopted WASPAS method for assessing human risks in an Indian dam. Nenadić [23] applied WASPAS method to identify the most dangerous road sections based on several quantitative and qualitative traffic safety criteria. Rajendra Prasad et al. [24] demonstrated the application of WASPAS method for parametric optimization of a modified air abrasive jet machining (AJM) operation on Nickel 233 alloy. Injection pressure, stand-off distance and abrasive mesh size were the input parameters, and taper angle, surface roughness and MRR were the responses. Vinchurkar & Samtani [25] considered four hydropower plants, i.e., Indira Sagar Project, Canal Head Power House, River Bed Power House and Lower Sileru Power House, and endeavored to rank them using WASPAS method. Its ranking performance was also contrasted with other popular MCDM techniques. Kumar et al. [26] considered 24 different models of five brands of hard disk drives available in the Indian market and attempted to single out the best hard disk drive using WASPAS method based on several evaluation criteria. Pathapalli et al. [27] applied WASPAS method as a multi-objective optimization tool to determine the best settings of cutting speed, feed rate, depth of cut and reinforcement weight percent of composite samples while machining metal-matrix composite materials. MRR, cutting force and surface roughness were treated as the responses.

4.2 ELECTROPLATING SYSTEM SELECTION PROBLEM

The electroplating operation is usually performed to alter characteristics of a surface in order to provide improved appearance, ability to withstand corrosive agents, resistance to abrasion or other desired properties [28]. Electroplating with various capabilities are now available for a wide-ranging application, like surface finishing, thickness maintaining, avoiding rusting, restoring dimension of under-size parts, aesthetic, etc. The use of an appropriate coating during electroplating operation can increase the life expectancy of a given product/component. Selection of the electroplating system to suit a specific application in manufacturing environment from a large number of available plating can often be treated as a typical MCDM problem. Kumar & Agrawal [28] considered an electroplating system selection problem, as provided in Table 4.1, where the best coating material needs to be chosen among seven candidate alternatives, i.e., silver, gold, lead, rhodium, nickel, chromium, and platinum, based on five evaluation criteria, i.e., hardness (H) (in HV), thickness (T) (in μm), aesthetic (AE), adhesion (AD) and cost (C). Thus, this decision matrix consists of both quantitative and qualitative criteria values which are linear normalized in Table 4.2. Using AHP method, the weights of the five criteria were evaluated as $w_H = 0.1761$, $w_T = 0.2042$, $w_{AE} = 0.2668$, $w_{AD} = 0.1243$, and $w_C = 0.2286$ [29]. In the subsequent WASPAS method-based analysis, these criteria weights are adopted. Applying an integrated AHP and graph theory and matrix approach (GTMA), Singh & Rao [29] obtained a rank ordering of the electroplating coating materials as 4-5-6-7-3-1-2, which identifies chromium and platinum as the two best choices. Rhodium became the worst chosen plating material. The results of WASPAS method-based analysis are provided in Table 4.2. Chromium emerges out as the best coating material, followed by platinum. On the other hand, rhodium is the least preferred electroplating material. Table 4.3 shows the effect of changing λ values on the ranking performance of WASPAS method. It is quite interesting to observe that for higher λ values, the rank orderings of the alternative electroplating materials exactly corroborate with those derived by Singh & Rao [29]. The influences of varying λ values on the ranking of the alternative electroplating materials are also graphically represented in Figure 4.1.

TABLE 4.1 Data for the Electroplating System Selection Problem [28]

Alternative	H	T	AE	AD	C
Silver	350	20	Good (4)	Good (4)	Medium (2)
Gold	250	25	Excellent (5)	Average (3)	High (3)
Lead	150	30	Average (3)	Poor (1)	Low (1)
Rhodium	400	20	Fair (2)	Average (3)	Medium (2)
Nickel	550	30	Poor (1)	Fair (2)	Low (1)
Chromium	600	35	Poor (1)	Excellent (5)	Low (1)
Platinum	580	30	Good (4)	Good (4)	High (3)

Source: Reprinted with permission from Ref. [28]. © 2021 Elsevier.

TABLE 4.2 Normalized Decision Matrix for the Electroplating System Selection Problem

Alternative	H	T	AE	AD	C	Q(1)	Q(2)	Q
Silver	0.5833	0.5714	0.8	0.8	0.5	0.6466	0.6345	0.6405
Gold	0.4167	0.7143	1	0.6	0.3333	0.6368	0.5842	0.6105
Lead	0.2500	0.8571	0.6	0.2	1	0.6326	0.5423	0.5874
Rhodium	0.6667	0.5714	0.4	0.6	0.5	0.5297	0.5209	0.5253
Nickel	0.9167	0.8571	0.2	0.4	1	0.6681	0.5543	0.6112
Chromium	1	1	0.2	1	1	0.7866	0.6509	0.7188
Platinum	0.9667	0.8571	0.8	0.8	0.3333	0.7343	0.6867	0.7105

TABLE 4.3 Effect of λ on Ranking Performance of WASPAS Method

$\lambda = 0$	$\lambda = 0.1$	$\lambda = 0.2$	$\lambda = 0.3$	$\lambda = 0.4$	$\lambda = 0.5$	$\lambda = 0.6$	$\lambda = 0.7$	$\lambda = 0.8$	$\lambda = 0.9$	$\lambda = 1.0$
0.6345 (3)	0.6357 (3)	0.6369 (3)	0.6381 (3)	0.6393 (3)	0.6405 (3)	0.6417 (3)	0.6430 (3)	0.6442 (4)	0.6454 (4)	0.6466 (4)
0.5842 (4)	0.5894 (4)	0.5947 (4)	0.6000 (4)	0.6052 (4)	0.6105 (5)	0.6158 (5)	0.6210 (5)	0.6263 (5)	0.6315 (5)	0.6368 (5)
0.5423 (6)	0.5513 (6)	0.5603 (6)	0.5694 (6)	0.5784 (6)	0.5874 (6)	0.5965 (6)	0.6055 (6)	0.6145 (6)	0.6236 (6)	0.6326 (6)
0.5209 (7)	0.5218 (7)	0.5227 (7)	0.5236 (7)	0.5244 (7)	0.5253 (7)	0.5262 (7)	0.5271 (7)	0.5279 (7)	0.5288 (7)	0.5297 (7)
0.5543 (5)	0.5657 (5)	0.5770 (5)	0.5884 (5)	0.5998 (5)	0.6112 (4)	0.6226 (4)	0.6340 (4)	0.6454 (3)	0.6567 (3)	0.6681 (3)
0.6509 (2)	0.6645 (2)	0.6780 (2)	0.6916 (2)	0.7052 (2)	0.7188 (1)	0.7323 (1)	0.7459 (1)	0.7594 (1)	0.7730 (1)	0.7866 (1)
0.6867 (1)	0.6915 (1)	0.6962 (1)	0.7010 (1)	0.7057 (1)	0.7105 (2)	0.7153 (2)	0.7200 (2)	0.7248 (2)	0.7296 (2)	0.7343 (2)

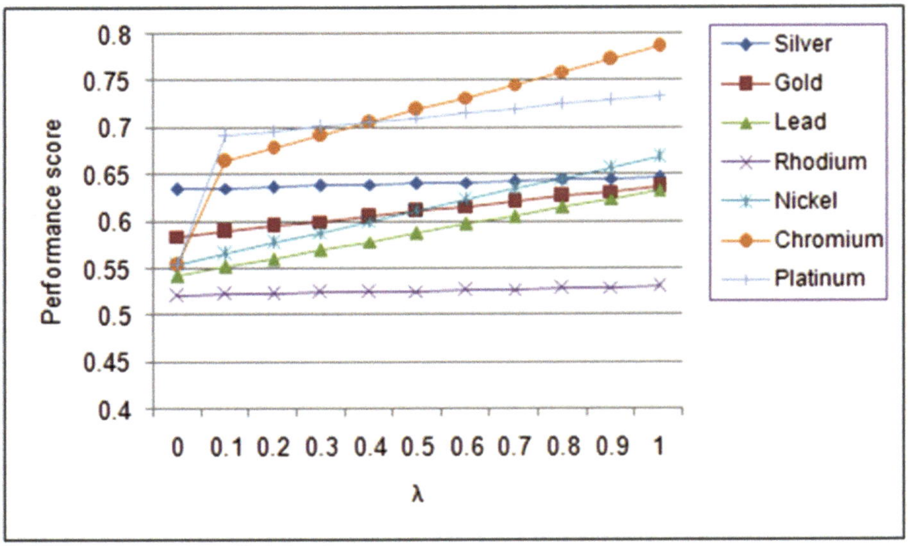

FIGURE 4.1 Variation of performance scores of electroplating materials with respect to λ.

KEYWORDS

- **aesthetic**
- **automated guided vehicle**
- **graph theory and matrix approach**
- **weighted aggregated sum product assessment**
- **weighted product model**
- **weighted sum model**

REFERENCES

1. MacCrimon, K. R., (1968). *Decision Making Among Multiple Attribute Alternatives: A Survey and Consolidated Approach.* Rand Memorandum, RM-4823-ARPA.
2. Triantaphyllou, E., & Mann, S. H., (1989). An examination of the effectiveness of multi-dimensional decision-making methods: A decision-making paradox. *Decision Support Systems, 5*(3), 303–312.
3. Miller, D. W., & Starr, M. K., (1969). *Executive Decisions and Operations Research.* Prentice-Hall, Inc., Englewood Cliffs, New Jersey.

4. Zavadskas, E. K., Antucheviciene, J., Šaparauskas, J., & Turskis, Z., (2013). Multi-criteria assessment of facades' alternatives: Peculiarities of ranking methodology. *Procedia Engineering, 57*, 107–112.
5. Zavadskas, E. K., Antucheviciene, J., Saparauskas, J., & Turskis, Z., (2013). MCDM methods WASPAS and MULTIMOORA: Verification of robustness of methods when assessing alternative solutions. *Economic Computation and Economic Cybernetics Studies and Research, 47*(2), 5–20.
6. Šaparauskas, J., Zavadskas, E. K., & Turskis, Z., (2011). Selection of facade's alternatives of commercial and public buildings based on multiple criteria. *International Journal of Strategic Property Management, 15*(2), 189–203.
7. Zavadskas, E. K., Turskis, Z., Antucheviciene, J., & Zakarevicius, A., (2012). Optimization of weighted aggregated sum product assessment. *Electronics and Electrical Engineering, 122*(6), 3–6.
8. Chakraborty, S., & Zavadskas, (2014). Applications of WASPAS method in manufacturing decision making. *Informatica, 25*(1), 1–20.
9. Madić, M., Gecevska, V., Radovanović, M., & Petković, D., (2014). Multi-criteria economic analysis of machining processes using the WASPAS method. *Journal of Production Engineering, 17*(2), 79–82.
10. Chakraborty, S., Zavadskas, E. K., & Antucheviene, J., (2015). Applications of WASPAS method as a multi-criteria decision-making tool. *Economic Computation and Economic Cybernetics Studies and Research, 49*(1), 5–22.
11. Chakraborty, S., Bhattacharyya, O., Zavadskas, E. K., & Antucheviciene, J., (2015). Application of WASPAS method as an optimization tool in non-traditional machining processes. *Information Technology and Control, 44*(1), 77–84.
12. Madić, M., Radovanović, M., Petković, D., & Nedić, B., (2015). Multi-criteria analysis of laser cut surface characteristics in CO_2 laser cutting of stainless steel. *Tribology in Industry, 37*(2), 236–243.
13. Petković, D., Madić, M., & Radenković, G., (2015). Selection of the most suitable non-conventional machining processes for ceramics machining by using MCDMs. *Science of Sintering, 47*, 229–235.
14. Karabašević, D., (2016). An approach to personnel selection based on SWARA and WASPAS methods. *Journal of Economics, Management and Informatics, 7*(1), 1–11.
15. Karande, P., Zavadskas, E. K., & Chakraborty, S., (2016). A study on the ranking performance of some MCDM methods for industrial robot selection problems. *International Journal of Industrial Engineering Computations, 7*, 399–422.
16. Madic, M., Antucheviciene, J., Radovanovic, M., & Petkovic, D., (2016). Determination of manufacturing process conditions by using MCDM methods: Application in laser cutting. *Engineering Economics, 27*(2), 144–150.
17. Mardani, A., Nilashi, M., Zakuan, N., Loganathan, N., Soheilirad, S., Saman, M. Z. M., & Ibrahim, O., (2017). A systematic review and meta-analysis of SWARA and WASPAS methods: Theory and applications with recent fuzzy developments. *Applied Soft Computing, 57*, 265–292.
18. Rudnik, K., (2017). Decision-making in a manufacturing system based on MADM methods. *International Scientific Journal "Industry, 4.0," II*(3), 104–107.
19. Suresh, R. K., Krishnaiah, G., & Venkataramaiah, P., (2017). Selection of best novel MCDM method during turning of hardened AISI D3 tool steel under minimum quantity

lubrication using bio-degradable oils as cutting fluids. *International Journal of Applied Engineering Research, 12*(19), 8082–8091.

20. Emovon, I., Norman, R. A., Murphy, A. J., & Okwu, M. O., (2018). Application of WASPAS in enhancing reliability centered maintenance for ship system maintenance. *Journal of Engineering and Technology, 9*(1), 1–16.
21. Stojić, G., Stević, Ž., Antuchevičienė, J., Pamučar, D., & Vasiljević, M., (2018). A novel rough WASPAS approach for supplier selection in a company manufacturing PVC carpentry products. *Information, 9*(121), 16. doi:10.3390/info9050121.
22. Bid, S., & Siddique, G., (2019). Human risk assessment of Panchet Dam in India using TOPSIS and WASPAS multi-criteria decision-making (MCDM) methods. *Heliyon, 5*, 01956.
23. Nenadić, D., (2019). Ranking dangerous sections of the road using the MCDM model. *Decision Making: Applications in Management and Engineering, 2*(1), 115–131.
24. Rajendra, P. S., Ravindranath, K., & Devakumar, M. L. S., (2019). Experimental study and optimization in modified air abrasive jet machining on nickel-233 alloy using MCDM techniques. *Manufacturing Technology, 19*(6), 1010–1019.
25. Vinchurkar, S. H., & Samtani, B. K., (2019). Performance evaluation of the hydropower plants using various multi-criteria decision-making techniques. *International Journal of Engineering and Advanced Technology, 8*(6), 2131–2138.
26. Kumar, R., Bhattacherjee, A., Singh, A. D., Singh, S., & Pruncu, C. I., (2020). Selection of portable hard disk drive based upon weighted aggregated sum product assessment method: A case of Indian market. *Measurement and Control, 53*(7, 8), 1218–1230.
27. Pathapalli, V. R., Basam, V. R., Gudimetta, S. K., & Koppula, M. R., (2020). Optimization of machining parameters using WASPAS and MOORA. *World Journal of Engineering, 17*(2), 237–246.
28. Kumar, A., & Agrawal, V. P., (2009). Attribute based specification, comparison and selection of electroplating system using MADM approach. *Expert Systems with Applications, 36*(8), 10815–10827.
29. Singh, D., & Rao, R. V., (2011). A hybrid multiple attribute decision making method for solving problems of industrial environment. *International Journal of Industrial Engineering Computations, 2*(3), 631–644.

CHAPTER 5

MULTI-OBJECTIVE OPTIMIZATION ON THE BASIS OF RATIO ANALYSIS (MOORA)

Multi-objective optimization, also known as multi-criteria or multi-attribute optimization, is the process of optimizing two or more competing objectives (attributes) under a set of constraints at the same time. Optimal judgments can be made in the context of tradeoffs between two or more challenging objectives in multi-objective optimization problems, which may be encountered in a variety of domains. Multi-objective optimization problems include maximization of profit, minimization of product cost, maximization of product performance, etc.

The presence of several decision makers with various interests and values complicates the decision-making process in a real-time manufacturing setting. It is assumed that for any decision-making problem, the objectives (attributes) are measurable and the corresponding outcomes are available for every decision alternative. Objective outcomes provide the basis of comparison of choices, consequently facilitating the selection of the best (satisfactory) option. As a result, multi-objective optimization techniques are the most suitable tools for identifying the best alternative from a collection of feasible options based on several, often contradictory benefit and cost criteria. The MOORA method, introduced by Brauers [1], is a multi-objective optimization technique that can be efficiently adopted to deal with various complex decision-making problems in real-time manufacturing environment.

The corresponding decision matrix (X), indicating performance scores of the alternatives in relation to the criteria under consideration, is the input to this approach for addressing any decision-making problem. After that, the original decision matrix is normalized to make all of its elements

Multi-Criteria Decision-Making Methods in Manufacturing Environments: Models and Applications.
Shankar Chakraborty, Prasenjit Chatterjee, & Partha Protim Das (Authors)

analogous and dimensionless. This normalization process adopts a ratio-based approach in which the performance of an alternative in relation to a criterion is compared to a denominator that represents the performance of all the alternatives for that criterion. The following equation is employed to normalize the decision matrix:

$$x_{ij}^{*} = \frac{x_{ij}}{\sum_{i=1}^{m} x_{ij}} \tag{5.1}$$

where; x_{ij}^{*} is a dimensionless number in [0,1] interval representing the normalized score of jth criterion for i^{th} alternative.

Brauers et al. [2] considered various ratio systems, e.g., total ratio, Schärlig ratio, Weitendorf ratio, Jüttler ratio, Stopp ratio, Körth ratio, etc., and concluded that for MOORA method, the best choice for the denominator is the square root of sum of squares of each alternative's score on each attribute. Thus, this ratio can be expressed as:

$$x_{ij}^{*} = x_{ij} \Big/ \left[\sum_{i=1}^{m} x_{ij}^{2}\right]^{1/2} \quad (j = 1, 2, \ldots, n) \tag{5.2}$$

It is important to note here that in all of the aforementioned normalization procedures, components of the decision matrix are typically normalized regardless of the type of criteria (beneficial or non-beneficial). It is occasionally noticed that in Eqn. (5.1), the normalized value for a decision matrix with a higher value for a specific criterion may surpass one. However, in Eqn. (5.2), it is observed that the maximum criterion value would be less than one. These normalized scores are added for beneficial criteria and deducted for non-beneficial criteria in MOORA method, resulting in the following equation:

$$y_i = \sum_{j=1}^{g} x_{ij}^{*} - \sum_{j=g+1}^{n} x_{ij}^{*} \tag{5.3}$$

where; g indicates the number of beneficial criteria; $(n - g)$ denotes the number of non-beneficial criteria; and y_i represents the evaluation score of ith alternative across all the criteria. The best alternative has the maximum assessment score when these scores are arranged in descending order. To determine the final ranking of the considered alternatives, an ordinal ranking of y_i values is recommended. Brauers & Zavadskas [3] proved that MOORA method is quite robust with respect to the judgments of the decision makers, their visions, and interactions between the alternatives and evaluation criteria.

In many of the decision-making problems, it is often observed that some criteria have more importance than the others. Thus, in order to emphasize the relative importance of those criteria, it is recommended to multiply the normalized elements of the decision matrix by the respective criteria weights [3]. When the criteria weights are considered in a decision-making problem, Eqn. (5.3) becomes as follows:

$$y_i = \sum_{j=1}^{g} w_j x_{ij}^* - \sum_{j=g+1}^{n} w_j x_{ij}^* \quad (j = 1, 2, \ldots, n) \tag{5.4}$$

where; w_j is the weight assigned to j^{th} criterion.

On the other hand, a maximum objective reference point is considered in the reference point approach [3]. As the coordinates (r_i) that are determined for the reference point are realized in one of the alternatives, the maximum objective reference point method is more realistic and non-subjective. The deviation of a criterion value from the established reference point (r_i) can be calculated using the normalized values of the decision matrix as ($r_i - x_{ij}^*$). An alternative would be considered to be the best when all of its beneficial criteria would attain the corresponding maximum values and non-beneficial criteria would have the minimum values. However, obtaining a single option with all of its beneficial criteria having maximum values and non-beneficial criteria having lowest values is not conceivable. As a result, there would be deviations for the beneficial criteria if the maximum values are not met, and also for the non-beneficial criteria if the minimum values are not attained.

In reference point approach, the performance index (P_i) estimates the total deviation for all the beneficial and non-beneficial criteria for i^{th} alternative and can be expressed as:

$$P_i = \underset{(i)}{\text{Min}} \left(\underset{(j)}{\text{Max}} \left| r_i - x_{ij}^* \right| \right) \tag{5.5}$$

The best alternative has the minimum total deviation (minimum P_i value) when all the beneficial and non-beneficial criteria are considered.

To develop the full multiplicative form of MOORA (MULTIMOORA) method from the mixed forms, Brauers & Zavadskas [5, 6] derived the following equation:

$$U_i = \frac{A_i}{B_i} \tag{5.6}$$

where; $A_i = \prod_{j=1}^{g} x_{ij}^*$, $B_i = \prod_{j=g+1}^{n} x_{ij}^*$ and U_i is the utility degree (overall performance score) for i^{th} alternative. In the above equation, the beneficial criteria are in the numerator, while the non-beneficial criteria are in the denominator [7]. If any of the x_{ij}^* values are 0, indicating the non-existence of a specific criterion in the decision matrix, Brauers & Zavadskas [5, 6] advised skipping the filtering stage or removing that criterion from the decision matrix.

From MULTIMOORA method, another variant in the form of multi-objective optimization on the basis of simple ratio analysis (MOOSRA) approach is evolved out. This method has been proved to be less sensitive to major changes in the criteria values. It considers a simple ratio of the sum of normalized performance values for beneficial criteria to the sum of normalized performance values for non-beneficial criteria to estimate the overall performance score of each alternative. The following equation is adopted to calculate the overall performance score (y_i):

$$y_i = \frac{\sum_{j=1}^{g} x_{ij}^*}{\sum_{j=g+1}^{n} x_{ij}^*} \quad (5.7)$$

Now, when the criteria weights are considered, Eqn. (5.7) becomes Eqn. (5.8):

$$y_i = \frac{\sum_{j=1}^{g} w_j x_{ij}^*}{\sum_{j=g+1}^{n} w_j x_{ij}^*} \quad (5.8)$$

5.1 LITERATURE SURVEY

Chakraborty [8] presented the application of MOORA method for solving several real-time manufacturing decisions making problems, like the selection of an industrial robot, an FMS, a computerized numerical control (CNC) machine, an NTM process for a given machining application, a rapid prototyping process (RP) and an automated inspection system. Its potentiality and flexibility while dealing with complex decision-making problems was also validated. Gadakh [8] considered six different milling processes and adopted MOORA method for their parametric optimization. Karande & Chakraborty [10] applied MOORA method for solving some of the industrial material selection problems. Considering gas tungsten arc welding (GTAW), submerged arc welding, CO_2 laser welding, gas metal arc welding (GMAW) and friction stir welding as the alternative welding processes, Gadakh et al.

[11] employed MOORA method to determine the most suitable welding parameters for a given joining operation. Stanujkic et al. [12] compared the performance and robustness of MOORA method against other popular MCDM techniques while ranking some of the Serbian banks. The reasons behind the discrepancies in the ranking results between the considered methods were also highlighted.

Applying MOOSRA method, Jagadish & Ray [13] solved a cutting fluid selection problem for gear hobbing process while minimizing environmental impact and production cost, and maximizing product quality. Achebo & Odinikuku [14] determined the optimal combinations of welding current, welding voltage, electrode diameter and welding speed for a welding process while proposing the integrated application of MOORA and standard deviation methods. Based on various evaluation criteria, like speed, cost, energy consumption, etc., Ertuğrul & Öztaş [15] adopted MOORA method for identification of the most apposite sewing machine in garment manufacturing. Khan & Maity [16] applied MOORA method for parametric optimization of nine NTM processes, like AJM, abrasive water jet machining (AWJM), electrochemical machining (ECM), electrochemical micro-machining (ECMM), laser beam machining (LBM), laser cutting (LC) process, plasma arc cutting (PAC), ultrasonic machining (USM) and wire electrical discharge machining (WEDM). Madić et al. [17] solved some NTM process selection problems using MOORA method and showed that its ranking performance would almost match with TOPSIS method. Mandal & Sarkar [18] considered three types of alternative manufacturing system, i.e., lean, agile, and le-agile, and applied MOORA method to identify the best system based on lead time, cost, quality, service level, leanness, and agility.

Patel & Maniya [19] determined the optimal parametric combination of a WEDM process using MOORA method, and also proved its applicability and potentiality as a multi-objective optimization tool. Based on MOORA and MOOSRA methods, Sarkar et al. [20] endeavored to develop a decision support method for non-traditional machine selection while taking into account several subjective and objective evaluation criteria. Karuppanna Prasad & Sekar [21] extensively reviewed the applications of MOORA, MULTIMOORA, and MOOSRA methods for selection of the best alternatives for various industrial applications. Employing MOORA method, Singarave et al. [22] determined the most suitable machining settings of a turning operation for having minimum surface roughness and micro-hardness, and maximum material removal rate (MRR). The weight of each of objective criteria was estimated by entropy method. Adalı & Işık [23] proposed the applications of MOORA, MULTIMOORA, and MOOSRA methods for solving a laptop selection problem. In a friction stir welding process of aluminum alloy AA2014-T6 material, Parikh et al. [24] studied the effects of rotational speed, welding speed and shoulder diameter on tensile strength, hardness in nugget zone and power consumption. The application of MOORA method was later explored as a multi-objective optimization tool for determining the optimal parametric intermix of the said process.

Singh [25] solved three real-time examples from Indian industries using MOORA method. Its application steps would consist of a) selection of the most suitable alternatives, b) ranking of internal assessment for benchmarking in an industry, and c) ranking of different types of benchmarking process depending on their usage. With an aim to fulfill customers' requirements and increased productivity, Dwivedi & Dwivedi [26] attempted to apply MOORA method to rank the competing suppliers in a manufacturing organization. Maghsoodi et al. [27] estimated weights of the considered criteria using Shannon's entropy method and later adopted MOORA approach for singling out the most appropriate performance appraisal method under dynamic and agile manufacturing environment. Mahale et al. [28] solved a friction material selection problem while adopting MOORA method. It ranking performance was also validated with respect to other MCDM techniques and extension evaluation method. Majumder & Saha [29] integrated MOORA method with principal component analysis to determine the optimal combination of spindle speed, depth of cut and feed rate while turning ASTM A588 mild steel material. The effectiveness of MOORA over TOPSIS method was also proved.

Jain & Ajmera [30] contrasted the ranking performance of MOORA and weighted Euclidean distance-based approach while evaluating flexibility in FMS for new part development. The weights of the considered criteria were determined using both entropy and AHP methods. Using MOORA method, Abhang et al. [31] optimized cutting speed, feed rate, depth of cut and tool nose radius during turning operation on alloy steel for achieving better values of chip-tool interface temperature, tool wear rate and cutting force. Kaladhar [32] proposed the application of MOORA method for determination of the optimal machining conditions, and predictive modeling for surface roughness and cutting tool flank wear while performing hard turning operation on AISI 4340 steel under dry environment. Kuriakose et al. [33] investigated the influences of feedstock extrusion parameters on the dimensional variation and surface roughness of the extruded multi-lumen tubes and applied the MOORA method to identify the optimal process conditions.

5.2 RAPID PROTOTYPING PROCESS SELECTION PROBLEM

The RP is special class of manufacturing process, used to quickly fabricate a scale model of a component or assembly using three-dimensional computer-aided design (CAD) data. As a visualization and testing tool, it has a wide range of applications. Male models for tooling, such as silicone rubber molds and investment castings, can be developed using RP models. The parts or components fabricated using RP technique can be employed for final use in some circumstances, however, the RP material is often not robust or precise enough.

Due to rapid development of RP technology, selection of the most apposite RP process to satisfy varying requirements of the customers from a pool of available processes has become increasingly important. This selection process also depends on some conflicting criteria. Furthermore, each of the RP processes has its own strengths, weaknesses, applications, utilities, and limitations. Using the modified TOPSIS method and based on six criteria, like accuracy (A), surface roughness (R), tensile strength (S), elongation (E), cost of the part (C), and build time (B), Byun & Lee [34] developed a decision support system for selection of a RP process. In this problem, cost of the manufactured part/component and build time are treated as linguistic variables and are subsequently converted into equivalent ranked value judgments using a fuzzy conversion scale [4]. The quantitative and qualitative data for this RP process selection problem are provided in Table 5.1. For this problem, S and E are considered as the beneficial criteria, whereas, A, R, C, and B are the non-beneficial criteria. Rao & Patel [4] determined the corresponding weights of those criteria as $w_A = 0.3185$, $w_R = 0.3185$, $w_S = 0.1291$, $w_E = 0.1291$, $w_C = 0.0524$, and $w_B = 0.0524$ using AHP method which are subsequently adopted here for the MOORA method-based analysis. Table 5.2 represents the MOORA method-based solution for this RP process selection problem. It can be observed from this table that SLA3500 emerges out as the best alternative, followed by Quadra. On the other hand, Z402 is identified to be the last choice. In Table 5.2, all the assessment values are observed to be negative because this problem consists of four non-beneficial criteria against two beneficial criteria. Both Byun & Lee [34]; Rao & Patel [4] derived the same observations while solving this RP process selection problem using the modified TOPSIS and PROMETHEE methods, respectively. Table 5.2 also exhibits the ranking results while solving this problem using reference point approach, MULTIMOORA and MOOSRA methods.

TABLE 5.1 Data for the RP Process Selection Problem [34]

RP Process	Accuracy (A)	Surface Roughness (R)	Tensile Strength (S)	Elongation (E)	Cost of the Part (C)	Build Time (B)
SLA3500	120	6.5	65	5.0	VH (0.745)	M (0.500)
SLS2500	150	12.5	40	8.5	VH (0.745)	M (0.500)
FDM8000	125	21.0	30	10.0	H (0.665)	VH (0.745)
LOM1015	185	20.0	25	10.0	SH (0.590)	SL (0.410)
Quadra	95	3.5	30	6.0	VH (0.745)	SL (0.410)
Z402	600	15.5	5	1.0	VVL (0.135)	VL (0.255)

Source: Reprinted with permission from Ref. [34]. © 2021 Springer.

TABLE 5.2 Assessment Values for the RP Process Selection Problem

RP Process	A	R	S	E	C	B	y_i	Rank	Reference Point	MULTIMOORA	MOOSRA
SLA3500	0.1777	0.1808	0.7145	0.2735	0.4737	0.4144	–0.0332	1	0.3878 (1/2/3)	30.9766 (2)	0.7935 (1)
SLS2500	0.2222	0.3477	0.4397	0.4650	0.4737	0.4144	–0.1113	3	0.3878 (1/2/3)	13.4810 (3)	0.5121 (3)
FDM8000	0.1851	0.5842	0.3298	0.5470	0.42287	0.6174	–0.1864	4	0.4869 (5)	6.3883 (5)	0.3779 (4)
LOM1015	0.2740	0.5564	0.2748	0.5470	0.3752	0.3398	–0.1959	5	0.459 (4)	7.7352 (4)	0.3514 (5)
Quadra	0.1407	0.0974	0.3298	0.3282	0.4737	0.3398	–0.0335	2	0.3878 (1/2/3)	49.0809 (1)	0.7171 (2)
Z402	0.8887	0.4312	0.0549	0.0547	0.0859	0.2114	–0.4218	6	0.748 (6)	0.4325 (6)	0.0325 (6)

KEYWORDS

- **abrasive jet machining**
- **abrasive water jet machining**
- **computerized numerical control**
- **electrochemical machining**
- **electrochemical micro-machining**
- **laser beam machining**

REFERENCES

1. Brauers, W. K. M., (2004). *Optimization Methods for a Stakeholder Society. A Revolution in Economic Thinking by Multiobjective Optimization.* Boston: Kluwer Academic Publishers.
2. Brauers, W. K. M., Zavadskas, E. K., Peldschus, F., & Turskis, Z., (2008). Multi-objective decision-making for road design. *Transport, 23,* 183–193.
3. Brauers, W. K. M., & Zavadskas, E. K., (2009). Robustness of the multi-objective MOORA method with a test for the facilities sector. *Technological and Economic Development of Economy: Baltic Journal on Sustainability, 15,* 352–375.
4. Rao, R. V., & Patel, B. K., (2009). Decision making in the manufacturing environment using an improved PROMETHEE method. *International Journal of Production Research, 48,* 4665–4682.
5. Brauers, W. K. M., & Zavadskas, E. K., (2011). MULTIMOORA optimization used to decide on a bank loan to buy property. *Technological and Economic Development of Economy, 17,* 174–188.
6. Brauers, W. K. M., & Zavadskas, E. K., (2010). Project management by MULTIMOORA as an instrument for transition economies. *Technological and Economic Development of Economy, 16,* 5–24.
7. Baležentis, A., Baležentis, T., & Valkauskas, R., (2010). Evaluating situation of Lithuania in the European Union: Structural indicators and MULTIMOORA method. *Technological and Economic Development of Economy, 16,* 578–602.
8. Chakraborty, S., (2011). Applications of the MOORA method for decision making in manufacturing environment. *International Journal of Advanced Manufacturing Technology, 54,* 1155–1166.
9. Gadakh, V. S., (2011). Application of MOORA method for parametric optimization of milling process. *International Journal of Applied Engineering Research, 1*(4), 743–758.
10. Karande, P., & Chakraborty, S., (2012). Application of multi-objective optimization on the basis of ratio analysis (MOORA) method for materials selection. *Materials and Design, 37*, 317–324.
11. Gadakh, V. S., Shinde, V. B., & Khemnar, N. S., (2013). Optimization of welding process parameters using MOORA method. *International Journal of Advanced Manufacturing Technology, 69*, 2031–2039.

12. Stanujkic, D., Đorđević, B., & Đorđević, M., (2013). Comparative analysis of some prominent MCDM methods: A case of ranking Serbian banks. *Serbian Journal of Management, 8*(2), 213–241.
13. Jagadish, & Ray, A., (2014). Green cutting fluid selection using MOOSRA method. *International Journal of Research in Engineering and Technology, 3*(3), 559–563.
14. Achebo, J., & Odinikuku, W. E., (2015). Optimization of gas metal arc welding process parameters using standard deviation (SDV) and multi-objective optimization on the basis of ratio analysis (MOORA). *Journal of Minerals and Materials Characterization and Engineering, 3*, 298–308.
15. Ertuğrul, İ., & Öztaş, T., (2015). The application of sewing machine selection with the multi-objective optimization on the basis of ratio analysis method (MOORA) in apparel sector. *Tekstilve Konfeksiyon, 25*(1), 80–85.
16. Khan, A., & Maity, K., (2015). Parametric optimization of some non-conventional machining processes using MOORA method. *International Journal of Engineering Research in Africa, 20*, 19–40.
17. Madić, M., Radovanović, M., & Petković, D., (2015). Non-traditional machining processes selection using multi-objective optimization on the basis of ratio analysis method. *Journal of Engineering Science and Technology, 10*(11), 1441–1452.
18. Mandal, U. K., & Sarkar, B., (2015). An enquiry of severe manufacturing system under utopian environment. *International Journal of Management and Applied Science, 1*(1), 26–34.
19. Patel, J. D., & Maniya, K. D., (2015). Application of AHP/MOORA method to select wire cut electrical discharge machining process parameter to cut EN31 alloys steel with brass wire. *Materials Today: Proceedings, 2*, 2496–2503.
20. Sarkar, A., Panja, S. C., Das, D., & Sarkar, B., (2015). Developing an efficient decision support system for non-traditional machine selection: An application of MOORA and MOOSRA. *Production & Manufacturing Research, 3*(1), 324–342.
21. Karuppanna, P. N., & Sekar, K., (2016). Optimal alternative selection using MOORA in industrial sector: A review. *International Journal of Fuzzy Logic Systems, 6*(2), 1–21.
22. Singarave, B., Selvaraj, T., & Vinodh, S., (2016). Multi-objective optimization of turning parameters using the combined MOORA and entropy method. *Transactions of the Canadian Society for Mechanical Engineering, 40*(1), 101–111.
23. Adalı, E. A., & Işık, A. T., (2017). The multi-objective decision making methods based on MULTIMOORA and MOOSRA for the laptop selection problem. *Journal of Industrial Engineering International, 13*, 229–237.
24. Parikh, V. K., Badgujar, A. D., & Ghetiya, N. D., (2017). Multi-objective optimization of FSW process parameter using MOORA method. *International Journal of Advances in Production and Mechanical Engineering, 3*(2), 19–24.
25. Singh, B., (2017). Applications of MOORA method for benchmarking decision in Indian industries. *International Journal of Advanced Operations Management, 9*(2), 88–105.
26. Dwivedi, S. K., & Dwivedi, A., (2018). Application of MOORA and WSM method for supplier selection in manufacturing. *International Journal of Advance Research and Development, 3*(7), 114–117.
27. Maghsoodi, A. I., Abouhamzeh, G., Khalilzadeh, M., & Zavadskas, E. K., (2018). Ranking and selecting the best performance appraisal method using the MULTIMOORA approach integrated Shannon's entropy. *Frontiers of Business Research in China, 12*(2), 1–21.

28. Mahale, V., Bijwe, J., & Sinha, S., (2018). Application and comparative study of new optimization method for performance ranking of friction materials. *Proceedings of Institution of Mechanical Engineers Part J: Journal of Engineering Tribology, 232*(2), 143–154.
29. Majumder, H., & Saha, A., (2018). Application of MCDM based hybrid optimization tool during turning of ASTM A588. *Decision Science Letters, 7*, 143–156.
30. Jain, V., & Ajmera, P., (2019). Application of MADM methods as MOORA and WEDBA for ranking of FMS flexibility. *International Journal of Data and Network Science, 3*, 119–136.
31. Abhang, L. B., Iqbal, M., & Hameedullah, M., (2020). Optimization of machining process parameters using Moora method. *Defect and Diffusion Forum, 402*, 81–89.
32. Kaladhar, M., (2020). Modeling and optimization for surface roughness and tool flank wear in hard turning of AISI 4340steel (35 HRC) using TiSiN-TiAlN nanolaminate coated insert. *Multidiscipline Modeling in Materials and Structures*. doi: 10.1108/MMMS-12-2019-0217.
33. Kuriakose, S., Parenti, P., & Annoni, M., (2020). Micro extrusion of high aspect ratio bi-lumen tubes using, 17-4PH stainless steel feedstock. *Journal of Manufacturing Processes, 58*, 443–457.
34. Byun, H. S., & Lee, K. S., (2004). A decision support system for the selection of a rapid prototyping process using the modified TOPSIS method. *International Journal of Advanced Manufacturing Technology, 26*, 1338–1347.

CHAPTER 6

TECHNIQUE FOR ORDER OF PREFERENCE BY SIMILARITY TO IDEAL SOLUTION (TOPSIS)

Hwang & Yoon [1] first developed this method, with subsequent refinements by Yoon [2] and Hwang et al. [3]. It is based on the assumption that the optimal alternative should be the one with the least Euclidean distance from the positive-ideal solution (PIS) and farthest away from the negative-ideal solution (NIS). The solution that maximizes the beneficial criteria while minimizing the non-beneficial criteria is known as a PIS. The NIS, on the other hand, maximizes the non-beneficial criteria while minimizing the beneficial criteria. In short, the PIS consists of all the best values that the considered criteria can have, whereas, the NIS is composed of all the worst values that the criteria can achieve [4].

In this method, the candidate alternatives are represented as points in an n-dimensional Euclidean space with every dimension corresponding to each criterion, as shown in Figure 6.1. It measures the distances of various alternatives (A_1, A_2, …, A_m) from the PIS (A^+) and the NIS (A^-). Based on these calculated Euclidean distances, the best alternative is selected. The procedural steps of TOPSIS method are enlisted as below:

- **Step 1:** Develop the corresponding decision matrix X with elements x_{ij} in which there are m alternatives and n criteria. The element x_{ij} indicates the performance of ith alternative in relation to j^{th} criterion. Thus, $X = (x_{ij})_{m \times n}$ ($i = 1, 2, \ldots, m; j = 1, 2, \ldots, n$).
- **Step 2:** In order to make all the elements of the initial decision matrix dimensionless and comparable, vector normalization is performed to obtain the normalized decision matrix, $R = (r_{ij})_{m \times n}$ using the following equation:

Multi-Criteria Decision-Making Methods in Manufacturing Environments: Models and Applications.
Shankar Chakraborty, Prasenjit Chatterjee, & Partha Protim Das (Authors)

$$r_{ij} = x_{ij} \Big/ \left[\sum_{i=1}^{m} x_{ij}^2 \right]^{1/2} \tag{6.1}$$

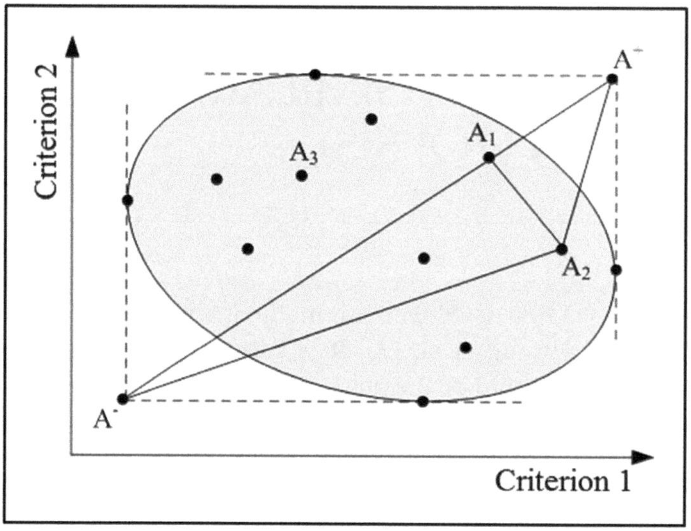

FIGURE 6.1 Fundamental of TOPSIS method.

- **Step 3:** Using the criteria weights, determine the weighted normalized matrix (V) while multiplying all the elements of the normalized decision matrix by the corresponding criteria weights.

$$v_{ij} = w_j \times r_j \tag{6.2}$$

- **Step 4:** Obtain the PIS (best) and the NIS (worst) using the following equations:

$$V^+ = \left\{ \left(\sum_i^{\max} v_{ij} \Big/ j \in J \right), \left(\sum_i^{\min} v_{ij} \Big/ j \in J' \right) \Big/ i = 1,2,\ldots,m \right\} = \left(V_1^+, V_2^+, V_3^+, \ldots, V_n^+ \right) \tag{6.3}$$

$$V^- = \left\{ \left(\sum_i^{\min} v_{ij} \Big/ j \in J \right), \left(\sum_i^{\max} v_{ij} \Big/ j \in J' \right) \Big/ i = 1,2,\ldots,m \right\} = \left(V_1^-, V_2^-, V_3^-, \ldots, V_n^- \right) \tag{6.4}$$

where; $J = (j = 1,2,\ldots,n)/j$ indicates beneficial criteria; and $J' = (j = 1,2,\ldots,n)/j$ represents non-beneficial criteria.

➢ **Step 5:** Determine the separation measures. The separations of each alternative from the PIS and the NIS are computed based on their Euclidean distances, as given in the following equations:

$$S_i^+ = \left\{ \sum_{j=1}^{n} \left(v_{ij} - V_j^+ \right)^2 \right\}^{0.5}, i = 1, 2, ..., m \tag{6.5}$$

$$S_i^- = \left\{ \sum_{j=1}^{n} \left(v_{ij} - V_j^- \right)^2 \right\}^{0.5}, i = 1, 2, ..., m \tag{6.6}$$

➢ **Step 6:** The following equation is applied to calculate the relative proximity of a given alternative to the ideal alternative (closeness coefficient):

$$C_i = S_i^- / \left(S_i^+ + S_i^- \right) \tag{6.7}$$

➢ **Step 7:** The most and the least favored alternatives are now identified when the considered alternatives are arranged in descending order of their C_i values. If $C_i = 1$, the alternative solution has the best condition, and if $C_i = 0$, the alternative solution has the worst condition. It signifies that the alternative which is distant from the NIS, becomes the best solution.

6.1 LITERATURE REVIEW

Byun & Lee [5] applied TOPSIS to identify the most appropriate RP technology while treating part cost, build time, strength, accuracy, roughness, and elongation as the most important evaluation criteria. Behzadian et al. [6] provided an exhaustive review of the applications of TOPSIS in diverse domains, including business and marketing management, supply chain management and logistics, engineering and manufacturing systems, human resources management, safety, and environmental management, energy management, water resources management, chemical engineering, etc. A framework for its future applications was also provided. Gadakh [7] presented the application of TOPSIS for solving multi-objective optimization problems in WEDM processes. Iç [8] endeavored to integrate TOPSIS and design of experiments (DoE) for solving four computer-integrated manufacturing technology selection problems, i.e., an industrial robot, a RP process, a CNC machine tool and a plant layout design. A regression metamodel was subsequently developed to reduce the computational steps of TOPSIS while providing more reliable results. Shahroudi & Tonekaboni [9] considered a supplier selection problem having both qualitative and quantitative criteria, and solved the same using TOPSIS. While applying TOPSIS, Vimal et al. [10] identified the most suitable supplier in the supply chain cycle of a manufacturing industry. Mansor et al. [11] considered three candidate thermoset materials, i.e., polyester, vinyl ester and epoxy matrix, and applied TOPSIS to identify the most appropriate material based on eight performance criteria for automotive bumper beam applications. Parida & Routara [12] applied TOPSIS for multi-response optimization of a dry turning process. Feed rate, cutting speed and depth of cut were considered as the input parameters, and surface roughness and material removal rate (MRR) were the process outputs. In order to achieve maximum MRR and minimum surface roughness, the ideal combination of the turning process parameters was finally identified. Employing TOPSIS, Senthil et al. [13] attempted to optimize the performance of an electrical discharge machining (EDM) process with Al-CuTiB_2 metal matrix composites as the work material. Discharge current, pulse-on time and pulse-off time were the input parameters, and MRR, tool wear rate and surface roughness were the process outputs.

Monjezi et al. [14] investigated the blasting operation in a mine and applied TOPSIS to identify the most suitable blasting pattern considering fragmentation, flyrock, and cost as the predominant evaluation criteria. Venkatesh et al. [15] applied TOPSIS to single out the best supplier for blood bags in a healthcare unit while validating its accuracy and effectiveness as a systematic decision-making tool. Yuvaraj & Pradeep Kumar [16] adopted TOPSIS as a multi-response optimization tool to identify the best parametric intermix of an AWJM process. During the process optimization, traverse rate, water jet pressure, flow rate of abrasives and stand-off distance were considered as the process parameters, whereas cutting rate, surface roughness, depth of penetration, taper cut ratio and top kerf width were the responses. Khan & Maity [17] applied TOPSIS to select the best modern manufacturing process from a pool of several alternatives, i.e., electrochemical honing (ECH), AWJM, AJM, LBM, PAC, LC, and EDM while recognizing the optimal combinations of the considered cutting parameters.

Applying TOPSIS, Khan & Maity [18] determined the most preferred combination of different process parameters in order to simultaneously optimize cutting force, surface roughness, machining temperature and MRR while turning CP-Ti Grade 2 pure titanium material. In a cryogenically cooled micro-electrical discharge drilling process, Manivannan & Pradeep Kumar [19] considered current, pulse-on duration, pulse-off duration and gap voltage as the controllable parameters, and taper angle, overcut, circularity at the entry and exit, MRR, tool wear rate and average roughness as the responses. The optimal parametric settings of the said drilling process were determined using TOPSIS. Shukla et al. [20] considered different manufacturing processes, like milling, drilling, turning, EDM, AJM, and micro-machining, and reviewed the applications of TOPSIS for parametric optimization of those processes. Asodariya et al. [21] employed TOPSIS for identification of the most apposite material for a flywheel design. The criteria weights were estimated using AHP and entropy methods, and the ranking performance of TOPSIS was later compared with that of COPRAS method. Genetic algorithm was finally employed for weight optimization of flywheel design with respect to a set of constraints. Durmusoglu [22] applied TOPSIS to evaluate alternative activity relationship charts based on environmental aspect, safety concerns, and manufacturing efficiency factors for arriving at the most suitable sustainable layout design. Khajanchee & Pradhan [23] compared the ranking performance of TOPSIS against multi-response signal-to-noise ratio, weighted signal-to-noise ratio and gray relational grade analysis methods while considering a case study involving laser welding of automotive gears made of 16MnCr5 alloy steel. Ma et al. [24] combined TOPSIS and information entropy technique for material selection considering life cycle assessment method. Information entropy technique was utilized to estimate the criteria weights, whereas, the alternative materials were ranked using TOPSIS. Yadav et al. [25] extensively reviewed the applications of TOPSIS in diverse domains, including electronics, information technology, manufacturing, biological science, etc., and demonstrated its procedural steps while solving a hypothetical example having six alternatives and five criteria.

Jiang et al. [26] conducted cutting operation on AISI 1045 steel, considering tradeoff among carbon emissions, surface roughness and processing time. The optimal cutting conditions were determined using non-dominated sorting genetic algorithm (NSGA-II) and were later validated using TOPSIS. Angelo et al. [27] optimized a fused deposition process in additive manufacturing using S-metric selection evolutionary multi-objective algorithm and verified the derived results while employing TOPSIS. Ersoy [28] considered cost, lifting capacity, height of lifting, lifting speed with load, speed of lowering with load and movement speed as the most important evaluation criteria, and proposed the application of TOPSIS for selection of the best fully automatic stacker machine for a manufacturing unit. Based on an orthogonal array design plan, Leo Princely et al. [29] performed robotic deburring operation on aluminum workpieces while considering deburring wheel diameter, tool rotational speed and tool feed rate as the input parameters, and deburring force and surface finish as the process responses. The TOPSIS was applied to determine the optimal parametric intermixes of the deburring operation for having better machining performance. Nguyen & Cao [30] integrated TOPSIS and principal component analysis to determine the optimal combination of different parameters of a burnishing process. Burnishing speed, feed, and depth were the controllable parameters, and energy consumption, power factor, surface roughness and surface hardness were the responses. In an EDM process, Sahu & Mahapatra [31] studied the influences of pulse-on time and discharge current on surface roughness, surface crack density, white layer thickness and micro-hardness of white layer. Finally, the optimal parametric combination for the said process was determined using TOPSIS in order to simultaneously optimize all the responses.

6.2 MATERIAL SELECTION FOR CONVEYOR BELT

A conveyor belt is a system of mechanical components that support and push a belt, which then transports the bulk material. The belt, which forms the moving and supporting surface on which the transported material rides, is one of the five essential components of this system. The other components are the idlers, that form the supports for the carrying and return strands of the belt; the pulleys to support and rotate the belt while controlling its tension; the drive to impart power to one or more pulley to move the belt and its load; and the structure to support and maintain alignment of the idlers and pulleys, while supporting the driving machinery. Besides being used for the transportation of industrial and agricultural materials, conveyor belts are also extensively utilized in self-unloading bulk freighters and live bottom trucks. This technology is commonly used in conveyor systems, such as moving sidewalks or escalators, and in industrial assembly lines. The conveyor belt dimensions are made flexible to match the user's specifications or application needs. The conveyor belts are usually made of a tough and strong material, making them capable of carrying heavy loads with ease. Its capacity may range from 500 to 5,000 m^3/hr. and more. The distance covered by a conveyor belt system may be as small as 50 m (for indoor work) to 5 km for transporting raw materials from mines [32]. While selecting a suitable belt material for a conveyor system, different important properties, like high strength, low elongation, high transverse flexibility and type of the cover layer are usually considered. Specially treated high strength fabric or cords can achieve higher tensile resistance with a minimum number of plies, thus reducing the weight and overall cost of the belt conveyor's structure and operation. This type of fabric can also achieve low elongation, requiring less readjustments while assuring higher service life. It can ensure high flexibility, allowing smaller pulleys and thus more compact structure and lower cost, and also increase transverse flexibility allowing high trough angle and thus higher bulk transport capacity. Synthetic rubber compounds for the cover layer of the belt conveyor can impart high resistance to diverse adverse working conditions and allow better resistance to heat, cold, oil, abrasion, static electricity build-up or chemical and bacteriological attack.

The wearing out of the conveyor belts is the most important factor to be considered by the designer. It mainly depends on the operating conditions of the belt and also on the type of the conveyed material. It has been noticed that about 80% of all the damages of a conveyor belt occur at its loading place. The impact force, which is one of the basic reasons for wearing out of the conveyor belt, rises at the input place. The impact of the sharp-edged pieces of the conveyed material on the belt causes the point load. If the impact energy is greater than the belt's capability, the conveyor belt damage occurs, mainly in the form of transverse and longitudinal scratches, perforations, and punctures at the top cover. The warp of the fabric or material of the cords defines the tensile strength of the conveyor belt. To enhance the desired characteristics of the belt, fabrics are frequently manufactured using a combination of different materials for the warp.

The fabrics or cords are impregnated with a special compound rubber to provide a permanent elastic bond with the surrounding layers. For selecting the right conveyor belt material for a particular industrial application, different mechanical properties of the belt material, and quality and gauge of the cover material are of prime importance. The materials that are most popularly used for conveyor belts are cotton, Kevlar, leather, aramid, neoprene, nylon, polyester, rubber, silicon, Teflon, and steel. Selection of the right conveyor belt material is based on the specific requirements of the conveyor system; the material being conveyed and the conditions under which it would operate as well as its cost-effectiveness.

In this demonstrative example, the most suitable conveyor belt material is selected for a given engineering application from the available list of aramid materials. Aramid material is as light as other synthetic fibers, like polyester and nylon, but as strong as steel. It has low elongation, low creep and excellent resistance against heat and chemicals. The aramid conveyor belts are fatigue-resistant throughout the entire lifetime. The reinforcement of aramid fibers does not corrode and has no effect on the wearing out of the belts. Different selection criteria for aramid belts with low elongation straight warp fabric are considered as the maximum working tension (MWT) (in (N/mm), carcass thickness (CT) (in mm), modulus of elasticity (ME) (in KN/mm), belt weight (BW) (kg/m^2) and cost per unit length (C) (in INR).

Among these five criteria, MWT and ME are the beneficial attributes as their higher values are desirable, whereas, CT, BW, and C are the non-beneficial attributes because their lower values are always preferable.

Selection of the most apposite conveyor belt material usually results in the lowest total cost per unit weight over its effective service life. For this, effective belt tension needs to be calculated while taking into account the tension required to move the empty belt, tension required to move the load horizontally and tension necessary to raise the load. Effective belt tension, also called allowable working tension, is the most critical factor for selection of the reinforcement ply combination for aramid fabric. The tensile strength of the belt material must be adequate to transmit power required in conveying the material over the distance involved.

Reinforcement or a carcass in the belt provides the required tensile strength and impact resistance to the belt, imparting rigidity for load support, and helping in joining the belt and force transmission. Carcass can be defined as a multi-element component of the conveyor belt connected with elastic material to ensure connection of the single components with one another and easy dispersal of energy in the conveyor belt while in operation. It also supports the load and provides structural strength to the conveyor belt. It consists of layers of aramid fabric or longitudinal cords, impregnated, or embedded in a specially treated elastic rubber compound. The breaker fabric, a specially woven polyamide fabric, is used on top of the carcass to add improved impact resistance. The belt carcass must have the essential qualities to withstand the weight, adapt to the curvature of the troughing idlers when empty, and bend well around the conveyor pulleys. The carcass is covered by a top protective edge, which resists it from mechanical damage caused by the delivered material, humidity, and chemical and thermal impacts on the conveyor belt. The bottom cover interacts with the conveyor's pulleys and idlers, protecting the carcass from any detrimental effect.

Cover thickness is also a function of the application, type of the material to be conveyed and frequency of the loading cycle. The quality and gauge of the cover material must be suitable to withstand the physical and chemical effects of the material conveyed. It also has a direct effect on the carcass thickness, BW and cost per unit length of the conveyor belt made of aramid fabric. The ME is a material's tendency to be deformed elastically (non-permanently) when a force is applied to it. The ME of a belt material measures its stiffness. Table 6.1 provides different selection criteria based on which the most suitable conveyor belt material is to be chosen from a set of aramid materials with low elongation straight warp fabric.

TABLE 6.1 Quantitative Data for Aramid Belt Materials

Designation of Belt	Belt Material	MWT (N/mm)	CT) (mm)	ME (KN/mm)	BW (kg/m²)	C (INR)
DP 500	M1	63	2.5	23	11.7	2,255
DP 630	M2	80	2.6	27	11.8	2,345
DP 800	M3	100	2.7	31	12.0	2,436
DP 1000	M4	125	3.5	37	12.8	3,157
DP 1250	M5	160	3.6	44	16.2	3,247
DP 1400	M6	175	3.7	49	16.4	3,338
DP 1600	M7	200	3.8	54	16.5	3,428
DP 1800	M8	225	4.2	60	16.9	3,789
DP 2000	M9	250	4.5	66	17.3	4,059
DP 2500	M10	315	5.0	80	17.9	4,510

As all the criteria considered for selecting the right aramid material for conveyor belt design are having different units and dimensions, their values are first vector normalized using Eqn. (6.1) and the corresponding normalized decision matrix is shown in Table 6.2.

TABLE 6.2 Normalized Decision Matrix

Designation of Belt	Belt Material	MWT	CT	ME	BW	C
DP 500	M1	0.1073	0.2139	0.1449	0.2443	0.2139
DP 630	M2	0.1363	0.2225	0.1701	0.2464	0.2225
DP 800	M3	0.1704	0.2311	0.1953	0.2506	0.2311
DP 1000	M4	0.2130	0.2995	0.2331	0.2673	0.2995
DP 1250	M5	0.2726	0.3081	0.2772	0.3383	0.3081
DP 1400	M6	0.2981	0.3167	0.3087	0.3425	0.3167
DP 1600	M7	0.3407	0.3252	0.3402	0.3446	0.3252
DP 1800	M8	0.3833	0.3595	0.3780	0.3529	0.3594
DP 2000	M9	0.4259	0.3851	0.4158	0.3613	0.3851
DP 2500	M10	0.5366	0.4279	0.5040	0.3738	0.4279

In order to determine the relative importance of each of the evaluation criteria affecting the conveyor belt material selection decision, a pair-wise comparison matrix, as shown in Table 6.3, is developed based on AHP method. The criteria weights are obtained as $w_{MWT} = 0.1450$, $w_{CT} = 0.1233$, $w_{ME} = 0.2737$, $w_{BW} = 0.2197$, and $w_C = 0.2383$. It is noticed that for this conveyor belt material selection problem, ME has the maximum importance, while CT is the least important selection criterion.

TABLE 6.3 Pair-Wise Comparison Matrix

Criteria	MWT	CT	ME	BW	C	Weight
MWT	1	1/2	1	1	1/2	0.1450
CT	2	1	1/3	1/2	1/3	0.1233
ME	1	3	1	1	2	0.2737
BW	1	2	1	1	1	0.2197
C	2	3	1/2	1	1	0.2383

Now, the values in the normalized decision matrix and the criteria weights are multiplied by applying Eqn. (6.2) to yield the weighted normalized decision matrix, as given in Table 6.4.

TABLE 6.4 Weighted Normalized Decision Matrix

Belt Material	MWT	CT	ME	BW	C
M1	0.0156	0.0264	0.0397	0.0537	0.0510
M2	0.0197	0.0274	0.0465	0.0541	0.0530
M3	0.0247	0.0285	0.0535	0.0551	0.0550
M4	0.0309	0.0369	0.0638	0.0587	0.0714
M5	0.0395	0.0380	0.0759	0.0743	0.0734
M6	0.0432	0.0390	0.0845	0.0753	0.0754
M7	0.0494	0.0401	0.0931	0.0757	0.0775
M8	0.0556	0.0443	0.1035	0.0775	0.0856
M9	0.0617	0.0475	0.1138	0.0794	0.0918
M10	0.0778	0.0527	0.1379	0.0821	0.1020

Based on Eqns. (6.3) and (6.4), the corresponding positive-ideal (best) and the negative-ideal (worst) solutions are estimated, as shown in Table 6.5.

TABLE 6.5 Positive-Ideal and Negative-Ideal Solutions

V^+	0.0778	0.0264	0.1379	0.0537	0.0510
V^-	0.0156	0.0527	0.0397	0.0821	0.1020

Using Eqns. (6.5) and (6.6), the separation measures of each alternative aramid material from the positive-ideal (best) and the negative-ideal (worst) solutions are computed, as shown in Table 6.6. Table 6.7 shows the relative closeness value of each alternative belt material, which is calculated using Eqn. (6.7).

TABLE 6.6 Separation Measures

Belt Material	M1	M2	M3	M4	M5	M6	M7	M8	M9	M10
S_i^+	0.1163	0.1083	0.0999	0.0908	0.0799	0.0726	0.0648	0.0615	0.0601	0.0641
S_i^-	0.0641	0.0623	0.0616	0.0505	0.0546	0.0609	0.0693	0.0777	0.0882	0.1163

TABLE 6.7 Relative Closeness Values

Belt Material	M1	M2	M3	M4	M5	M6	M7	M8	M9	M10
C_i	0.3551	0.3653	0.3814	0.3573	0.4059	0.4562	0.5170	0.5582	0.5947	0.6449

Now, the alternative aramid materials for conveyor belt design are arranged in descending order according to their relative closeness values. From Table 6.7, it can be propounded that belt material M10 (DP 2500) is the best choice for designing the conveyor belt for the given industrial application. The final ranking of the aramid materials is M10-M9-M8-M7-M6-M5-M3-M2-M4-M1. Belt material M10 (DP 2500) outperforms other alternatives due to its high working tension and ME, although its thickness, weight, and cost per unit length are higher than the others to some extent.

KEYWORDS

- **electrical discharge machining**
- **electrochemical honing**
- **laser cutting**
- **maximum working tension**
- **modulus of elasticity**
- **non-dominated sorting genetic algorithm**

REFERENCES

1. Hwang, C. L., & Yoon, K., (1981). *Multiple Attribute Decision Making: Methods and Applications.* Springer-Verlag, New York.
2. Yoon, K., (1987). A reconciliation among discrete compromise situations. *Journal of the Operational Research Society, 38,* 277–286.
3. Hwang, C. L., Lai, Y. J., & Liu, T. Y., (1993). A new approach for multiple objective decision making. *Computers and Operational Research, 20,* 889–899.
4. Çelikbilek, Y., & Tüysüz, F., (2020). An in-depth review of theory of the TOPSIS method: An experimental analysis. *Journal of Management Analytics, 7*, 281–300.
5. Byun, H. S., & Lee, K. H., (2005). A decision support system for the selection of a rapid prototyping process using the modified TOPSIS method. *International Journal of Advanced Manufacturing Technology, 26,* 1338–1347.
6. Behzadian, M., Otaghsara, S. K., Yazdani, M., & Ignatius, J., (2012). A state-of-the-art survey of TOPSIS applications. *Expert Systems with Applications, 39*, 13051–13069.
7. Gadakh, V. S., (2012). Parametric optimization of wire electrical machining using TOPSIS method. *Advances in Production Engineering & Management, 7*(3), 157–164.
8. Iç, Y. T., (2012). An experimental design approach using TOPSIS method for the selection of computer-integrated manufacturing technologies. *Robotics and Computer-Integrated Manufacturing, 28*, 245–256.
9. Shahroudi, K., & Tonekaboni, S. M. S., (2012). Application of TOPSIS method to supplier selection in Iran auto supply chain. *Journal of Global Strategic Management, 6*(2), 123–131.
10. Vimal, J., Chaturvedi, V., & Dubey, A. K., (2012). Application of TOPSIS method for supplier selection in manufacturing industry. *International Journal of Research in Engineering & Applied Sciences, 2*(5), 25–35.
11. Mansor, M. R., Sapuan, S. M., Hambali, A., Zainudin, E. S., & Nuraini, A. A., (2014). Materials selection of hybrid bio-composites thermoset matrix for automotive bumper beam application using topsis method. *Advances in Environmental Biology, 8*(8), 3138–3142.
12. Parida, A. K., & Routara, B. C., (2014). Multiresponse optimization of process parameters in turning of GFRP using TOPSIS method. *International Scholarly Research Notices* (p. 10). Article ID: 905828. http://dx.doi.org/10.1155/2014/905828.
13. Senthil, P., Vinodh, S., & Singh, A. K., (2014). Parametric optimization of EDM on Al-Cu/TiB_2 in-situ metal matrix composites using TOPSIS method. *International Journal of Machining and Machinability of Materials, 16*(1), 80–94.
14. Monjezi, M., Dehghani, H., Singh, T. N., Sayadi, A. R., & Gholinejad, A., (2015). Application of TOPSIS method for selecting the most appropriate blast design. *Arabian Journal of Geosciences, 5*, 95–101.
15. Venkatesh, V. G., Dubey, R., Joy, P., Thomas, M., Vijeesh, V., & Moosa, A., (2015). Supplier selection in blood bags manufacturing industry using TOPSIS model. *International Journal of Operational Research, 24*(4), 461–488.
16. Yuvaraj, N., & Pradeep, K. M., (2015). Multiresponse optimization of abrasive water jet cutting process parameters using TOPSIS approach. *Materials and Manufacturing Processes, 30*(7), 882–889.
17. Khan, A., & Maity, K. P., (2016). Application of MCDM-based TOPSIS method for the optimization of multi quality characteristics of modern manufacturing processes. *International Journal of Engineering Research in Africa, 23*, 33–51.

18. Khan, A., & Maity, K., (2017). Application of MCDM-based TOPSIS method for the selection of optimal process parameter in turning of pure titanium. *Benchmarking: An International Journal, 24*(7), 2009–2021.
19. Manivannan, R., & Pradeep, K. M., (2017). Multi-attribute decision-making of cryogenically cooled micro-EDM drilling process parameters using TOPSIS method. *Materials and Manufacturing Processes, 32*(2), 209–215.
20. Shukla, A., Agarwal, P., Rana, R. S., & Purohit, R., (2017). Applications of TOPSIS algorithm on various manufacturing processes: A review. M*aterials Today: Proceedings, 4,* 5320–5329.
21. Asodariya, H., Patel, H. V., Babariya, D., & Maniya, K. D., (2018). Application of multi-criteria decision making method to select and validate the material of a flywheel design. *Materials Today: Proceedings, 5,* 17147–17155.
22. Durmusoglu, Z. D. U., (2018). A TOPSIS-based approach for sustainable layout design: Activity relation chart evaluation. *Kybernetes, 47*(10), 2012–2024.
23. Khajanchee, A., & Pradhan, S. K., (2018). Comparison and identification of suitable multi-response optimization technique for laser welding process parameters. *Materials Today: Proceedings, 5,* 12197–12208.
24. Ma, F., Zhao, Y., Pu, Y., & Li, J., (2018). A comprehensive multi-criteria decision making model for sustainable material selection considering life cycle assessment method. *IEEE Access, 6,* 58338–58354.
25. Yadav, S. K., Joseph, D., & Jigeesh, N., (2018). A review on industrial applications of TOPSIS approach. *International Journal of Services and Operations Management, 30*(1), 23–38.
26. Jiang, Z., Gao, D., Lu, Y., & Liu, X., (2019). Optimization of cutting parameters for trade-off among carbon emissions, surface roughness, and processing time. *Chinese Journal of Mechanical Engineering, 32*, 94.
27. Angelo, L. D., Stefano, P. D., Dolatnezhadsomarin, A., Guardiani, E., & Khorram, E., (2020). A reliable build orientation optimization method in additive manufacturing: The application to FDM technology. *International Journal of Advanced Manufacturing Technology, 108*, 263–276.
28. Ersoy, Y., (2020). Green machine selection in a manufacturing company using TOPSIS method. *Academia Journal of Nature and Human Sciences, 6*(1), 45–56.
29. Leo, P. F., Senthil, P., & Selvaraj, T., (2020). Application of TOPSIS method for optimization of process parameters in robotic deburring. *Materials Today: Proceedings, 27*, 2137–2141.
30. Nguyen, T. T., & Cao, L. H., (2020). Optimization of the burnishing process for energy responses and surface properties. *International Journal of Precision Engineering and Manufacturing, 21*, 1143–1152.
31. Sahu, A. K., & Mahapatra, S. S., (2020). Surface characteristics of EDMed titanium alloy and AISI 1040 steel workpieces using rapid tool electrode. *Arabian Journal for Science and Engineering, 45*, 699–718.
32. Colijn, H., (1985). *Mechanical Conveyors for Bulk Solids*. Elsevier Science Publishers, USA.

CHAPTER 7

VIEKRITERIJUMSKO KOMPROMISNO RANGIRANJE (VIKOR)

Usually, the decision-making problems in the manufacturing environment are often characterized by several non-commensurable and conflicting criteria, and there may be no solution satisfying all the criteria simultaneously. Therefore, the solution is a set of non-inferior solutions or a compromise solution according to the decision maker's preference. A compromise solution for a problem with conflicting criteria can allow the decision maker to arrive at a final decision. The VIKOR (the Serbian name is 'VIekriterijumsko KOmpromisno Rangiranje' which means multi-criteria optimization and compromise solution) method was established by Zeleny [1], and later advocated by Opricovic & Tzeng [2, 3]. It is developed to solve MCDM problems with conflicting and non-commensurable attributes (attributes having different units), assuming that compromise can be acceptable for conflict resolution, when the decision maker wants a solution that is the closest to the positive-ideal solution (PIS) and the alternatives can be evaluated with respect to all the established attributes. It focuses on ranking and selecting the best alternative from a finite set of alternatives with conflicting criteria, and on proposing the compromise solution (one or more). The compromise solution is a feasible solution, which is the closest to the PIS and farthest from the negative-ideal solution (NIS), and a compromise means an agreement established by mutual concessions made between the alternatives.

The basic concept of the VIKOR method lies in defining the positive- and the negative-ideal points in the solution space. It evaluates a multi-criterion ranking index based on the closeness to the PIS. When each alternative is evaluated with respect to each criterion, the compromise ranking can be obtained while comparing the relative closeness measure to the positive-ideal

Multi-Criteria Decision-Making Methods in Manufacturing Environments: Models and Applications.
Shankar Chakraborty, Prasenjit Chatterjee, & Partha Protim Das (Authors)

alternative. The VIKOR method starts with an L_p-metric, which is used as an aggregating function in the compromise programming method and is developed into the multi-criteria measure for compromise ranking [4].

The L_p-metric is formulated as follows:

$$L_{p,i} = \left\{ \sum_{j=1}^{n} \left(w_j \left[\left(x_{ij}\right)_{max} - x_{ij} \right] / \left[\left(x_{ij}\right)_{max} - \left(x_{ij}\right)_{min} \right] \right)^p \right\}^{1/p} \quad 1 \le p \le \infty;\ i = 1, 2, ..., m \quad (7.1)$$

where; x_{ij} is the performance of ith alternative with respect to j^{th} criterion, m is the number of feasible alternatives, n is the number of criteria and w_j is the weight assigned to the jth criterion. The VIKOR method is not only developed with the above form of L_p-metric, but it also uses $L_{1,i}$ and $L_{\infty,i}$ values to formulate the ranking measure. The steps of VIKOR method are explained as below:

- **Step 1:** Identify the corresponding MCDM problem along with the major evaluation criteria and shortlist the candidate alternatives on the basis of the considered criteria satisfying the requirements. A quantitative or qualitative value is assigned to each identified criterion to construct the related decision matrix.
- **Step 2:** After shortlisting the alternatives and formation of the decision matrix, determine the best, $(x_{ij})_{max}$ and the worst, $(x_{ij})_{min}$ values for all the criteria.
- **Step 3:** The weights (relative importance) of the criteria under consideration are estimated using AHP or entropy method.
- **Step 4:** Calculate S_i and R_i values.

$$S_i = L_{1,i} = \sum_{j=1}^{n} w_j \left[\left(x_{ij}\right)_{max} - x_{ij} \right] / \left[\left(x_{ij}\right)_{max} - \left(x_{ij}\right)_{min} \right] \quad (7.2)$$

$$R_i = L_{\infty,i} = \text{Max of } \left\{ w_j \left[\left(x_{ij}\right)_{max} - x_{ij} \right] / \left[\left(x_{ij}\right)_{max} - \left(x_{ij}\right)_{min} \right] \right\} \quad (j = 1, 2, ..., n) \quad (7.3)$$

It is worthwhile to mention here that Eqn. (7.2) is only applicable to beneficial criteria. For non-beneficial criteria, the term $[(x_{ij})_{max} - x_{ij}]$ in Eqn. (7.2), is to be replaced by $[x_{ij} - (x_{ij})_{min}]$. Hence, for non-beneficial criteria, Eqn. (7.2) can be reformulated as:

$$S_i = L_{1,i} = \sum_{j=1}^{n} w_j \left[\left(x_{ij}\right) - \left(x_{ij}\right)_{min} \right] / \left[\left(x_{ij}\right)_{max} - \left(x_{ij}\right)_{min} \right] \quad (7.4)$$

When the value of p is small, the group utility is emphasized and as p increases, the individual regrets receive more weight. In addition, the compromise solution, $\min_i L_{p,i}$ would be chosen because its value is closest to the positive-ideal level. Therefore, between $\min_i S_i$ and $\min_i R_i$, $\min_i S_i$ expresses to minimize the sum of the individual regrets, while $\min_i R_i$ attempts to minimize the maximum individual regret. Thus, $\min_i S_i$ denotes the maximum group utility, whereas, $\min_i R_i$ identifies the minimum among the maximum individual regrets.

- **Step 5:** Calculate Q_i value.

$$Q_i = v\left((S_i - S_{i\min})/(S_{i\max} - S_{i\min})\right) + (1-v)\left((R_i - R_{i\min})/(R_{i\max} - R_{i\min})\right) \quad (7.5)$$

where; $S_{i\max}$ and $S_{i\min}$ are the maximum and minimum values of S_i, respectively; and $R_{i\max}$ and $R_{i\min}$ are the maximum and minimum values of R_i, respectively. The term v is introduced as a weight for the strategy of maximum group utility, whereas, $(1 - v)$ is the weight of the individual regret. The value of v lies between 0 and 1, with the most preferred value of v taken as 0.5. Thus, when $v > 0.5$, it represents a decision-making process that can adopt the strategy of maximum group utility (i.e., if v is large, group utility is emphasized), or by consensus when $v \approx 0.5$, or with veto when $v < 0.5$.

- **Step 6:** Arrange the alternatives in ascending order, according to the values of *S*, *R*, and *Q*. It provides three ranking lists. In the ranking list derived based on the ascending values of *Q*, $a(1)$ is the compromise solution (with minimum *Q* value) if the following two conditions are satisfied:
  - **C1:** 'Acceptable advantage'

$$Q(a(2)) - Q(a(1)) \geq DQ$$

where; $a(2)$ is the second-best alternative in the ranking list by *Q* and $DQ = 1/(m - 1)$.
  - **C2:** 'Acceptable stability in decision-making'

Alternative $a(1)$ must also be the best ranked by *S* or/and *R*. If any one of the above-mentioned conditions is not satisfied, a set of compromise solutions is then proposed as follows:

- Alternatives $a(1)$ and $a(2)$ if only condition C2 is not satisfied, or
- Alternatives $a(1)$, $a(2)$, ..., $a(M)$ if condition C1 is not satisfied, and $a(M)$ is determined by the relation $Q(a(M)) - Q(a(1)) < DQ$ for maximum M (the positions of these alternatives are 'in closeness').

The VIKOR method is an effective MCDM tool, particularly when the decision maker is not able, or does not know to express his/her preference at the beginning of the decision-making process. The obtained compromise solution can be accepted because it provides a maximum 'group utility' of the 'majority,' and a minimum of the individual regret of the 'opponent.' The compromise solutions can be the basis for negotiations, involving the decision maker's preference on criteria weights. In the VIKOR method, the performance ratings of the alternatives with respect to a set of criteria are quantified as crisp values. But, under any circumstances, crisp data are inadequate to model real-life situations. In addition, in case of conflicting situations or criteria, the decision maker must also consider imprecise or ambiguous data.

7.1 SURVEY OF THE LITERATURE

Tong et al. [5] attempted to adopt the VIKOR method as a multi-response optimization tool for identifying the most suitable parametric settings for plasma-enhanced chemical vapor deposition and copper chemical mechanical polishing processes. Chatterjee et al. [6] employed both VIKOR and ELECTRE methods to solve two material selection problems for mechanical components. The ranking performance of the considered methods was also contrasted. Chatterjee et al. [7] endeavored to solve two real-time industrial robot selection problems using VIKOR and ELECTRE methods. The applicability and potentiality of VIKOR as a compromise ranking method was also proved. Using VIKOR method, Ghaleb et al. [8] identified the most apposite manufacturing process with respect to productivity, accuracy, complexity, flexibility, material utilization, quality, and operation cost. Its performance was also appraised based on the number of alternatives and criteria, agility of decision-making, computational complexity, adequacy in supporting group decision-making and rank reversal. Tavakkoli-Moghaddam [9] integrated AHP and VIKOR methods for identification of the best plant location for a specific manufacturing organization. The criteria weights were estimated using AHP method, while VIKOR method helped in compromised ranking of the candidate alternatives.

Athawale et al. [10] employed VIKOR method as a compromise ranking tool to select the most appropriate robot for an industrial application while achieving the desired performance with minimum cost. Treating distance between the nozzle and rotor, type of extractive nozzle and type of draw-off tube as the control parameters in a rotor spinning process, Fallahpour & Moghassem [11] applied VIKOR method to determine their optimal operating levels for having maximum spinning performance. Dağ & Önder [12] considered 10 evaluation criteria, like supply of raw material, proximity to customer, proximity to airport, proximity to harbor, transportation cost, availability of skilled labor, labor cost, proximity to industrial zone, government facilities and investment cost for identification of the best facility location using VIKOR method. Jain & Raj [13] combined AHP and VIKOR methods for measuring flexibility in FMS based on 15 evaluation criteria. It was observed that amongst the candidate alternatives, production flexibility and program flexibility had the maximum and minimum impacts, respectively. Mohanty & Mahapatra [14] proposed the application of VIKOR method as an MCDM tool to identify an optimal set of design parameters for a given product so as to meet the customers' requirements.

While attempting to review the applications of VIKOR method, Yazdani & Graeml [15] categorized its applications in different domains, i.e., (a) design and manufacturing management; (b) business and marketing management; (c) supply chain and logistics management; (d) environmental resources and energy management; (e) construction management; (f) education management; (g) healthcare and risk management; and (h) tourism management. Koyee et al. [16] integrated Taguchi methodology and VIKOR method to determine the optimal input parameters while simulating the machining operation of standard duplex EN 1.4462 and super duplex EN 1.4410 stainless steel materials. Chatterjee & Chakraborty [17] compared the ranking performance of VIKOR method with its other variants, like comprehensive VIKOR, fuzzy VIKOR, regret theory based VIKOR, modified VIKOR and interval VIKOR methods. The application of the traditional VIKOR method was recommended for uncomplicated decision-making problems. Mardani et al. [18] performed a systematic review of VIKOR method in several application areas of engineering and management decision-making. Venkateswarlu & Dattatraya Sarma [19] solved a supplier selection problem using simple additive weighting (SAW) and VIKOR methods for a manufacturing organization.

Kumar et al. [20] considered three responses, i.e., material removal rate (MRR), surface roughness, and dimensional deviation during end milling of EN 31 steel material, and later applied the VIKOR method for parametric optimization of the said process. Sindhwani & Malhotra [21] stressed on the determination of certain facilitators and criteria that would help in effective and smooth implementation of the agile manufacturing system. The criteria weights were determined, and the considered facilitators were evaluated using the VIKOR method. Based on VIKOR method, Jing et al. [22] developed a tool selection model for lean management. It was claimed that the derived results would be more stable, not depending on the weights of the evaluation criteria. Baloyi & Meyer [23] applied the VIKOR method to identify the most appropriate mining method in the coal mining industry, and later compared its performance with other popular MCDM techniques. In a piston material selection problem, Dev et al. [24] first determined the weights of various material properties of the considered alternatives, and later applied the VIKOR method to single out the most appropriate piston material. Divya et al. [25] employed VIKOR method for parametric optimization of a turning process and also compared its performance with other MCDM tools. Using VIKOR method, Sindhwani et al. [26] identified the most important barrier to be considered for effective implementation of lean-green manufacturing system in an organization.

7.2 MATERIAL SELECTION FOR A FLYWHEEL

It has already been well established that materials play an important role in engineering design. For the design and manufacturing of different mechanical components, knowledge of material properties, cost, design considerations and their influences are mandatory. A large number of available alternative materials, having complex relationships with various selection parameters (mechanical properties), make the material selection process a challenging task. Selection of the most suitable material involves the study of a large number of factors, like mechanical, electrical, and physical properties, and cost considerations of the materials. The most important mechanical properties usually encountered in the material selection process are strength, stiffness, toughness, hardness, density, and creep resistance. Hence, there must be some methodological approach so as to assist the decision maker in selecting the most appropriate material for a given mechanical component. Perhaps the first step in the material selection process is to specify the performance requirements of the component, and broadly outline the main material characteristics and processing requirements. The main design task lies in comparing the properties of a finite set of materials and selecting the best one out of this set. While choosing a new material for a mechanical component, the decision maker usually applies trial and error methods or employs his/her knowledge from previous experimentations.

This illustrative example deals with the selection of the most appropriate material for the design of a flywheel [27] which is a device to store kinetic energy as used in automobiles, urban subway trains, mass transit buses, wind-power generators, etc. Despite its many advantages, relatively poor energy storage and the chance of a catastrophic failure hinder its applications. The main requirements in a flywheel design are to store the maximum amount of kinetic energy per unit mass and ensure against premature failure due to fatigue or brittle fracture. If the nature of failure is fatigue, the performance index would be σ_{limit}/ρ (where σ_{limit} is the fatigue limit of the material and ρ is the material density). This signifies that higher is the value of σ_{limit}/ρ, lower is the weight of the material for a given fatigue strength, and consequently, the kinetic energy per unit mass of the flywheel would be higher. For failure due to brittle fracture, fracture toughness (K_{IC}) of the material would be the performance measure. If the flywheel breaks into small pieces at final failure, the hazard would be much reduced. Thus, fragmentability of the flywheel material is an important property from the safety point of view. Hence, four criteria are considered here, i.e., fatigue limit (σ_{limit}/ρ), fracture toughness (K_{IC}/ρ), price per unit mass and fragmentability. Among these four criteria, fatigue limit, fracture toughness (FT) and fragmentability are beneficial where higher values are desirable, and price/mass is a non-beneficial attribute where smaller value is always preferable. Around 10 materials comprising four metals and six unidirectional fiber-reinforced-epoxy composites are considered as the alternatives [27]. Thus, the MCDM problem consists of 10 alternative materials and 4 material selection criteria, as shown in Table 7.1.

TABLE 7.1 Quantitative Data for the Flywheel Material Selection Problem [27]

SL. No.	Material	σ_{limit}/ρ	K_{IC}/ρ	Price/Mass	Fragment Ability
1.	300M	100	8.6125	4,200	Poor (3)
2.	2024-T3	49.6454	13.4752	2,100	Poor (3)
3.	7050-T73651	78.0142	12.5532	2,100	Poor (3)
4.	Ti-6Al-4V	108.8795	26.0042	10,500	Poor (3)
5.	E glass-epoxy FRP	70	10	2,735	Excellent (9)
6.	S glass-epoxy FRP	165	25	4,095	Excellent (9)
7.	Carbon-epoxy FRP	440.2516	22.0126	35,470	Fairly good (7)
8.	Kevlar 29-epoxy FRP	242.8571	28.5714	11,000	Fairly good (7)
9.	Kevlar 49-epoxy FRP	616.4384	34.2466	25,000	Fairly good (7)
10.	Boron-epoxy FRP	500	23	3,15,000	Good (5)

Source: Reprinted with permission from Ref. [27]. © 2021 Elsevier.

In order to solve this flywheel material selection problem using VIKOR method, the best and the worst values of all the criteria are first identified. Using entropy method, Jee & Kang [27] determined weights of the considered criteria as $w_1 = 0.4$, $w_2 = 0.3$, $w_3 = 0.2$, and $w_4 = 0.1$, which are also employed here for VIKOR method-based analysis. Now, the values of S_i and R_i are calculated using Eqns. (7.2) or (7.4) and (7.3), respectively, as given in Table 7.2. Table 7.2 also provides the values of Q_i for $v = 0.5$ and the compromise ranking list of the candidate materials. These materials are finally arranged in ascending order, according to the values of Q_i. The best choice of material for flywheel design is material 9 (Kevlar 49-epoxy FRP). Carbon-epoxy FRP is the second choice, and the last choice is material 2 (2024-T3). Jee & Kang [27] obtained a ranking of the alternative materials as 5-9-7-6-8-3-4-2-1-10 (for case 1), whereas, using the VIKOR method, the compromise ranking of materials is 9-10-8-6-7-5-2-4-1-3. It is observed that in the VIKOR method, the best choice of material for flywheel design remains the same.

While calculating Q_i values, the value of v is usually taken as 0.5, but actually, its value lies between 0 and 1. Table 7.3 exhibits the compromise rankings of the alternative materials for flywheel design for two extreme values of $v = 0.1$ and $v = 0.9$. In both these cases, the best choice of material (Kevlar 49-epoxy FRP) does not alter, although the rankings of the intermediate alternative materials change slightly.

TABLE 7.2 S_i, R_i, and Q_i Values

Material	S_i	R_i	Q_i	Rank
300M	0.76580	0.3644	0.9515	9
2024-T3	0.74311	0.4000	0.9844	10
7050-T73651	0.7339	0.3799	0.9507	8
Ti-6Al-4V	0.5599	0.3582	0.7356	6
E glass-epoxy FRP	0.6698	0.3856	0.9138	7
S glass-epoxy FRP	0.4280	0.3186	0.6538	5
Carbon-epoxy FRP	0.3221	0.1431	0.3408	2
Kevlar 29-epoxy FRP	0.3690	0.2636	0.5377	4
Kevlar 49-epoxy FRP	0.0479	0.0333	0	1
Boron-epoxy FRP	0.4804	0.2000	0.5287	3

TABLE 7.3 Ranking of Materials for Varying of v

Q_i ($v = 0.1$)	Material	Q_i ($v = 0.9$)	Material
0.9128	300M (7)	0.9903	300M (10)
0.9969	2024-T3 (10)	0.9720	2024-T3 (9)
0.9464	7050-T73651 (8)	0.9549	7050-T73651 (8)
0.8559	Ti-6Al-4V (6)	0.6152	Ti-6Al-4V (6)
0.9514	E glass-epoxy FRP (9)	0.8761	E glass-epoxy FRP (7)
0.7531	S glass-epoxy FRP (5)	0.5544	S glass-epoxy FRP (4)
0.3077	Carbon-epoxy FRP (2)	0.3738	Carbon-epoxy FRP (2)
0.6100	Kevlar 29-epoxy FRP (4)	0.4654	Kevlar 29-epoxy FRP (3)
0	Kevlar 49-epoxy FRP (1)	0	Kevlar 49-epoxy FRP (1)
0.4693	Boron-epoxy FRP (3)	0.5878	Boron-epoxy FRP (5)

KEYWORDS

- **boron-epoxy FRP**
- **entropy method**
- **mechanical components**
- **simple additive weighting**
- **tourism management**
- **Viekriterijumsko Kompromisno Rangiranje**

REFERENCES

1. Zeleny, M., (2002). *Multiple Criteria Decision Making*. McGraw Hill, New York.
2. Opricovic, S., & Tzeng, G. H., (2004). Compromise solution by MCDM methods: A comparative analysis of VIKOR and TOPSIS. *European Journal of Operational Research, 156*(2), 445–455.
3. Opricovic, S., & Tzeng, G. H., (2007). Extended VIKOR method in comparison with outranking methods. *European Journal of Operational Research, 178*(2), 514–529.
4. Rao, R. V., (2007). *Decision Making in the Manufacturing Environment Using Graph Theory and Fuzzy Multiple Attribute Decision Making Methods*. Springer Verlag, London.

5. Tong, L. I., Chen, C. C., & Wang, C. H., (2007). Optimization of multi-response processes using the VIKOR method. *International Journal of Advanced Manufacturing Technology, 31*, 1049–1057.
6. Chatterjee, P., Athawale, V. M., & Chakraborty, S., (2009). Selection of materials using compromise ranking and outranking methods. *Materials and Design, 30*, 4043–4053.
7. Chatterjee, P., Athawale, V. M., & Chakraborty, S., (2010). Selection of industrial robots using compromise ranking and outranking methods. *Robotics and Computer-Integrated Manufacturing, 26,* 483–489.
8. Ghaleb, A. M., Kaid, H., Alsamhan, A., Mian, S. H., & Hidri, L., (2010). Assessment and comparison of various MCDM approaches in the selection of manufacturing process. *Advances in Materials Science and Engineering* (p. 16). Article ID: 4039253. https://doi.org/10.1155/2020/4039253.
9. Tavakkoli-Moghaddam, R., Mousavi, S. M., & Heydar, M., (2011). An integrated AHP-VIKOR methodology for plant location. *International Journal of Engineering Transactions B: Applications, 24*(2), 127–137.
10. Athawale, V. M., Chatterjee, P., & Chakraborty, S., (2012). Selection of industrial robots using compromise ranking method. *International Journal of Industrial and Systems Engineering, 11*(1, 2), 3–15.
11. Fallahpour, A. R., & Moghassem, A. R., (2012). Evaluating applicability of VIKOR method of multi-criteria decision making for parameters selection problem in rotor spinning. *Fibers and Polymers, 13*(6), 802–808.
12. Dağ, S., & Önder, E., (2013). Decision-making for facility location using VIKOR method. *Journal of International Scientific Publication: Economy & Business, 7*(1), 308–330.
13. Jain, V., & Raj, T., (2014). Evaluation of flexibility in FMS by VIKOR methodology. *International Journal of Industrial and Systems Engineering, 18*(4), 483–498.
14. Mohanty, P. P., & Mahapatra, S. S., (2014). A compromise solution by VIKOR method for ergonomically designed product with optimal set of design characteristics. *Procedia Materials Science, 6*, 633–640.
15. Yazdani, M., & Graeml, F. R., (2014). VIKOR and its applications: A state-of-the-art survey. *International Journal of Strategic Decision Sciences, 5*(2), 28–45.
16. Koyee, R. D., Schmauder, S., Heisel, U., & Eisseler, R., (2015). Numerical modeling and optimization of machining duplex stainless steels. *Production & Manufacturing Research, 3*(1), 36–83.
17. Chatterjee, P., & Chakraborty, S., (2016). A comparative analysis of VIKOR method and its variants. *Decision Science Letters, 5*, 469–486.
18. Mardani, A., Zavadskas, E. K., Govindan, K., Senin, A. A., & Jusoh, A., (2016). VIKOR technique: A systematic review of the state-of-the-art literature on methodologies and applications. *Sustainability, 8*, 21. doi: 10.3390/su8010037.
19. Venkateswarlu, P., & Dattatraya, S. B., (2016). Selection of supplier by using saw and Vikor methods. *International Journal of Engineering Research and Application, 6*(9), 80–88.
20. Kumar, G. V. A., Reddy, D. V., & Nagaraju, N., (2018). Multi-objective optimization of end milling process parameters in machining of EN 31 steel: Application of AHP embedded with VIKOR and WASPAS methods. *Journal of Mechanical Engineering, 8*(4), 39–46.
21. Sindhwani, R., & Malhotra, V., (2018). An integrated approach for implementation of agile manufacturing system in an Indian manufacturing industry. *Benchmarking: An International Journal, 25*(4), 1106–1120.

22. Jing, S., Niu, Z., & Chang, P. C., (2019). The application of VIKOR for the tool selection in lean management. *Intelligent Manufacturing, 30*, 2901–2912.
23. Baloyi, V. D., & Meyer, L. D., (2020). The development of a mining method selection model through a detailed assessment of multi-criteria decision methods. *Results in Engineering, 8*, 100172.
24. Dev, S., Aherwar, A., & Patnaik, A., (2020). Material selection for automotive piston component using entropy-VIKOR method. *Silicon, 12*, 155–169.
25. Divya, C., Suvarna, R. L., & Singaravel, B., (2020). Application of MCDM methods for process parameter optimization in turning process: A review. In: Narasimham, G. S. V. L., Veeresh B. A., S Sreenatha, R. S., & Dhanasekaran, R., (eds.), *Recent Trends in Mechanical Engineering* (pp. 199–207). Springer Singapore.
26. Sindhwani, R., Singh, P. L., Kaushik, V., Sharma, S., & Phanden, R. K., (2020). Analysis of barriers to lean-green manufacturing system (LGMS): A multi-criteria decision-making approach. In: *Advances in Intelligent Manufacturing* (pp. 181–188). Springer Nature Switzerland.
27. Jee, D. H., & Kang, K. J., (2000). A method for optimal material selection aided with decision making theory. *Materials & Design, 21*, 199–206.

CHAPTER 8

ELIMINATION ET CHOICE TRANSLATING REALITY (ELECTRE)

Roy [1] developed the ELimination Et Choice Translating REality (ELECTRE) method, which is based on multi-attribute utility theory (MAUT), with the goal of increasing efficiency while taking into account less information [2]. It gradually decreases the number of candidate alternatives presented to the decision-maker in a set of non-dominated alternatives. The idea of an outranking relationship S is presented as a binary relation in an alternative set of A. For two specific alternatives A_i and A_k, A_i outranks A_k (A_iSA_k) if there are sufficient arguments to conclude that A_i is at least as good as A_k based on all the available information about these two alternatives. The purpose of this outranking strategy is to identify all the alternatives that outrank other alternatives, but they cannot be dominated by any other alternative in the considered set of feasible solutions. The ELECTRE method also requires knowledge about the criteria weights in order to select the best or optimal alternative.

Each criterion $C_j \in C$ is assigned a subjective weight, w_j (sum of weights of all the criteria equals to 1), and every pair of alternatives A_i and A_k is assigned a concordance index, $c(i,k)$, as given below:

$$c(i,k) = \sum_{g_j(i) \geq g_j(k)} w_j, \quad i,k = 1,2,...,m, i \neq k \tag{8.1}$$

where; $g_j(i)$ and $g_j(k)$ indicate the normalized performance scores of ith and k^{th} alternatives, respectively in the decision matrix with regard to the jth criterion. The concordance index $c(i,k)$ is the sum of all the weights for those criteria where A_i is at least as good as A_k in terms of the performance for an ordered pair of alternatives (A_i,A_k). The concordance index ranges between 0 and 1. Additionally, a discordance index, $d(i,k)$, can be estimated using the following equation:

Multi-Criteria Decision-Making Methods in Manufacturing Environments: Models and Applications.
Shankar Chakraborty, Prasenjit Chatterjee, & Partha Protim Das (Authors)

$$d(i,k) = 0, \quad \text{if } g_j(i) \geq g_j(k),\ j = 1,2,\ldots,n$$

$$= \frac{\max\limits_{g_j(k) > g_j(i)} \left(g_j(k) - g_j(i)\right)}{\max\limits_{j=1,\ldots,n} \left(\left|g_j(k) - g_j(i)\right|\right)} \quad \text{otherwise}, i,k = 1,2,\ldots,m, i \neq k \tag{8.2}$$

Once these two indices are computed, an outranking relation S can be expressed as below:

$$A_i S A_k \text{ when } c(i,k) \geq \hat{c} \text{ and} \tag{8.3}$$
$$\mathrm{d(i,k)} \leq \hat{d}$$

where; $\hat{c}$ and $\hat{d}$ are the threshold values decided by the concerned decision maker. It would become more difficult to verify the tests if the threshold values are set too high (usually, $\hat{c}$ = 0.7 and $\hat{d}$ = 0.3) [3]. Both the concordance and discordance indices must not violate their respective threshold values for an outranking relation to be considered true. When these two tests are performed for all ordered alternative pairs, the preferred alternatives are those which outrank more than being outranked.

The procedural steps of ELECTRE method are given as follows:

- **Step 1:** From the initial decision matrix, derive the normalized criteria values.
- **Step 2:** Using the concordance and discordance definitions, construct the outranking relations and associated digraph to depict the dominance relations among the alternatives. If alternative A_i outranks alternative A_k in this digraph, there should be a directed arc from A_i to A_k.
- **Step 3:** Using the lowest concordance and highest discordance indices, develop a minimum dominating subset.
- **Step 4:** Choose the final alternative if the subset contains only one element or is small enough to apply value judgment. Otherwise, repeat Steps (2) through (4) until a single element is achieved.
- **Step 5:** If it is required to derive a complete ranking of the alternatives, apply ELECTRE II method, which is an extension of ELECTRE. For this purpose, calculate the following two indices:

$$\text{Pure concordance index } (C_i) = \sum_{k=1}^{m} c(i,k) - \sum_{i=1}^{m} c(k,i) \qquad (i \neq k) \tag{8.4}$$

$$\text{Pure discordance index } (D_i) = \sum_{k=1}^{m} d(i,k) - \sum_{i=1}^{m} d(k,i) \qquad (i \neq k) \tag{8.5}$$

- Once these two indices are determined, derive two different rankings based on them;

- From these two ranks, calculate an average rank;
- Finally, the alternative with the maximum average rank is selected.

While interactively varying the threshold values in ELECTRE method, the size of the subset consisting of the promising pair of alternatives can be changed. The ELECTRE I method is employed to develop a partial ranking and identify a set of the promising alternatives. On the other hand, ELECTRE II is adopted for ranking all the alternatives under consideration. ELECTRE I, the initial version of ELECTRE method for selecting the optimal alternative from a list of candidate choices, was proposed in 1965. Then onwards, it has become a popular outranking method successfully applied in diverse fields of engineering and management decision-making. To account for the concept of a veto threshold, ELECTRE Iv was later proposed. Subsequently, a further version, known as ELECTRE IS, appeared to model situations with imperfect data. This version of ELECTRE method is currently adopted for choice problematic. Afterwards, ELECTRE II was developed for deriving the complete ranking of the considered alternatives from the best to the worst. On the other hand, ELECTRE III method can effectively consider pseudo-criteria along with fuzzy binary outranking relations. A further extension, called ELECTRE IV, was presented to rank alternatives without using the values of criteria weights. Nowadays, a more simple and accurate method (ELECTRE TRI) is available for solving sorting problems in decision-making.

8.1 LITERATURE SURVEY

Applying both VIKOR and ELECTRE methods, Chatterjee et al. [4] solved two material selection problems for mechanical components, and proved the effectiveness and ranking stability of the considered MCDM approaches. Under a group decision-making environment, Afshari et al. [5] proved the potentiality of ELECTRE method for dealing with complex real-time personnel selection problems. Cavallaro [6] presented the application of ELECTRE III method for identification of the best production process of thin-film solar technology to be used in the renewable energy sector. Chatterjee et al. [7] compared the performance of VIKOR method as a compromise ranking tool and ELECTRE as an outranking approach in an attempt to solve two robot selection examples. Marzouk [8] employed the ELECTRE III method in the context of value engineering while explaining its various procedural steps, like calculation of concordance indices, estimation of discordance indices, computation of credibility scores and development of complete ranking of the alternatives. Vinodh & Girubha [9] explored the application of ELECTRE II method for sustainable concept selection based on three evaluation criteria, i.e., material, product, and process.

Ashari & Parsaei [10] demonstrated the application potentiality of ELECTRE III as an effective outranking tool for weapon (infantry rifle) selection based on several conflicting criteria. Chatterjee et al. [11] contrasted the ranking performance of different variants of ELECTRE method, i.e., ELECTRE II, IS, III, IV, and TRI in a discrete manufacturing context, while considering an automated inspection device selection example. Genc [12] employed two outranking methods in the form of ELECTRE III and PROMETHEE II to identify the best military tank using several conflicting evaluation criteria. Sabaei et al. [13] reviewed the applications of ELECTRE method along with other MCDM techniques and compared their performance from a maintenance management point of view. Govindan & Jepsen [14] performed a comprehensive literature survey of ELECTRE methods with respect to their mathematical formulations, applications, modifications, and comparisons with other MCDM methods. Hodgett [15] developed a software framework employing AHP, multi-attribute range evaluations and ELECTRE III methods for solving an equipment selection problem in a chemical manufacturing process. Kadziński & Ciomek [16] proposed a framework integrating ELECTRE and PROMETHEE methods for preference modeling and robustness analysis in outranking-based multi-criteria decision-making (MCDM). Supraja & Kousalya [17] applied AHP and ELECTRE methods for identification of the best PC framework from a pool of three alternatives based on four evaluation criteria.

Considering six factoring companies and financial statement ratios as the evaluation criteria, Alper & Başdar [18] contrasted the application potentialities of ELECTRE and TOPSIS methods as effective MCDM techniques. Öztürk et al. [19] adopted the ELECTRE method to identify the best supplier for a cable industry based on product quality, cost, customer satisfaction and production flexibility. In a printing industry, Bathrinath et al. [20] identified the critical risk factors in waste management based on the application of ELECTRE I method. Using productivity, flexibility, and quality as the main indicators, Jain & Ajmera [21] integrated AHP, combinatorial mathematics-based approach and ELECTRE methods to assess the performance of some competing FMSs. Considering surface roughness and energy consumption as the responses, Tiwari et al. [22] applied ELECTRE I and II methods for parametric optimization of an EDM process.

8.2 MATERIAL HANDLING EQUIPMENT SELECTION PROBLEM

Material handling is defined as an activity that employs suitable technique to deliver the appropriate amount of material at the proper time, in the correct sequence, at the intended location, with minimum cost involvement [23]. The role of a material handling equipment in the manufacturing industry is critical. Tompkins et al. [24] estimated that in a typical manufacturing system, 25% of the number of employees, 55% of all plant area and 87% of production time are assigned to material handling, and it accounts for between 15% and 70% of the total product cost. An efficient material handling equipment can improve competitiveness of a manufacturing organization while reducing handling cost, enhancing the manufacturing process by increasing production and system flexibility, increasing efficiency of material flow, providing effective utilization of manpower, improving facility utilization, reducing maintenance and operation cost, and decreasing manufacturing lead time. Therefore, selection of the most appropriate material handling equipment for a given handling application is of immense importance for reduced costs and increased profits.

This illustrative example is taken from Kulak [25] where the most suitable conveyor is to be selected from four competing alternatives based on six evaluation criteria. Conveyors are fixed and portable devices used for transporting materials between two fixed points, through intermittent or continuous movement. They are usually employed where the workflow is relatively constant. Industrial conveyors can be classified into several categories, such as bulk material and unit handling conveyors, accumulating and non-accumulating conveyors, and fixed and portable conveyors. Bulk conveyors are employed to handle coal, ashes, cement grains, sand, and almost any material that is not packaged, crated, or otherwise handled in unit loads. The unit conveyors are designed to handle small pieces in containers weighing from a few kilograms to large units, such as filled pallets, crates or assemblies weighing tons. Non-accumulating conveyors stop moving whenever loading or unloading occurs, while accumulating conveyors work continuously, and thus, contribute to the accumulation of materials at the workstation. Fixed conveyors are inflexible, as they cannot accommodate alterations in location, shape, capacity, or type of material handled. On the other hand, portable conveyors can be used for temporary as well as permanent arrangements.

While solving the material handling equipment (conveyor) selection problems, sometimes non-compensatory aggregation procedures are required where compensation of the loss on a given selection criterion by a gain in another one cannot be accepted by the decision maker. This type of problem can be well dealt with while applying ELECTRE methods which are proved to be quite flexible based on simple logic having the strength to solve problems in the presence of incomparability.

Table 8.1 exhibits the detailed objective and subjective information for the considered case study [25] where the six criteria/attributes are fixed cost per hour (FC) (in €/hr), variable cost per hour (VC) (in €/hr), speed of the conveyor (SC) (in m/min), item width (IW) (in cm), item weight (W) (in kg) and flexibility (F). The flexibility criterion is defined subjectively. Hence, using the corresponding conversion of linguistic terms into fuzzy score values [26], the appropriate objective values are assigned to this criterion, as shown in Table 8.2. The chosen conveyor should have low fixed and variable costs, faster speed with the capacity to handle large items with greater flexibility.

TABLE 8.1 Data for the Conveyor Selection Problem [25]

Conveyor	FC	VC	SC	IW	W	F
1	2.00	0.45	12	15	10	Very good
2	2.30	0.44	13	20	10	Excellent
3	2.25	0.45	11	30	20	Excellent
4	2.40	0.46	10	25	15	Very good

Source: Reprinted with permission from Ref. [25]. © 2021 Elsevier.

TABLE 8.2 Objective Information for the Conveyor Selection Problem

Conveyor	FC	VC	SC	IW	W	F
1	2.00	0.45	12	15	10	0.745
2	2.30	0.44	13	20	10	0.955
3	2.25	0.45	11	30	20	0.955
4	2.40	0.46	10	25	15	0.745

In this problem, FC and VC are non-beneficial criteria, whereas, the remaining four criteria are beneficial in nature. The criteria values, as given in Table 8.2, are first normalized using Eqn. (8.6), and the weighted normalized matrix is obtained by applying Eqn. (8.7), as given in Table 8.3. Rao [26] assigned equal weights to all the considered criteria ($w_j = 0.1667$ for $j = 1, 2, \ldots, 6$) while solving the same problem using SAW, WPM, AHP, TOPSIS, modified TOPSIS and GTMA methods. These same criteria weights are also employed here for ELECTRE II method-based analysis.

$$r_{ij} = x_{ij} \Big/ \sqrt{\sum_{i=1}^{m} x_{ij}^2} \quad (j = 1, 2, ..., n) \tag{8.6}$$

where; x_{ij} represents the ith alternative's performance on the jth criterion.

Now, the weighted normalized value, v_{ij}, is calculated as follows:

$$v_{ij} = w_j \times r_{ij} \tag{8.7}$$

where; w_j is the weight of *the* j^{th} criterion.

TABLE 8.3 Weighted Normalized Decision Matrix

Conveyor	FC	VC	SC	IW	W	F
1	0.0743	0.0833	0.0866	0.0539	0.0580	0.0725
2	0.0855	0.0815	0.0938	0.0719	0.0580	0.0929
3	0.0836	0.0833	0.0794	0.1079	0.1161	0.0929
4	0.0892	0.0852	0.0721	0.0899	0.0871	0.0725

Now, the concordance index values for the first alternative conveyor are calculated as follows:

- $c(1,2) = 0.1667 + 0.5 \times 0.1667 = 0.2501$;
- $c(1,3) = 0.1667 + 0.5 \times 0.1667 + 0.1667 = 0.4168$;
- $c(1,4) = 0.1667 + 0.1667 + 0.1667 + 0.5 \times 0.1667 = 0.5835$.

The same procedure is also adopted for determining the concordance indices of the remaining alternative conveyors. Table 8.4 depicts the concordance matrix developed for this conveyor selection problem.

TABLE 8.4 Concordance Matrix

Conveyor	1	2	3	4
1	–	0.2501	0.4168	0.5835
2	0.7500	–	0.4168	0.6668
3	0.5833	0.5833	–	1.0000
4	0.4166	0.3332	0	–

While computing the concordance index values, if there is a tie between the alternatives with respect to a particular criterion, one-half of that criterion weight should be considered. As the decision matrix of Table 8.2 is normalized using vector normalization procedure, Eqn. (8.1) is applicable for calculating the concordance index values for the beneficial criteria. For non-beneficial criteria, their lower values are taken into account while calculating the concordance indices.

Now, using Eqn. (8.2), the discordance indices for this MCDM problem are calculated as given below:

- $d(1,2) = \max(0.0112, 0.0072, 0.0180, 0.0204)/$
- $\max|(0.0112, -0.0019, 0.0072, 0.0180, 0, 0.0204)| = 0.0204/0.0204 = 1$
- $d(1,3) = \max(0.0093, 0.0539, 0.0580, 0.0204)/$
- $\max|(0.0093, 0, -0.0072, 0.0539, 0.0580, 0.0204)| = 0.0580/0.0580 = 1$
- $d(1,4) = \max(0.0149, 0.0019, 0.0360, 0.0290)/ \max|(0.0149, 0.0019, -0.0144, 0.0360, 0.0290, 0)| = 0.0360/0.0360 = 1$

Following the same mathematical steps, the discordance indices for all the alternative conveyors are estimated, as shown in Table 8.5.

TABLE 8.5 Discordance Matrix

Conveyor	1	2	3	4
1	–	1.0000	1.0000	1.0000
2	0.0906	–	1.0000	1.0000
3	0.1243	0.2486	–	0.1921
4	0.4013	0.7458	1.0000	–

Now, the pure concordance and pure discordance indices for the alternative conveyors are estimated while employing Eqns. (8.4) and (8.5), respectively. The average rankings are subsequently determined, on the basis of which the final rankings of the alternatives are obtained, as given in Table 8.6. From the rankings of the candidate conveyors, it can be observed that conveyor 3 is the best choice, followed by conveyor 2. Conveyor 1 is the last choice. Rao [26] also derived the best and the worst choices as conveyor 3 and conveyor 1, respectively, while applying SAW, WPM, AHP, GTMA, TOPSIS, and modified TOPSIS methods to the same problem. Thus, the rankings of the conveyors under consideration as obtained using the ELECTRE II method exactly match with those suggested by Rao [26], proving the applicability of the ELECTRE II method for solving complex material handling equipment selection problems. Figure 8.1 shows the outranking graph for this conveyor selection problem.

TABLE 8.6 Rankings of the Alternative Conveyors

Conveyor	Pure Concordance Index	Initial Rank	Pure Discordance Index	Initial Rank	Average Rank	Final Rank
1	–0.4995	3	2.3838	4	3.5	4
2	0.6670	2	0.0963	3	2.5	2
3	1.3330	1	–2.4350	1	1	1
4	–1.5005	4	–0.0451	2	3	3

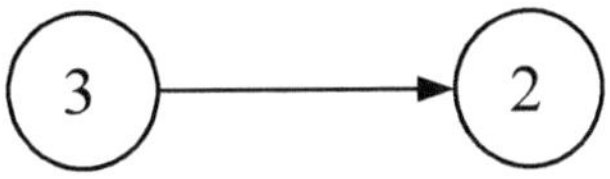

FIGURE 8.1 Outranking graph for conveyor selection problem.

KEYWORDS

- **ELECTRE II method**
- **elimination et choice translating reality**
- **item width**
- **literature survey**
- **multi-attribute utility theory**
- **threshold values**

REFERENCES

1. Roy, B., & Vincke, P., (1981). Multi-criteria analysis: Survey and new directions. *European Journal of Operational Research, 8*(3), 207–218.
2. Cho, K. T., (2003). Multicriteria decision methods: An attempt to evaluate and unify. *Mathematics and Computer Modeling, 37*(9, 10), 1099–1119.
3. Almeida, A. T., (2005). Multicriteria modeling of repair contract based on utility and ELECTRE I method with dependability and service quality criteria. *Annals of Operations Research, 138*, 113–126.
4. Chatterjee, P., Athawale, V. M., & Chakraborty, S., (2009). Selection of materials using compromise ranking and outranking methods. *Materials and Design, 30*, 4043–4053.
5. Afshari, A. R., Mojahed, M., Yusuff, R. M., Hong, T. S., & Ismail, M. Y., (2010). Personnel selection using ELECTRE. *Journal of Applied Sciences, 10*(23), 3068–3075.

6. Cavallaro, F., (2010). A comparative assessment of thin-film photovoltaic production processes using the ELECTRE III method. *Energy Policy, 38*, 463–474.
7. Chatterjee, P., Athawale, V. M., & Chakraborty, S., (2010). Selection of industrial robots using compromise ranking and outranking methods. *Robotics and Computer-Integrated Manufacturing, 26*, 483–489.
8. Marzouk, M. M., (2011). ELECTRE III model for value engineering applications. *Automation in Construction, 20*, 596–600.
9. Vinodh, S., & Girubha, R. J., (2012). Sustainable concept selection using ELECTRE. *Clean Technologies and Environmental Policy, 14*, 651–656.
10. Ashari, H. E., & Parsaei, M., (2014). Application of the multi-criteria decision method ELECTRE III for the weapon selection. *Decision Science Letters, 3*, 511–522.
11. Chatterjee, P., Mondal, S., & Chakraborty, S., (2014). A comprehensive solution to automated inspective device selection problems using ELECTRE methods. *International Journal of Technology, 2*, 193–208.
12. Genc, T., (2015). Application of ELECTRE III and PROMETHEE II in evaluating military tanks. *International Journal of Procurement Management, 8*(4), 457–475.
13. Sabaei, D., Erkoyuncu, J., & Roy, R., (2015). A review of multi-criteria decision making methods for enhanced maintenance delivery. *Procedia CIRP, 37*, 30–35.
14. Govindan, K., & Jepsen, M. B., (2016). ELECTRE: A comprehensive literature review on methodologies and applications. *European Journal of Operational Research, 250*, 1–29.
15. Hodgett, R. E., (2016). Comparison of multi-criteria decision-making methods for equipment selection. *International Journal of Advanced Manufacturing Technology, 85*, 1145–1157.
16. Kadziński, M., & Ciomek, K., (2016). Integrated framework for preference modeling and robustness analysis for outranking-based multiple criteria sorting with ELECTRE and PROMETHEE. *Information Sciences, 352, 353*, 167–187.
17. Supraja, S., & Kousalya, P., (2016). ELECTRE method for the selection of best computer system. *Indian Journal of Science and Technology, 9*(39), 1–5.
18. Alper, D., & Başdar, C., (2017). A comparison of TOPSIS and ELECTRE methods: An application on the factoring industry. *Business and Economics Research Journal, 8*(3), 627–646.
19. Öztürk, H., Pekel, E., & Elevli, B., (2018). Using ANP and ELECTRE methods for supplier selection: Cable industry application. *Sakarya University Journal of Science, 22*(5), 1190–1198.
20. Bathrinath, S., Vignesh, A. K. N., Venkatesh, S., & Satheshkumar, R., (2019). Examination of waste management in printing industries using electre-I method. *International Journal of Recent Technology and Engineering, 8*(4S2), 284–288.
21. Jain, V., & Ajmera, P., (2019). Evaluation of performance factors of FMS by combined decision making methods as AHP, CMBA and ELECTRE methodology. *Management Science Letters, 9*, 519–534.
22. Tiwari, R., Agrawal, S., & Kasdekar, D. K., (2020). Application of ELECTRE-I, II methods for EDM performance measures in manufacturing decision making. *IOP Conference Series: Materials Science and Engineering, 748*, 012015.
23. Apple, J. M., (1977). *Plant Layout and Material Handling*. John Wiley & Sons, USA.

24. Tompkins, A., White, J. A., Bozer, Y. A., Frazelle, E. H., Tanchoco, J. M. A., & Trevino, J., (1996). *Facilities Planning*. Wiley, USA.
25. Kulak, O., (2005). A decision support system for fuzzy multi-attribute selection of material handling equipments. *Expert Systems with Applications, 29*(2), 310–319.
26. Rao, R. V., (2007). *Decision Making in the Manufacturing Environment Using Graph Theory and Fuzzy Multiple Attribute Decision Making Methods*. Springer-Verlag, London.

CHAPTER 9

PREFERENCE RANKING ORGANIZATION METHOD FOR ENRICHMENT EVALUATION (PROMETHEE)

The PROMETHEE technique, proposed by Brans et al. [1, 2], is a well-accepted MCDM tool being augmented to solve various complex engineering and managerial decision-making problems [3]. It can bc classified as an outranking method allowing for a finite number of alternatives to be ranked based on a finite number of criteria, which are often conflicting. The PROMETHEE I define a partial ranking list of the selected alternatives; on the other hand, a complete ranking for the alternatives can be obtained using PROMETHEE II. Brans et al. [2] later proposed two extensions of this method, i.e., PROMETHEE III for providing an interval order emphasizing indifference and PROMETHEE IV for dealing with continuous sets of possible alternatives. Around the mid-90s, two more extensions of this method in the form of PROMETHEE V with segmentation constraints and PROMETHEE VI to represent the human brain were also developed [4]. The PROMETHEE has a very strong mathematical foundation and has already been proved to be quite robust under changing decision-making scenario [5, 6]. The PROMETHEE II has become quite popular among the decision-making community due to its simple mathematical steps while providing a complete ranking for a set of candidate alternatives from the most preferred to the least preferred. The ranking for the alternatives is derived based on their pair-wise comparisons with respect to each criterion using preference functions. These preference functions are later aggregated using criteria weights to obtain net outranking flows while providing a complete ranking of the alternatives.

Multi-Criteria Decision-Making Methods in Manufacturing Environments: Models and Applications.
Shankar Chakraborty, Prasenjit Chatterjee, & Partha Protim Das (Authors)

Each criterion used within a PROMETHEE model to help rank alternatives is assigned a preference function by the decision maker. The preference function translates the difference (either positive or negative) in the value of a criterion between two alternatives in a pair-wise comparison into a preference degree in a range from 0 to 1. To estimate the preference degree between two alternatives (a and b) with respect to j^{th} criterion, i.e., $P_j(a,b)$, six types of generalized criterion [2], as provided in Figure 9.1, can be employed. In this figure, s and r are the two parameters to be set by the concerned decision maker. Parameter s signifies the insignificance threshold where the largest difference is observed to be negligible when comparing two alternatives, and parameter r represents the least difference with a strict preference to one between the two alternatives. The details of these six generalized criteria are provided as below:

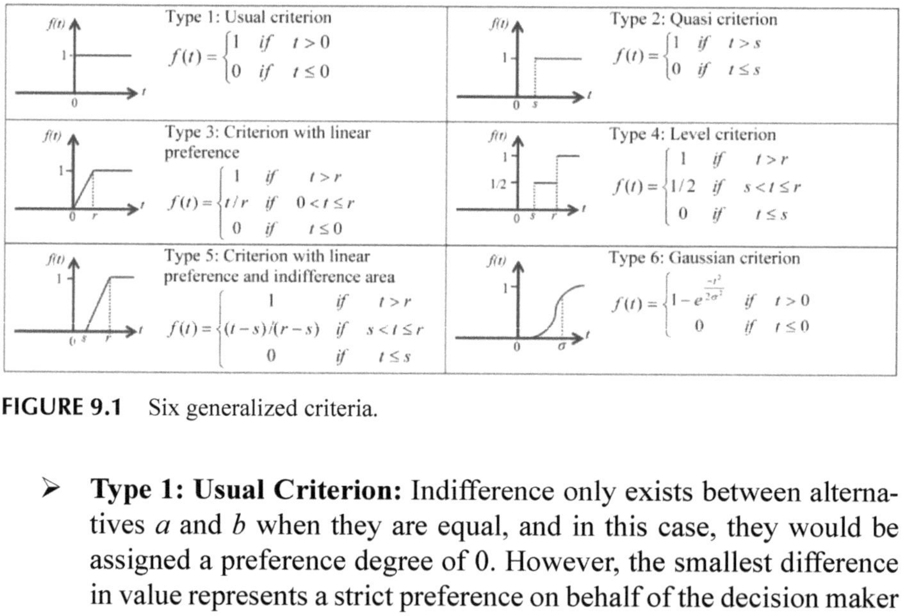

FIGURE 9.1 Six generalized criteria.

- **Type 1: Usual Criterion:** Indifference only exists between alternatives a and b when they are equal, and in this case, they would be assigned a preference degree of 0. However, the smallest difference in value represents a strict preference on behalf of the decision maker and is assigned a preference degree of 1.
- **Type 2: Quasi Criterion:** In this case, alternatives a and b are assumed to be indifferent within a range defined by the decision maker and would be assigned a preference degree of 0. Beyond that area of indifference, the preference becomes strict, with a preference degree of 1 being assigned.
- **Type 3: Criterion with Linear Preference:** Extending the usual criterion, in this case, the decision maker's preference increases

for progressively larger deviations between alternatives *a* and *b*. When *a* equals *b*, the alternatives are indifferent and are assigned a preference degree of 0. From that point, the intensity of preference increases linearly to a threshold value defined by the decision maker at which point and beyond which the preference is strict and assigned a preference degree of 1.

- **Type 4: Level Criterion:** An extension to quasi criterion, in this case, alternatives *a* and *b* are considered to be indifferent within a range defined by the decision maker and would be assigned a preference degree of 0. Beyond that, a second range defined by the decision maker provides a weak preference where a preference degree of 0.5 would be assigned. Beyond that range, a strict preference with a preference degree of 1 is assigned.
- **Type 5: Criterion with Linear Preference and Indifference Area:** A combination of criteria types 2 and 3, in this case, alternatives *a* and *b* are considered to be indifferent within a range defined by the decision maker and would be assigned a preference degree of 0. Beyond this range, the decision maker's preference increases progressively for larger deviations between alternatives *a* and *b*. The intensity of preference increases linearly to a threshold value defined by the decision maker at which point and beyond which the preference is strict and assigned a preference degree of 1.
- **Type 6: Gaussian Criteria:** Similar to criterion type 3, if a particular criterion is of the Gaussian type, the preference of the decision maker would still grow with increasing deviation between alternatives *a* and *b*, but the relationship is not linear. Through the decision maker providing a value of σ, representing the distance between the origin and point of inflection, the relationship between the difference in alternatives *a* and *b* and the preference degree is given. The preference degree would vary from 0 where *a* equals *b* to approaching 1 where the difference is very large.

The procedural steps of PROMETHEE method for solving MCDM problems are presented as below:

- **Step 1: Development of the Decision Matrix:** Let $e_1, e_2, \ldots, e_m$ be m number of alternatives and $q_1, q_2, \ldots, q_n$ be n number of criteria. Each element $q_j(e_i)$ represents the performance measure for i^{th} alternative against j^{th} criterion. Thus, the following decision matrix can be formulated:

$$D = \begin{bmatrix} q_1(e_1) & \cdots & q_j(e_1) & \cdots & q_n(e_1) \\ \cdots & \cdots & \cdots & \cdots & \cdots \\ q_1(e_i) & \cdots & q_j(e_i) & \cdots & q_n(e_i) \\ \cdots & \cdots & \cdots & \cdots & \cdots \\ q_1(e_m) & \cdots & q_j(e_m) & \cdots & q_n(e_m) \end{bmatrix}$$

- **Step 2: Estimation of the Preference Function:** The preference function, $P_j(a,b)$ in PROMETHEE is based on a preference structure being estimated using pair-wise comparisons between the criteria values of two alternatives a and b with respect to a specific performance measure. Here, the deviations between two criteria values are taken into account. A higher deviation indicates greater preference, while smaller deviation signifies lower preference which can be ignored. The value of $P_j(a,b)$ typically ranges from 0 to 1 with $P_j(a,b) = 1$, signifying alternative a to be strictly preferred to alternative b, while $P_j(a,b) = 0$ indicates that both the alternatives are indifferent to each other with respect to a particular criterion. Thus, the preference function can be estimated as follows:

$$P_j(a,b) = f_j[t_j(a,b)] \quad \forall a,b \in D \tag{9.1}$$

where; $t_j(a,b) = [q_j(a) - q_j(b)]$, $0 \leq P_j(a,b) \leq 1$ and f_j is a non-decreasing function with respect to the deviation $t_j(a,b)$ between alternatives a and b.

For beneficial criterion, this function provides the preference of alternative a over alternative b with respect to a particular criterion $q(.)$. The preference function is presumed to be zero if it takes a negative value. On the other hand, for non-beneficial criterion, this preference function should be reversed as follows:

$$P_j(a,b) = f_j[-t_j(a,b)] \tag{9.2}$$

- **Step 3: Computation of the Overall Preference Function:** The overall degree of preference by which alternative a is preferred to alternative b can now be obtained as follows:

$$\pi(a,b) = \sum_{j=1}^{n} w_j P_j(a,b) \tag{9.3}$$

Similarly, the degree by which alternative b is preferred to alternative a is given by:

$$\pi(b,a) = \sum_{j=1}^{n} w_j P_j(b,a) \tag{9.4}$$

where; w_j is the relative weight assigned to *the* j^{th} criterion.

- **Step 4: Calculation of the Positive Outranking Flow and the Negative Outranking Flow:** Based on the calculated overall preference functions, the positive (leaving) outranking flow, $\varphi^+(a)$ and the negative (entering) outranking flow, $\varphi^-(a)$ with respect to alternative a can be obtained as below:

$$\varphi^+(a) = \frac{1}{m-1}\sum_{b \in D} \pi(a,b), \text{ where } a \neq b \tag{9.5}$$

$$\varphi^-(a) = \frac{1}{m-1}\sum_{b \in D} \pi(b,a), \text{ where } a \neq b \tag{9.6}$$

- **Step 5: Partial Ranking of the Alternatives:** In this step, the partial ranking of the candidate alternatives is derived using the PROMETHEE I method based on the estimated values of the positive and the negative outranking flows. When a and b represent two alternatives in the alternative set, the following conditions may arise.
  - When one of the followings is ensured, a is preferred to b:

    $\varphi^+(a) > \varphi^+(b)$ or $\varphi^-(a) < \varphi^-(b)$
    $\varphi^+(a) > \varphi^+(b)$ or $\varphi^-(a) = \varphi^-(b)$
    $\varphi^+(a) = \varphi^+(b)$ or $\varphi^-(a) < \varphi^-(b)$

  - When the following is true, alternatives a and b are identical:

    $\varphi^+(a) = \varphi^+(b)$ or $\varphi^-(a) = \varphi^-(b)$

  - When one of the followings is ensured, alternatives a and b cannot be compared:

    $\varphi^+(a) > \varphi^+(b)$ or $\varphi^-(a) > \varphi^-(b)$
    $\varphi^+(a) < \varphi^+(b)$ or $\varphi^-(a) < \varphi^-(b)$

- **Step 6: Estimation of the Net Outranking Flow:** Based on the net outranking preference flows, the complete ranking of the alternatives is obtained using PROMETHEE II.

$$\text{Net outranking flow, } \varphi(a) = \varphi^+(a) - \varphi^-(a) \tag{9.7}$$

The alternative having the highest net outranking flow is selected to be the best among all the available options. Thus, the following ranking is decided depending on the net outranking flow.

When $\varphi(a) > \varphi(b)$, alternative a outranks alternative b.
When $\varphi(a) = \varphi(b)$, alternatives a and b are identical.

The PROMETHEE II method has significant advantages over the other MCDM approaches. Unlike the other methods, it can classify the alternatives which are difficult to be compared because of a tradeoff relation of evaluation standards as non-comparable alternatives. It is also different from AHP technique since it is not essential to perform pair-wise comparisons again when comparative alternatives are added or deleted.

9.1 PROMETHEE-GAIA METHOD

The geometrical analysis for interactive aid (GAIA) is developed as an extension of PROMETHEE method, which would act as a visual decision aid to the decision maker to arrive at the most accurate and viable selection decision. Based on the positive and the negative outranking flows of PROMETHEE, the net outranking flow value can be reformulated, as follows:

$$\varphi(a) = \varphi^{+}(a) - \varphi^{-}(a) = \frac{1}{m-1}\sum_{j=1}^{n}\sum_{x\in D} w_j[P_j(a,x) - P_j(x,a)] \tag{9.8}$$

Thus,

$$\varphi(a) = \sum_{j=1}^{n} w_j \varphi_j(a) \tag{9.9}$$

$$\text{if}\varphi_j(a) = \frac{1}{m-1}\sum_{x\in D}[P_j(a,x) - P_j(x,a)] \tag{9.10}$$

where; $\varphi_j(a)$ is the single criterion net flow obtained for j^{th} criterion. It can be noticed that the scalar product of the vector of the criteria weights and profile vector of an alternative results in the value of the global net flow of that alternative. This property is primarily used for developing the corresponding GAIA plane. Now, the matrix $M_{(m\times n)}$ is defined based on the single criterion net flows of all the alternatives.

$$M = \begin{bmatrix} \varphi_1(a_1) & \varphi_2(a_1) & \dots & \varphi_j(a_1) & \dots & \varphi_n(a_1) \\ \varphi_1(a_2) & \varphi_2(a_2) & \dots & \varphi_j(a_2) & \dots & \varphi_n(a_2) \\ \dots & \dots & \dots & \dots & \dots & \dots \\ \varphi_1(a_i) & \varphi_2(a_i) & \dots & \varphi_j(a_i) & \dots & \varphi_n(a_i) \\ \dots & \dots & \dots & \dots & \dots & \dots \\ \varphi_1(a_m) & \varphi_2(a_m) & \dots & \varphi_j(a_m) & \dots & \varphi_n(a_m) \end{bmatrix} \tag{9.11}$$

Matrix M is more useful than the initial decision/evaluation matrix because the degrees of preference provided by the generalized criteria are considered in this matrix. Moreover, in the decision matrix, $q_j(e_i)$ values are expressed on their own scale, while in matrix M, $\varphi_j(a_i)$ values are dimensionless. In addition, matrix M does not depend on criteria weights.

The set of m alternatives can be shown in the form of a cluster of m points in an n-dimensional space. When the number of criteria becomes larger than two, it is difficult to obtain a clear view of the relative positions of the alternatives with regard to the criteria. The GAIA plane is developed by projecting these points on a plane such that very little information is lost. In this plane, alternatives (a_1, a_2, ..., a_m) are denoted by points and the criteria (c_1, c_2, ..., c_n) are represented by axes. From Eqn. (9.9), it can be said that the net flow a_i is the vector's projection of the single criterion net flows on w. Consequently, the relative positions of the projections of all the alternatives on w yield the PROMETHEE-based ranking for the alternatives. Clearly, the vector w plays an important role. It is illustrated in GAIA plane by the projection of the unit vector of the weights. This projection is referred to as the PROMETHEE decision axis (π). This axis directs towards the compromise solution resulting from the weights allocated to the criteria. If a particular criterion has the maximum relative importance (weight), the π axis would coincide with the axis of that criterion in GAIA plane. When the weights are distributed among all the criteria, the π axis acts as the weighted resultant of all the criteria axes.

A long π axis signifies a strong decision power, inviting the decision maker to identify suitable alternatives as far as possible on its direction. On the other hand, a short π axis has no strong decision power, the criteria are strongly conflicting, and selection of a good compromise solution is a hard problem. In the GAIA plane, the criteria which are oriented in the same direction have similar preferences on the evaluation data, while, the criteria positioned in opposite directions are usually conflicting in nature. With variations in the criteria weights, the positions of the criteria and alternatives would remain unchanged in GAIA plane. The weight vector acts as a decision stick that the decision-maker can control with respect to the preference of a particular criterion. While performing a sensitivity analysis by altering weights of the criteria, the PROMETHEE decision stick (w) in addition to the π axis moves in a way such that the outcomes for the decision analysis can easily be visualized in the GAIA plane.

9.2 REVIEW OF THE LITERATURE

For a small and medium-sized enterprise (SME), Anand & Kodali [7] demonstrated the applicability of an extended version of PROMETHEE in the form of PROMETHEE V while identifying the most suitable lean manufacturing system. Dağdeviren [8] considered an equipment selection problem for a manufacturing organization and applied PROMETHEE to determine the complete ranking list of the candidate options. A sensitivity analysis was also performed while changing the criteria weights to prove the robustness of the adopted technique. Behzadian et al. [3] critically reviewed the applications of PROMETHEE in diverse areas, like chemistry, environment management, manufacturing and assembly, logistics and transportation, hydrology and water management, energy management, business, and financial management, and social. Rao & Patel [9] proved the application potentiality of PROMETHEE while solving four real-time manufacturing-related decision-making problems, i.e., selection of (a) cutting fluid, (b) environmentally conscious manufacturing program, (c) best end-of-life scenario, and (d) RP process. Based on several economic, environmental, technical, and social criteria, Turcksin et al. [10] proposed the combined application of AHP and PROMETHEE techniques for evaluation of the shortlisted policy scenarios for final selection of a clean vehicle fleet. Bogdanovic et al. [11] applied AHP technique to frame the hierarchical structure for selection of the best-suited mining method and PROMETHEE was later employed to rank the considered alternative mining methods. Karande & Chakraborty [12] presented the integrated application of PROMETHEE and GAIA techniques while solving four NTM process selection problems. For a supplier selection problem, Safari et al. [13] determined the criteria weights using entropy method and ranked the competing alternative suppliers while employing PROMETHEE-GAIA technique. Vinodh & Jeya Girubha [14] applied PROMETHEE to identify the best sustainable concept in a manufacturing organization based on social, economical, and natural perspectives. It was concluded that material oriented sustainable concept would be provided the maximum priority.

Maity & Chakraborty [15] demonstrated the simultaneous application of PROMETHEE and GAIA approaches while solving two gear material selection problems. Nitralloy 135M and Nylon glass fiber reinforced 6/6 were respectively identified as the best metallic and non-metallic gear materials. Meysam Mousavi et al. [16] combined Delphi, AHP, and PROMETHEE techniques for solving a plant location selection problem. The most influential evaluation criterion was first identified by the Delphi method, the corresponding criteria weights were estimated using AHP method and the location alternatives were finally ranked employing PROMETHEE. Öztürk et al. [17] considered personnel selection for an organization as an MCDM problem and adopted PROMETHEE for choosing the best personnel in an accommodation enterprise. Balali et al. [18] applied PROMETHEE for simultaneous selection of the most suitable material, construction procedure and structural system of bridges with respect to a set of evaluation criteria. Balali et al. [19] compared the ranking performance and consistency of AHP and PROMETHEE techniques while solving a building structural system selection problem. It was observed that 3D panels consisting of light walls in building frames were the most apposite choice. Murali et al. [20] solved a supplier selection problem using PROMETHEE and contrasted its performance against TOPSIS method.

While identifying the most appropriate enterprise resource planning (ERP) system for an SME, Kilic et al. [21] first employed ANP method to determine the weights and interrelationships of the considered criteria, and then proposed the application of PROMETHEE for optimal ranking of the candidate alternatives. Sen et al. [22] explored the application of PROMETHEE II method as an effective decision-making tool for identification of the most apposite industrial robot based on a set of quantitative evaluation data. Bogdanović et al. [23] integrated PROMETHEE and GAIA techniques while identifying the best benchmarking partner in building a CHP plant for biomass processing. Using the group technology concept, Abdelhadi [24] implemented PROMETHEE as a strategic decision-making technique to select clusters of maintainable machines and schedule their maintenance activities according to the nature of failure in order to minimize the total maintenance cost. Using PROMETHEE, Bottero et al. [25] developed a model to evaluate the candidate urban regeneration processes. The strengths-weaknesses-opportunities-threats (SWOT) analysis and stakeholders' analysis were also integrated with PROMETHEE to ensure accurate ranking results. Using PROMETHEE, Peterková & Franek [26] proposed a decision-making model to aid the policy planners in selecting the most suitable innovation management concept. Amongst the candidate alternatives, value analysis was identified as the best option. Sapkota et al. [27] presented the combined application of PROMETHEE-GAIA technique for evaluation of a number of alternative hybrid water supply scenarios. The ranking stability was also validated based on sensitivity analysis. It was noticed that the centralized system with stormwater harvesting and centralized system with treated wastewater and rainwater tanks occupied the top two positions, respectively. On the other hand, the centralized water supply system with treated wastewater was the least preferred hybrid water supply system. Considering 27 lean six sigma barriers and 22 alternatives, Yadav et al. [28] proposed a hybrid framework integrating fuzzy AHP and PROMETHEE techniques to prioritize lean six sigma adoption in a manufacturing industry. Considering AISI 304, 310, 316, 321, and 430 stainless steels as the candidate alternatives, Zindani & Kumar [29] applied PROMETHEE to single out the most apposite material for turbine seal strips based on six criteria, like creep strength, resistance to oxidation, coefficient of thermal expansion, yield strength, toughness, and limit strain.

Abdullah et al. [30] solved a green supplier selection problem consisting of seven evaluation attributes, five decision-makers and four suppliers, and exhibited the effects of different generalized criteria of PROMETHEE on the derived ranking results. Crnjac et al. [31] selected the most suitable table design for a CNC milling machine using PROMETHEE based on the chosen preferences and criteria, and later applied Taguchi methodology to propose the most suitable parameters for the identified design. Considering 24 Indian steel companies as the alternatives and 15 financial ratio indicators as the criteria, Raikar [32] adopted PROMETHEE to rank the considered alternatives from the best to the worst. Yougbaré [33] first determined weights of various evaluation criteria and applied PROMETHEE to compute preference indices for ranking of different districts with respect to risk of flood and runoff. Baloyi & Meyer [34] presented the application of PROMETHEE as a systematic MCDM approach for identifying the most applicable mining method. Its superiority over the other existing methods was also validated. Nasrollahi et al. [35] estimated the relative importance of various criteria using fuzzy BWM and then ranked the competing industrial robots using PROMETHEE. With an aim to solve an assignment place selection problem, Soba et al. [36] first determined the corresponding criteria weights and employed PROMETHEE to grade and rank the alternative assignment locations.

9.3 FACILITY LAYOUT SELECTION PROBLEM

In order to demonstrate and validate the application of PROMETHEE II method in solving MCDM problems, the following facility layout selection is considered here. Chakraborty & Banik [37] identified six important criteria which would directly affect the facility layout selection decision-making problem. These are material flow (C_1), information flow (C_2), equipment flow (C_3), maintenance (C_4), shape ratio (C_5), and adjacency (C_6). The quantitative measures of these six criteria for 10 alternative layouts are provided in Table 9.1. Material flow is defined as the path traveled by the materials during the production process and can be measured by calculating the total path flow distance between centroids of the departments through which the materials are processed. Information flow deals with the transfer of information between various departments of an organization and can be estimated as the linear distance between the administrative office and other organizational departments. Equipment flow involves the movement of different material handling equipment while transferring materials from one place to another within the organization. Maintenance criterion deals with the distance traveled by the maintenance personnel and maintenance tools/equipment for smooth and trouble-free maintenance actions. Shape ratio can be measured as the maximum value between the depth-to-width ratio and width-to-depth ratio of the smallest rectangle that entirely covers a department. Adjacency can be achieved by maximizing the closeness between various departments within an organization. Among these six criteria, C_1, C_2, C_3, C_4, and C_5 are beneficial in nature, whereas, C_6 is the only non-beneficial criterion. Using AHP techniques, the corresponding criteria weights are determined as $w_{C1} = 0.4524$, $w_{C2} = 0.2475$, $w_{C3} = 0.1308$, $w_{C4} = 0.0750$, $w_{C5} = 0.0587$, and $w_{C6} = 0.0356$, which are subsequently employed for the PROMETHEE II method-based analysis.

TABLE 9.1 Quantitative Data for Alternative Layouts

Layout	C_1	C_2	C_3	C_4	C_5	C_6
A_1	90.31	240.71	111.65	91.94	1.81	188.3
A_2	90.57	227.14	112.72	88.32	2.83	186.38
A_3	73.84	153.99	106.05	75.67	1.25	161.5
A_4	73.84	200.25	106.05	75.67	1.25	164.87
A_5	72.07	179.5	108.07	96.62	1.25	161.21
A_6	106.85	134.87	127.58	105.41	1.25	145.39
A_7	70.34	207.41	109.34	75.42	1.55	162.89
A_8	77.73	140.1	113.73	102.28	1.25	151.07
A_9	81.6	169.55	117.6	75.67	2.83	163.33
A_{10}	74.83	178.11	113.27	75.67	2.83	146.06

Depending on the type of the evaluation criteria considered, the original decision matrix is first normalized to make all of its elements dimensionless and comparable. Table 9.2 shows the corresponding normalized decision matrix. Now, the preference functions are calculated for all the pairs of alternative facility layouts based on Eqns. (9.1) and (9.2) and are given in Table 9.3. Table 9.4 provides the aggregated preference function values for all the paired alternatives, calculated using Eqns. (9.3) and (9.4). The positive and the negative outranking flows for different alternative layouts are then estimated using Eqns. (9.5) and (9.6), respectively, and are exhibited in Table 9.5.

TABLE 9.2 Normalized Decision Matrix for the Layout Selection Problem

Layout	C_1	C_2	C_3	C_4	C_5	C_6
A_1	0.5469	1	0.2601	0.5508	0.3526	0
A_2	0.5541	0.8718	0.3095	0.4301	1	0.0447
A_3	0.0958	0.1806	0	0.0085	0	0.6244
A_4	0.0958	0.6178	0	0.0085	0	0.5459
A_5	0.0475	0.4217	0.0937	0.7069	0	0.6314
A_6	1	0	1	1	0	1
A_7	0	0.6854	0.1524	0	0.1908	0.5923
A_8	0.2024	0.0494	0.3564	0.8957	0	0.8677
A_9	0.3084	0.3277	0.5363	0.0085	1	0.5818
A_{10}	0.123	0.4085	0.3353	0.0085	1	0.9845

TABLE 9.3 Preference Function Values for the Layout Selection Problem

Layout pair	C_1	C_2	C_3	C_4	C_5	C_6
(A_1,A_2)	0	0.1282	0	0.1206	0	0
(A_1,A_3)	0.4511	0.8194	0.2601	0.5423	0.3526	0
(A_1,A_4)	0.4511	0.3822	0.2601	0.5423	0.3526	0
(A_1,A_5)	0.4994	0.5783	0.1664	0	0.3526	0
(A_1,A_6)	0	1	0	0	0.3526	0
(A_1,A_7)	0.5469	0.3146	0.1077	0.5508	0.1618	0
(A_1,A_8)	0.3445	0.9506	0	0	0.3526	0
(A_1,A_9)	0.2385	0.6723	0	0.5423	0	0
(A_1,A_{10})	0.4239	0.5915	0	0.5423	0	0
(A_2,A_1)	0.0072	0	0.0494	0	0.6474	0.0447
(A_2,A_3)	0.4583	0.6911	0.3095	0.4216	1	0
(A_2,A_4)	0.4583	0.254	0.3095	0.4216	1	0
(A_2,A_5)	0.5066	0.4501	0.2158	0	1	0
(A_2,A_6)	0	0.8718	0	0	1	0

TABLE 9.3 *(Continued)*

Layout pair	C_1	C_2	C_3	C_4	C_5	C_6
(A_2,A_7)	0.5541	0.1864	0.1571	0.4301	0.8092	0
(A_2,A_8)	0.3518	0.8224	0	0	1	0
(A_2,A_9)	0.2457	0.5441	0	0.4216	0	0
(A_2,A_{10})	0.4312	0.4632	0	0.4216	0	0
(A_3,A_1)	0	0	0	0	0	0.6244
(A_3,A_2)	0	0	0	0	0	0.5797
(A_3,A_4)	0	0	0	0	0	0.0785
(A_3,A_5)	0.0483	0	0	0	0	0
(A_3,A_6)	0	0.1806	0	0	0	0
(A_3,A_7)	0.0958	0	0	0.0085	0	0.0321
(A_3,A_8)	0	0.1312	0	0	0	0
$(A3,A_9)$	0	0	0	0	0	0.0426
(A_3,A_{10})	0	0	0	0	0	0
(A_4,A_1)	0	0	0	0	0	0.5459
(A_4,A_2)	0	0	0	0	0	0.5012
(A_4,A_3)	0	0.4371	0	0	0	0
(A_4,A_5)	0.0483	0.1961	0	0	0	0
(A_4,A_6)	0	0.6178	0	0	0	0
(A_4,A_7)	0.0958	0	0	0.0085	0	0
(A_4,A_8)	0	0.5684	0	0	0	0
(A_4,A_9)	0	0.2901	0	0	0	0
(A_4,A_{10})	0	0.2092	0	0	0	0
(A_5,A_1)	0	0	0	0.1561	0	0.6314
(A_5,A_2)	0	0	0	0.2768	0	0.5867
(A_5,A_3)	0	0.241	0.0937	0.6984	0	0.007
(A_5,A_4)	0	0	0.0937	0.6984	0	0.0855
(A_5,A_6)	0	0.4217	0	0	0	0
(A_5,A_7)	0.0475	0	0	0.7069	0	0.0391
(A_5,A_8)	0	0.3723	0	0	0	0
(A_5,A_9)	0	0.094	0	0.6984	0	0.0496
(A_5,A_{10})	0	0.0131	0	0.6984	0	0
(A_6,A_1)	0.4531	0	0.7399	0.4492	0	1
(A_6,A_2)	0.4459	0	0.6905	0.5699	0	0.9553
(A_6,A_3)	0.9042	0	1	0.9915	0	0.3756
(A_6,A_4)	0.9042	0	1	0.9915	0	0.4541
(A_6,A_5)	0.9525	0	0.9063	0.2931	0	0.3686
(A_6,A_7)	1	0	0.8476	1	0	0.4077
(A_6,A_8)	0.7976	0	0.6436	0.1043	0	0.1323

TABLE 9.3 *(Continued)*

Layout pair	C_1	C_2	C_3	C_4	C_5	C_6
(A_6,A_9)	0.6916	0	0.4637	0.9915	0	0.4182
(A_6,A_{10})	0.877	0	0.6647	0.9915	0	0.0155
(A_7,A_1)	0	0	0	0	0	0.5923
(A_7,A_2)	0	0	0	0	0	0.5475
(A_7,A_3)	0	0.5047	0.1524	0	0.1908	0
(A_7,A_4)	0	0.0676	0.1524	0	0.1908	0.0463
(A_7,A_5)	0	0.2637	0.0587	0	0.1908	0
(A_7,A_6)	0	0.6854	0	0	0.1908	0
(A_7,A_8)	0	0.636	0	0	0.1908	0
(A_7,A_9)	0	0.3577	0	0	0	0.0104
(A_7,A_{10})	0	0.2768	0	0	0	0
(A_8,A_1)	0	0	0.0963	0.345	0	0.8677
(A_8,A_2)	0	0	0.0469	0.4656	0	0.8229
(A_8,A_3)	0.1065	0	0.3564	0.8872	0	0.2432
(A_8,A_4)	0.1065	0	0.3564	0.8872	0	0.3217
(A_8,A_5)	0.1548	0	0.2627	0.1888	0	0.2363
(A_8,A_6)	0	0.0494	0	0	0	0
(A_8,A_7)	0.2024	0	0.204	0.8957	0	0.2754
(A_8,A_9)	0	0	0	0.8872	0	0.2858
(A_8,A_{10})	0.0794	0	0.0211	0.8872	0	0
(A_9,A_1)	0	0	0.2762	0	0.6474	0.5818
(A_9,A_2)	0	0	0.2268	0	0	0.5371
(A_9,A_3)	0.2126	0.1471	0.5363	0	1	0
(A_9,A_4)	0.2126	0	0.5363	0	1	0.0359
(A_9,A_5)	0.2609	0	0.4426	0	1	0
(A_9,A_6)	0	0.3277	0	0	1	0
(A_9,A_7)	0.3084	0	0.3839	0.0085	0.8092	0
(A_9,A_8)	0.106	0.2783	0.1799	0	1	0
(A_9,A_{10})	0.1855	0	0.201	0	0	0
(A_{10},A_1)	0	0	0.0752	0	0.6474	0.9845
(A_{10},A_2)	0	0	0.0258	0	0	0.9397
(A_{10},A_3)	0.0271	0.2279	0.3353	0	1	0.36
(A_{10},A_4)	0.0271	0	0.3353	0	1	0.4385
(A_{10},A_5)	0.0754	0	0.2416	0	1	0.3531
(A_{10},A_6)	0	0.4085	0	0	1	0
(A_{10},A_7)	0.123	0	0.1829	0.0085	0.8092	0.3922
(A_{10},A_8)	0	0.3591	0	0	1	0.1168
(A_{10},A_9)	0	0.0808	0	0	0	0.4026

TABLE 9.4 Aggregated Preference Function Values for the Layout Selection Problem

Layout	A_1	A_2	A_3	A_4	A_5	A_6	A_7	A_8	A_9	A_{10}
A_1	–	0.0408	0.5023	0.3941	0.4115	0.2682	0.3902	0.4118	0.315	0.3789
A_2	0.0493	–	0.5092	0.401	0.4275	0.2745	0.3971	0.4214	0.2774	0.3413
A_3	0.0222	0.0206	–	0.0028	0.0218	0.0447	0.0451	0.0325	0.0015	0
A_4	0.0194	0.0178	0.1082	–	0.0704	0.1529	0.044	0.1407	0.0718	0.0518
A_5	0.0342	0.0416	0.1245	0.0677	–	0.1044	0.0759	0.0921	0.0774	0.0556
A_6	0.371	0.3688	0.6276	0.6304	0.5845	–	0.6528	0.4576	0.4628	0.5586
A_7	0.0211	0.0195	0.1561	0.0495	0.0841	0.1808	–	0.1686	0.0889	0.0685
A_8	0.0694	0.0703	0.17	0.1728	0.127	0.0122	0.1952	–	0.0767	0.1052
A_9	0.0948	0.0488	0.2614	0.2263	0.2346	0.1398	0.2379	0.1991	–	0.1102
A_{10}	0.0829	0.0368	0.1841	0.1304	0.137	0.1598	0.1417	0.1517	0.0343	–

TABLE 9.5 Positive, Negative, Net Outranking Flows, and Ranks for the Layout Selection Problem

Layout	Positive Flow	Negative Flow	Net Flow	Rank
A_1	0.3459	0.0849	0.2609	3
A_2	0.3443	0.0739	0.2704	2
A_3	0.0212	0.2937	–0.2725	10
A_4	0.0752	0.2306	–0.1553	8
A_5	0.0748	0.2332	–0.1583	9
A_6	0.5238	0.1486	0.3752	1
A_7	0.093	0.2422	–0.1492	7
A_8	0.111	0.2306	–0.1196	6
A_9	0.1725	0.1562	0.0163	4
A_{10}	0.1176	0.1856	–0.0679	5

The values of the net outranking flow for all the alternative layouts are now computed using Eqn. (9.7) and based on these values, the alternative layouts are ranked in descending order, as shown in Table 9.5. The best choice of layout for the given industrial application is layout A_6. Layout A_2 is the second choice, and the last choice is layout A_3.

9.4 NON-TRADITIONAL MACHINING (NTM) PROCESS SELECTION PROBLEM

The non-traditional machining (NTM) methods belong to a group of machining processes where the main purpose is to remove material from the workpiece surface using direct energy sources in the form of thermal, mechanical, chemical, electrical, or a combination of them. These NTM processes have now become indispensable to machine newly developed hard-to-machine metals and non-metals with excellent surface finish, minimum surface damage and complex part geometry [38]. In conventional machining processes, the material is removed in the form of chips by applying shear forces with the help of a sharp-edged cutting tool on the workpiece surface under the specific machining condition. The application of such force develops plastic deformation of the workpiece resulting in shear deformation and chip formation. Using conventional methods to machine harder metals and alloys incurs increased energy and machining cost; and in some cases, a conventional material removal process may not be suitable for a given machining application. Conventional machining also results in tool wear and loss of product quality due to induced residual stresses. Development and continuous use of newer exotic work materials with intricate features have been putting pressure on the capabilities of the conventional machining processes to generate complex shape geometries with the desired tolerances economically. To overcome these problems, in many modern-day industries, NTM processes are now being recognized as efficient and economic replacements to conventional machining processes. It is also a known fact that probably the NTM processes may never fully replace the conventional machining processes presently used in many of the industries. But these new machining processes ensure an increasingly important role because of their steadily improving capabilities [39]. Due to the presence of complex physical characteristics of various NTM processes along with a dearth of experts in this field, it has become a challenging task for a process engineer to identify the most suitable NTM process for a required machining application. A NTM process found suitable under the given circumstances may not be equally applicable to other situations. Thus, a careful selection of the most suitable NTM process for a given machining application is necessary while taking into consideration the following important attributes:

- Ability of the NTM processes in machining different shapes on the work materials;
- Physical and operational features of the NTM processes;
- Capability of those processes in machining different types of work material; and
- Economic aspects of different NTM processes.

In this example, pockets having length 16.5 mm, width 8.9 mm and depth 2.3 mm are generated on 4140 die steel within a tolerance of 0.13

mm. Based on these machining requirements, Yurdakul & Cogun [40] eliminated water jet machining (WJM), electrochemical grinding (ECG), electron beam machining (EBM), electrochemical honing (ECH), LBM, WEDM, and plasma arc machining (PAM) processes (on the basis of shape applications), and also USM and chemical machining (CHM) processes (on the basis of process capabilities), and identified the pertinent attributes as surface finish (SF) (in μm), surface damage (SD) (in mm), corner radii (CR) (in mm), taper (in mm/mm), material removal rate (MRR) (in mg/min), work material (WM) and cost, affecting the selection decision of the NTM process. Based on the recommendation of Yurdakul & Cogun [40], the decision matrix of Table 9.6 is developed here consisting of only three NTM processes, i.e., AJM, ECM, and EDM as the feasible alternatives to generate pockets of the desired dimensions on die steel material. Using AHP method, the criteria weights are determined as $w_{SF} = 0.27$, $w_{SD} = 0.05$, $w_{CR} = 0.05$, $w_{Taper} = 0.05$, $w_{MRR} = 0.27$, $w_{WM} = 0.05$ and $w_{Cost} = 0.27$, which are subsequently used in the PROMETHEE-GAIA method-based analysis. In this example, SF, SD, CR, taper, and cost are non-beneficial criteria, and MRR and WM are beneficial criteria. The initial input window of PROMETHEE-GAIA software is shown in Figure 9.2 where the end user can easily set the number of alternatives, number of criteria, number of scenarios for the group decision-making problem, type of each criterion (beneficial or non-beneficial), weights for the criteria and unit for each criterion. The PROMETHEE-GAIA software is available to be downloaded from www.promethee-gaia.net/software.html website. Among the six possible preference functions for PROMETHEE method, the simplest one (usual criterion) is chosen here. If the user selects other preference functions, the corresponding threshold values (*s* or *r*) need to be specified by the decision maker in the input window.

TABLE 9.6 Decision Matrix for the NTM Process Selection Problem

NTM Process	SF	SD	CR	Taper	MRR	WM	Cost
ECM	0.9	0	0.18	0.0009	2,200	4	3
EDM	1.8	18	0.36	0.0009	880	4	5
AJM	0.54	2.25	0.27	0.0045	55	3	3

When this problem of NTM process selection for pocketing operation on die steel is solved using PROMETHEE-GAIA software, ECM emerges out as the most suitable NTM process, as displayed in Figure 9.3. The EDM process is the worst choice and the capability of AJM process falls in between ECM and EDM processes. The developed GAIA plane is shown in Figure 9.4, where the decision axis (π) points towards the ECM process, indicating its suitability to generate pockets of the required dimensions on die steel. It can be noted that MRR and WM, and SF, SD, CR, taper, and cost are the conflicting criteria as they are oriented in opposite directions. In GAIA plane, criteria SD and CR are superimposed on each other, suggesting that these two criteria may have similar influences on the NTM process selection decision. It signifies that these two criteria may be correlated. Criteria MRR and cost, and criteria WM and cost are not related to each other with respect to their preferences as they are represented by the orthogonal axes. It can also be observed that this NTM process selection problem has a solution accuracy (reliability) of 100%. Figure 9.5 exhibits the walking weight diagram for this NTM process selection problem where the process engineer can easily modify the criteria weights to visualize the effects of those changing weights on the ranking of the candidate NTM processes. It mainly performs as a sensitivity analysis tool. Using hybrid AHP and TOPSIS methods, Yurdakul & Cogun [40] also recommended ECM as the best NTM process for this machining application.

Scenario1	SF	SD	CR	Taper	MRR	WM	Cost
Unit	Microns	mm	mm	mm	mg/min		
Cluster/Group							
Preferences							
Min/Max	min	min	min	min	max	max	min
Weight	0.27	0.05	0.05	0.05	0.27	0.05	0.27
Preference Fn.	Usual	Usual	Usual	Usual	Usual	Usual	Usual
Thresholds	absolute	absolute	absolute	absolute	absolute	absolute	absolute
- Q: Indifference	n/a	n/a	n/a	n/a	n/a	n/a	n/a
- P: Preference	n/a	n/a	n/a	n/a	n/a	n/a	n/a
- S: Gaussian	n/a	n/a	n/a	n/a	n/a	n/a	n/a
Statistics							
Minimum	0.54	0.0	0.2	0.001	55	3	3
Maximum	1.80	18.0	0.4	0.004	2200	4	7
Average	1.08	6.8	0.3	0.002	1045	4	5
Standard Dev.	0.53	8.0	0.1	0.002	883	0	2
Evaluations							
ECM	0.90	0.0	0.2	0.001	2200	4	3
EDM	1.80	18.0	0.4	0.001	880	4	5
AJM	0.54	2.3	0.3	0.004	55	3	3

FIGURE 9.2 Input window for NTM process selection problem.

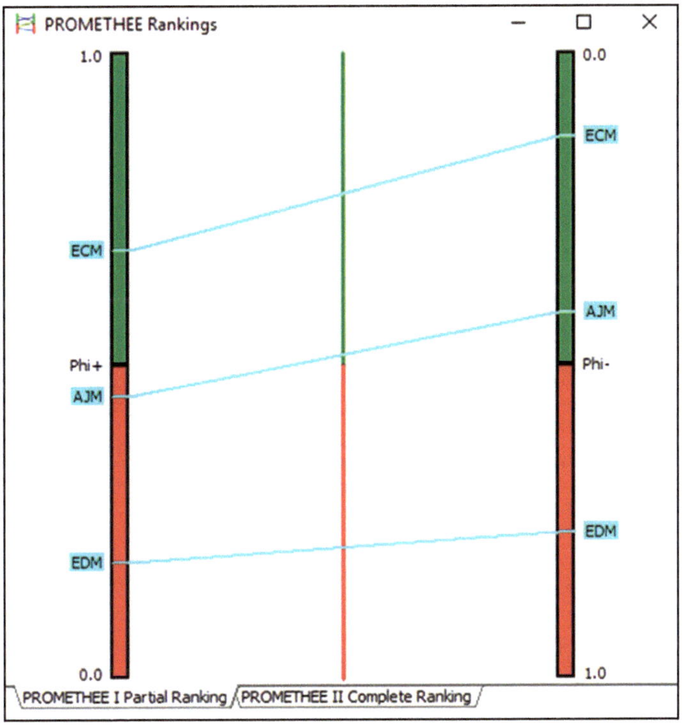

FIGURE 9.3 PROMETHEE II-based ranking of the NTM processes.

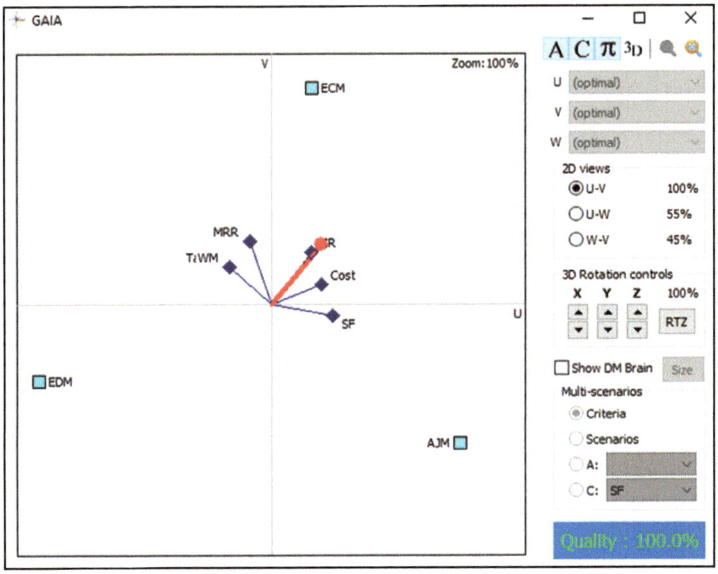

FIGURE 9.4 GAIA plane for NTM process selection problem.

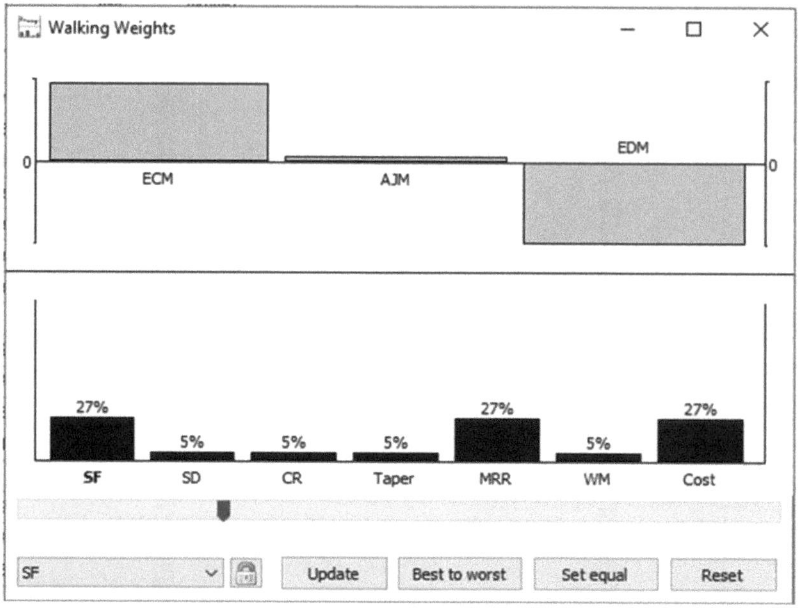

FIGURE 9.5 Walking weights for NTM process selection problem.

KEYWORDS

- **electrochemical grinding**
- **enterprise resource planning**
- **geometrical analysis for interactive aid**
- **preference ranking organization method**
- **small- and medium-sized enterprise**
- **strengths-weaknesses-opportunities-threats**
- **water jet machining**

REFERENCES

1. Brans, J. P., & Vincke, P., (1985). A preference ranking organization method. *Management Science, 31*(6), 647–656.
2. Brans, J. P., Vincke, P., & Mareschal, B., (1986). How to select and how to rank projects: The PROMETHEE method. *European Journal of Operational Research, 24*(2), 228–238.

3. Behzadian, M., Kazemzadeh, R. B., Albadvi, A., & Aghdasi, M., (2010). PROMETHEE: A comprehensive literature review on methodologies and applications. *European Journal of Operational Research, 200*(1), 198–215.
4. Brans, J. P., De Smet, Y., (2016). PROMETHEE methods. In: *Multiple Criteria Decision Analysis* (pp. 187–219). Springer, New York.
5. Abdullah, L., Chan, W., & Afshari, A., (2019). Application of PROMETHEE method for green supplier selection: A comparative result based on preference functions. *Journal of Industrial Engineering International, 15*(2), 271–285.
6. Singh, A., Gupta, A., & Mehra, A., (2020). Best criteria selection based PROMETHEE II method. *OPSEARCH, 58*(1), 160–180.
7. Anand, G., & Kodali, R., (2008). Selection of lean manufacturing systems using the PROMETHEE. *Journal of Modeling in Management, 3*(1), 40–70.
8. Dağdeviren, M., (2008). Decision making in equipment selection: An integrated approach with AHP and PROMETHEE. *Journal of Intelligent Manufacturing, 19*, 397–406.
9. Rao, R. V., & Patel, B. K., (2010). Decision making in the manufacturing environment using an improved PROMETHEE method. *International Journal of Production Research, 48*(16), 4665–4682.
10. Turcksin, L., Bernardini, A., & Macharis, C., (2011). A combined AHP-PROMETHEE approach for selecting the most appropriate policy scenario to stimulate a clean vehicle fleet. *Procedia Social and Behavioral Sciences, 20*, 954–965.
11. Bogdanovic, D., Nikolic, D., & Ilic, I., (2012). Mining method selection by integrated AHP and PROMETHEE method. *Annals of the Brazilian Academy of Sciences, 84*(1), 1–10.
12. Karande, P., & Chakraborty, S., (2012). Application of PROMETHEE-GAIA method for non-traditional machining processes selection. *Management Science Letters, 2*, 2049–2060.
13. Safari, H., Fagheyi, M. S., Ahangari, S. S., & Fathi, M. R., (2012). Applying PROMETHEE method based on entropy weight for supplier selection. *Business Management and Strategy, 3*(1), 97–106.
14. Vinodh, S., & Jeya, G. R., (2012). PROMETHEE based sustainable concept selection. *Applied Mathematical Modeling, 36*, 5301–5308.
15. Maity, S. R., & Chakraborty, S., (2013). A visual decision aid for gear materials selection. *Journal of the Institution of Engineers (India) Series C, 94*, 199–212.
16. Meysam, M. S., Tavakkoli-Moghaddam, R., Heydar, M., & Ebrahimnejad, S., (2013). Multi-criteria decision making for plant location selection: An integrated Delphi-AHP-PROMETHEE methodology. *Arabian Journal of Science and Engineering, 38*, 1255–1268.
17. Öztürk, A., Kecek, G., & Söylemez, E. Y., (2013). Personnel selection in an accommodation enterprise by method. *International Journal of Business and Commerce, 5*(3), 1–19.
18. Balali, V., Mottaghi, A., Shoghli, O., & Golabchi, M., (2014). Selection of appropriate material, construction technique, and structural system of bridges by use of multicriteria decision-making method. *Journal of the Transportation Research Board, 3*, 79–87.
19. Balali, V., Zahraie, B., & Roozbahani, A., (2014). A comparison of AHP and PROMETHEE family decision making methods for selection of building structural system. *American Journal of Civil Engineering and Architecture, 2*(5), 149–159.
20. Murali, P., Diwakar, R. V., & Naga, P. A., (2014). Supplier selection by using multi-criteria decision-making methods. *International Journal of Engineering Research and General Science, 2*(6), 533–539.

21. Kilic, H. S., Zaim, S., & Delen, D., (2015). Selecting "the best" ERP system for SMEs using a combination of ANP and PROMETHEE methods. *Expert Systems with Applications, 42*, 2343–2352.
22. Sen, D. K., Datta, S., Patel, S. K., & Mahapatra, S. S., (2015). Multi-criteria decision making towards selection of industrial robot exploration of PROMETHEE II method. *Benchmarking: An International Journal, 22*(3), 465–487.
23. Bogdanović, D., Vučković, D., Pajović, M., & Gubijan, N., (2017). The implementation of the PROMETHEE-GAIA method in the process of determining the benchmarking partner in building a CHP plant. *Engineering Management, 3*(2), 36–42.
24. Abdelhadi, A., (2018). Maintenance scheduling based on PROMETHEE method in conjunction with group technology philosophy. *International Journal of Quality & Reliability Management, 35*(7), 1423–1444.
25. Bottero, M., D'Alpaos, C., & Oppio, A., (2018). Multicriteria evaluation of urban regeneration processes: An application of PROMETHEE method in Northern Italy. *Advances in Operations Research* (p. 12). Article ID: 9276075. https://doi.org/10.1155/2018/9276075.
26. Peterková, J., & Franek, J., (2018). Decision making support for managers in innovation management: A PROMETHEE approach. *International Journal of Innovation, 6*(3), 256–274.
27. Sapkota, M., Arora, M., Malano, H., Sharma, A., & Moglia, M., (2018). Integrated evaluation of hybrid water supply systems using a PROMETHEE-GAIA approach. *Water, 10*, 15. doi: 10.3390/w10050610.
28. Yadav, G., Seth, D., & Desai, T. N., (2018). Application of hybrid framework to facilitate lean six sigma implementation: A manufacturing company case experience. *Production Planning & Control, 29*(3), 185–201.
29. Zindani, D., & Kumar, K., (2018). Material selection for turbine seal strips using PROMETHEE-GAIA method. *Materials Today: Proceedings, 5*, 17533–17539.
30. Abdullah, L., Chan, W., & Afshari, A., (2019). Application of PROMETHEE method for green supplier selection: A comparative result based on preference functions. *Journal of Industrial Engineering International, 15*, 271–285.
31. Crnjac, M., Aljinovic, A., Gjeldum, N., & Mladineo, M., (2019). Two-stage product design selection by using PROMETHEE and Taguchi method: A case study. *Advances in Production Engineering & Management, 14*(1), 39–50.
32. Raikar, A. V., (2019). An assessment of performance of steel companies in India using PROMETHEE method of multi criteria decision technique. *International Journal of Management Studies, VI*(4), 46–56.
33. Yougbaré, J. W., (2019). The ranking of districts in Ouagadougou by the risk of flood and runoff using the PROMETHEE. *European Journal of Pure and Applied Mathematics, 12*(4), 1731–1743.
34. Baloyi, V. D., & Meyer, L. D., (2020). The development of a mining method selection model through a detailed assessment of multi-criteria decision methods. *Results in Engineering, 8*, 100172.
35. Nasrollahi, M., Ramezani, J., & Sadraei, M., (2020). A FBWM-PROMETHEE approach for industrial robot selection. *Heliyon, 6*, e03859.
36. Soba, M., Ersoy, Y., Altınay, A. T., Erkan, B., & Şik, E., (2020). Application of multiple criteria decision-making methods in assignment place selection. *Mathematical Problems in Engineering* (p. 13). Article ID: 6748342. https://doi.org/10.1155/2020/6748342.

37. Chakraborty, S., & Banik, B., (2007). An analytic hierarchy process (AHP) based approach for optimal facility layout design. *Journal of the Institution of Engineers, Part PR: Production Engineering, 88*, 12–18.
38. Pandey, P. C., & Shan, H. S., (2005). *Modern Machining Processes*. Tata MacGraw-Hill Publishing Company Limited, New Delhi.
39. Jain, V. K., (2005). *Advanced Machining Processes*. Allied Publishers Pvt. Limited, New Delhi.
40. Yurdakul, M., & Cogun, C., (2003). Development of a multi-attribute selection procedure for non-traditional machining processes. *Proc. IMechE Part B: Journal of Engineering Manufacture, 217*, 993–1009.

CHAPTER 10

COMPLEX PROPORTIONAL ASSESSMENT (COPRAS) METHOD

The COmplex PRoportional ASsessment (COPRAS) method, developed by Zavadskas et al. [1], assumes direct and proportional dependences of the significance and utility degree of the available alternatives under the presence of mutually conflicting criteria. It takes into account the performance of the alternatives with respect to different criteria and the corresponding criteria weights. This method selects the best decision considering both the ideal and the ideal-worst solutions. The COPRAS method adopts a step-wise ranking and evaluating procedure of the alternatives in terms of their significance and utility degree. This method has already been successfully applied to solve various problems in the field of construction, property management, economics, etc. [1–5]. The procedural steps of the COPRAS method are presented as below:

- **Step 1:** Develop the decision matrix, X.

$$X = [x_{ij}]_{m \times n} = \begin{bmatrix} x_{11} & x_{12} & \cdots & x_{1n} \\ x_{21} & x_{22} & \cdots & x_{2n} \\ \cdots & \cdots & \cdots & \cdots \\ x_{m1} & x_{m2} & \cdots & x_{mn} \end{bmatrix} \quad (10.1)$$

  where; x_{ij} is the performance value of ith alternative on j^{th} criterion, m is the number of alternatives and n is the number of criteria.

- **Step 2:** Normalize the decision matrix using the following equation to obtain dimensionless values of different criteria so that all of them can be compared.

$$R = [r_{ij}]_{m \times n} = \frac{x_{ij}}{\sum_{i=1}^{m} x_{ij}} \quad (10.2)$$

Multi-Criteria Decision-Making Methods in Manufacturing Environments: Models and Applications.
Shankar Chakraborty, Prasenjit Chatterjee, & Partha Protim Das (Authors)

- **Step 3:** Determine the weighted normalized decision matrix, *D*.

$$D = [y_{ij}]_{m \times n} = r_{ij} \times w_j \quad (i = 1,2,\ldots,m; j = 1,2,\ldots,n) \tag{10.3}$$

where; r_{ij} is the normalized performance value of ith alternative on j^{th} criterion and w_j is the weight of jth criterion. The sum of dimensionless weighted normalized values of each criterion is always equal to the weight for that criterion.

$$\sum_{i=1}^{m} y_{ij} = w_j \tag{10.4}$$

In other words, it can be said that the weight, w_j of the investigated criterion is proportionally distributed among all the alternatives according to their weighted normalized value, y_{ij}.

- **Step 4:** Sums of the weighted normalized values are calculated for both beneficial and non-beneficial criteria.
These sums are calculated using the following equations:

$$S_{+i} = \sum_{j=1}^{n} y_{+ij} \tag{10.5}$$

$$S_{-i} = \sum_{j=1}^{n} y_{-ij} \tag{10.6}$$

where; y_{+ij} and y_{-ij} are the weighted normalized values for beneficial and non-beneficial criteria, respectively.
The greater the value of S_{+i}, the better is the alternative, and the lower the value of S_{-i}, the better is the alternative. The S_{+i} and S_{-i} values express the degree of goals attained by each alternative. In any case, the sums of 'pluses' S_{+i} and 'minuses' S_{-i} of the alternatives are always respectively equal to the sums of weights for the beneficial and non-beneficial criteria, as expressed by the following equations:

$$S_{+} = \sum_{i=1}^{m} S_{+i} = \sum_{i=1}^{m}\sum_{j=1}^{n} y_{+ij} \tag{10.7}$$

$$S_{-} = \sum_{i=1}^{m} S_{-i} = \sum_{i=1}^{m}\sum_{j=1}^{n} y_{-ij} \tag{10.8}$$

In this way, Eqns. (10.7) and (10.8) can be used to verify the calculations.

- **Step 5:** Determine the significances of the alternatives on the basis of defining the positive alternatives S_{+i} and negative alternatives S_{-i} characteristics.

- **Step 6:** Determine the relative significances or priorities of the alternatives. The priorities of the candidate alternatives are calculated on the basis of Q_i. The greater the value of Q_i, the higher is the priority of the alternative. The relative significance value of an alternative shows the degree of satisfaction attained by that alternative. The alternative with the highest relative significance value (Q_{max}) is the best choice among the candidate alternatives.
  The relative significance value (priority), Q_i of ith alternative can be obtained as below:

$$Q_i = S_{+i} + \frac{S_{-\min}\sum_{i=1}^{m} S_{-i}}{S_{-i}\sum_{i=1}^{m}(S_{-\min}/S_{-i})} \quad (i = 1,2,\ldots,m) \tag{10.9}$$

  where; S_{-min} is the minimum value of S_{-i}.
- **Step 7:** Calculate the quantitative utility (U_i) for i^{th} alternative. The degree of an alternative's utility is directly associated with its relative significance value (Q_i). The degree of an alternative's utility, leading to a complete ranking of the candidate alternatives, is determined by comparing the priorities of all the alternatives with the most efficient one and can be denoted as below:

$$U_i = \left[\frac{Q_i}{Q_{\max}}\right] \text{x} 100\% \tag{10.10}$$

  where; Q_{max} is the maximum relative significance value. With the increase or decrease in the value of the relative significance for an alternative, it is observed that its degree of utility also increases or decreases [1]. These utility values of the candidate alternatives range from 0% to 100%. Thus, this approach allows for evaluating the direct and proportional dependence of significance and utility degree of the considered alternatives in a decision-making problem involving multiple criteria, their weights and performance values of the alternatives with respect to all the criteria.

10.1 SURVEY OF THE LITERATURE

Kaklauskas et al. [6] proposed the application of COPRAS method as an MCDM tool for selection of the most suitable contractor for replacement of windows in a public building. Zavadskas et al. [5] demonstrated the concept of the application of COPRAS method while identifying the best road design alternative. Applying COPRAS method, Chatterjee et al. [7] solved two real-time material selection problems for mechanical components and derived the ranking lists of the alternative materials from the best to the worst. Chatterjee & Chakraborty [8] proved the application potentiality of both COPRAS and gray COPRAS (COPRAS-G) methods for efficiently solving material selection problems for mechanical components. Popović et al. [9] explored the application potentialities of COPRAS and COPRAS-G methods for the selection of investment projects based on several financial analysis criteria and imprecise data. Chatterjee & Chakraborty [10] contrasted the applications of COPRAS and ARAS methods while solving a gear material selection problem in a given manufacturing environment.

While employing COPRAS method, Gadakh [11] proposed an effective decision-making framework for solving vendor selection problems. Its efficacy was later proved using two demonstrative examples. Xia et al. [12] combined COPRAS method with coefficient of variation approach and presented its application for solving an industrial material selection problem. Kouchaksaraei et al. [13] developed a hybrid framework integrating SWARA and COPRAS methods for identification of the best location for a glasshouse in Iran. Makhesana [14] adopted COPRAS method for identification of the most appropriate RP process and compared its ranking performance with other MCDM techniques. Petković et al. [15] considered 10 NTM processes and identified the most suitable process for machining of ceramic materials based on 10 criteria using COPRAS method. Stefano et al. [16] presented an exhaustive review of COPRAS method and its applications in diverse domains. Adalı & Işik [17] selected the most appropriate air conditioner based on several conflicting criteria while applying both COPRAS and ARAS methods. The ranking results obtained using those methods were finally compared. Kundakcı & Işık [18] first applied the MACBETH method for the determination of the criteria weights based on subjective assessment and later adopted the COPRAS method for ranking of the alternative air compressors in a textile spinning mill.

While solving a material selection problem for flywheel design, Asodariya et al. [19] first applied AHP and entropy methods to estimate weights of the subjective and objective criteria under consideration, and then employed both COPRAS and TOPSIS methods for identifying the most apposite material. Borkar et al. [20] stressed on the application of COPRAS method for singling out the best alternative from a pool of available options. Applying COPRAS method, Goran et al. [21] solved a problematic ranking problem of cadastral municipalities where land consolidation would be performed based on effective land management approach. George & Badoniya [22] combined entropy and COPRAS methods for identifying the best supplier in a manufacturing organization while resolving the uncertainty involved in the supplier selection problems. Taking into account 10 alternative mobile phone models from different brands, Goswami & Mitra [23] proposed the applications of both COPRAS and ARAS methods for finding out the best mobile based on price, internal storage, RAM, and brand. The weights of the considered evaluation criteria were estimated using AHP method. The ranking performance of those two MCDM tools was observed to be almost similar with slight variations in the intermediate rankings of the mobiles. Pamučar & Savin [24] presented the integrated application of BWM and COPRAS for selection of the optimal off-road vehicle to fulfill the requirements of the Serbian Armed Force. The derived results were also validated with respect to other MCDM tools. Sriram & Vinodh [25] analyzed and determined the priority ranking of 15 readiness factors for Industry 4.0 implementation in SMEs based on COPRAS method. Jain & Chand [26] applied COPRAS method for ranking flexibility and performance indicators of an FMS. It was noticed that product flexibility (PF) for new part configuration had the maximum importance and productivity was the primary performance indicator of an FMS.

10.2 MATERIAL SELECTION FOR A NAVAL CRAFT

This demonstrative example deals with the selection of the most appropriate material for a light-weight high-speed naval craft [27]. The high-speed naval craft material selection problem consists of six alternative materials, i.e., Grade A Steel (M_1), Single Skin Aluminum (A5086-H34) (M_2), Aluminum Sandwich (honeycomb core) (M_3), LASCOR Steel (M_4), Composite (CFRP) Carbon w/Vinyl ester resin (M_5) and DUCTAL (UHP2C) (M_6). The key requirements in designing a high-speed naval craft are yield strength (YS) (in MPa), Young's modulus (YM) (in GPa), fire resistance (FR), repairability (RY), resistance to corrosion (RC), fabrication cost (FC), risk (R), density (D) (in kg/m^3) and overall potential for weight savings (WS). Among these nine criteria, YS, YM, FR, RY, RC, and WS are the beneficial attributes, whereas, FC, R, and D are the non-beneficial attributes. Thus, the light-weight high-speed naval craft material selection problem consists of six alternative materials and nine material selection criteria, as shown in Table 10.1.

TABLE 10.1 Quantitative Data for the Naval Craft Material Selection Problem [27]

Material	YS	YM	FR	RY	RC	FC	R	D	WS
M_1	234.4	204.1	0.6818	0.7727	0.3182	0.5	0.3182	7,800	0
M_2	137.9	67	0.3182	0.6818	0.6818	0.3182	0.5	2,700	0.6818
M_3	268.9	67	0.5	0.5	0.6818	0.5	0.5	1,800	0.7727
M_4	379.2	204.1	0.6818	0.5	0.6818	0.7727	0.6818	5,200	0.6818
M_5	1496.2	227.5	0.3182	0.5	0.7727	0.7727	0.5	1,800	0.7727
M_6	220.6	53.9	0.7727	0.7727	0.7727	0.3182	0.7727	2,500	0.6818

Source: Reprinted with permission from Ref. [27].

While applying COPRAS method for solving this material selection problem, the evaluation/decision matrix, as shown in Table 10.1, is first normalized using Eqn. (10.2) and the corresponding weighted normalized decision matrix is developed, as given in Table 10.2. As mentioned earlier, the purpose of normalization is to obtain dimensionless values of different material selection criteria so that all of them can be compared. Rao & Patel [27] employed an integrated weighting method to determine the corresponding criteria weights and took into account seven cases of the integrated weightings for the criteria. The second case, where $w_{YS} = 0.3914$, $w_{YM} = 0.1265$, $w_{FR} = 0.0540$, $w_{RY} = 0.0291$, $w_{RC} = 0.0332$, $w_{FC} = 0.0676$, $w_R = 0.0399$, $w_D = 0.1549$, and $w_{WS} = 0.1036$, is adopted here for the COPRAS method-based analysis. Now, using Eqns. (10.5) and (10.6), the sums of the weighted normalized values are calculated for both the beneficial criteria (S_{+i}) and non-beneficial criteria (S_{-i}), as given in Table 10.3. Then, applying Eqn. (10.9), the relative significance or priority value (Q_i) for each candidate material is determined, as shown in Table 10.4.

TABLE 10.2 Weighted Normalized Decision Matrix

Material	YS	YM	FR	RY	RC	FC	R	D	WS
M_1	0.0335	0.0313	0.0112	0.0060	0.0027	0.0000	0.0106	0.0039	0.0554
M_2	0.0197	0.0103	0.0053	0.0053	0.0058	0.0197	0.0068	0.0061	0.0192
M_3	0.0385	0.0103	0.0083	0.0039	0.0058	0.0223	0.0106	0.0061	0.0128
M_4	0.0542	0.0313	0.0112	0.0039	0.0058	0.0197	0.0164	0.0083	0.0369
M_5	0.2139	0.0349	0.0053	0.0039	0.0066	0.0223	0.0164	0.0061	0.0128
M_6	0.0315	0.0083	0.0127	0.0060	0.0066	0.0197	0.0068	0.0094	0.0178

TABLE 10.3 Sums of the Weighted Normalized Values

Material	S_{+i}	Value	S_{-i}	Value
M_1	S_{+1}	0.0849	S_{-1}	0.0699
M_2	S_{+2}	0.0660	S_{-2}	0.0320
M_3	S_{+3}	0.0890	S_{-3}	0.0295
M_4	S_{+4}	0.1262	S_{-4}	0.0617
M_5	S_{+5}	0.2869	S_{-5}	0.0353
M_6	S_{+6}	0.0848	S_{-6}	0.0339

Table 10.4 also exhibits the value of quantitative utility (U_i) for each alternative on the basis of which the complete ranking of the alternative materials is obtained. The candidate materials for high-speed naval craft are then arranged in descending order of U_i value.

TABLE 10.4 Q_i and U_i Values

Material	Q_i	U_i	Rank
M_1	0.1093	32.5967	6
M_2	0.1194	35.6136	5
M_3	0.1469	43.8188	3
M_4	0.1539	45.8980	2
M_5	0.3354	100.0000	1
M_6	0.1352	40.3252	4

The best choice of material for high-speed naval craft is Composite (CFRP) Carbon w/Vinyl ester resin. LASCOR Steel is the second choice and the last choice is Grade A Steel. Rao & Patel [27] obtained a ranking of the alternative materials as M_5-M_3-M_4-M_6-M_2-M_1 (for case 2), whereas, using COPRAS method, the ranking of candidate materials is derived as M_5-M_4-M_3-M_6-M_2-M_1. It can be noticed that the rankings of the best and the worst alternative materials are exactly the same, which suggests that the COPRAS method can be successfully applied for dealing with complex decision-making problems.

KEYWORDS

- **complex proportional assessment**
- **decision-making problems**
- **repairability**
- **resistance to corrosion**
- **weight savings**
- **Young's modulus**

REFERENCES

1. Zavadskas, E. K., Kaklauskas, A., Turskis, Z., & Tamošaitien, J., (2008). Selection of the effective dwelling house walls by applying attributes values determined at intervals. *Journal of Civil Engineering and Management, 14,* 85–93.
2. Kaklauskas, A., Zavadskas, E. K., & Trinkunas, V., (2007). A multiple criteria decision support online system for construction. *Engineering Applications of Artificial Intelligence, 20,* 163–175.
3. Zavadskas, E. K., Turskis, Z., Tamošaitienė, J., & Marina, V., (2008). Multicriteria selection of project managers by applying grey criteria. *Technological and Economic Development of Economy, Baltic Journal of Sustainability, 14,* 462–477.
4. Kaklauskas, A., Zavadskas, E. K., Raslanas, S., Ginevicius, R., Komka, A., & Malinauskas, P., (2006). Selection of low e-windows in retrofit of public buildings by applying multiple criteria method COPRAS: A Lithuanian case. *Energy and Buildings, 38,* 454–462.
5. Zavadskas, E. K., Kaklauskas, A., Peldschus, F., & Turskis, Z., (2007). Multi-attribute assessment of road design solutions by using the COPRAS method. *Baltic Journal of Road and Bridge Engineering, 2*(4), 195–203.
6. Kaklauskas, A., Zavadskas, E. K., Raslanas, S., Ginevicius, R., Komka, A., & Malinauskas, P., (2006). Selection of low-e windows in retrofit of public buildings by applying multiple criteria method COPRAS: A Lithuanian case. *Energy and Buildings, 38,* 454–462.
7. Chatterjee, P., Athawale, V. M., & Chakraborty, S., (2011). Materials selection using complex proportional assessment and evaluation of mixed data methods. *Materials and Design, 32,* 851–860.
8. Chatterjee, P., & Chakraborty, S., (2012). Materials selection using COPRAS and COPRAS-G methods. *International Journal of Materials and Structural Integrity, 6*(2, 3, 4), 111–133.
9. Popović, G., Stanujkić, D., & Stojanović, S., (2012). Investment project selection by applying COPRAS method and imprecise data. *Serbian Journal of Management, 7*(2), 257–269.

10. Chatterjee, P., & Chakraborty, S., (2013). Gear material selection using complex proportional assessment and additive ratio assessment-based approaches: A comparative study. *International Journal of Materials Science and Engineering, 1*(2), 104–111.
11. Gadakh, V. S., (2014). Application of complex proportional assessment method for vendor selection. *International Journal of Logistics Research and Applications, 17*(1), 23–34.
12. Xia, F., Wei, H., & Yang, L. W., (2014). Improved COPRAS method and application in material selection problem. *Applied Mechanics and Materials, 707*, 505–508.
13. Kouchaksaraei, R. H., Zolfani, S. H., & Golabchi, M., (2015). Glasshouse locating based on SWARA-COPRAS approach. *International Journal of Strategic Property Management, 19*(2), 111–122.
14. Makhesana, M. A., (2015). Application of improved complex proportional assessment (COPRAS) method for rapid prototyping system selection. *Rapid Prototyping Journal, 21*(6), 671–674.
15. Petković, D., Madić, M., & Radenković, G., (2015). Selection of the most suitable non-conventional machining processes for ceramics machining by using MCDMs. *Science of Sintering, 47*, 229–235.
16. Stefano, N. M., Casarotto, F. N., Vergara, L. G. L., & Rocha, R. U. G., (2015). COPRAS (complex proportional assessment): State of the art research and its applications. *IEEE Latin America Transactions, 13*(12), 3899–3906.
17. Adalı, E. A., & Işık, A. T., (2016). Air conditioner selection problem with COPRAS and ARAS methods. *MANAS Journal of Social Studies, 5*(2), 124–138.
18. Kundakcı, N., & Işık, A. T., (2016). Integration of MACBETH and COPRAS methods to select air compressor for a textile company. *Decision Science Letters, 5*(2016). 381–394.
19. Asodariya, H., Patel, H. V., Babariya, D., & Maniya, K. D., (2018). Application of multi-criteria decision making method to select and validate the material of a flywheel design. M*aterials Today: Proceedings, 5,* 17147–17155.
20. Borkar, P., Jaiswal, S., Dhawan, P., & Saraf, P., (2018). Decision making using COPRAS and ELECTRE technique. *Helix, 8*(5), 4039–4043.
21. Goran, M., Jelena, L., Ilija, G., & Zoran, I., (2018). Application of COPRAS method for land consolidation projects ranking. *Archives for Technical Sciences, 19*(1), 35–44.
22. George, J., & Badoniya, P., (2019). A decision support system for selecting the best supplier for multiple criteria decision making problems using entropy weight method and COPRAS method. *National Journal of Advanced Research, 5*(3), 1–3.
23. Goswami, S. S., & Mitra, S., (2020). Selecting the best mobile model by applying AHP-COPRAS and AHP-ARAS decision making methodology. *International Journal of Data and Network Science, 4*, 27–42.
24. Pamučar, D. S., & Savin, L. M., (2020). Multiple-criteria model for optimal off-road vehicle selection for passenger transportation: BWM-COPRAS model. *Military Technical Courier, 68*(1), 28–64.
25. Sriram, R. M., & Vinodh, S., (2020). Analysis of readiness factors for Industry, 4.0 implementation in SMEs using COPRAS. *International Journal of Quality & Reliability Management, 38*(5), 1178–1192.
26. Jain, V., & Chand, M., (2021). Decision making in FMS by COPRAS approach. *International Journal of Business Performance Management, 22*(1), 75–92.
27. Rao, R. V., & Patel, B. K., (2010). A subjective and objective integrated multiple attribute decision making method for material selection. *Materials and Design, 31*(11), 4738–4747.

CHAPTER 11

ORGANIZATION, RANGEMENT ET SYNTHESE DE DONNES RELATIONNELLES (ORESTE) METHOD

The organization, Rangement Et Synthese De Donnes RElationnelles (ORESTE) is an effective MCDM method based on ordinal outranking relations, which require only the ordinal data and ranking of the criteria based on their relative importance. Basically, it is a compensatory preference ranking approach, as it uses the differences between the ranks of pairs of actions based on their evaluations [1, 2]. In many real-time decision-making situations involving both cardinal and ordinal criteria, this method is particularly appropriate where no quantitative data are available, and the decision maker is not able to provide any information for determining criteria weights or formulation of the corresponding preference function. It deals with the situation where an alternative A_i ($i = 1, 2, \ldots, m$) is ranked according to criterion C_j ($j = 1, 2, \ldots, n$), and the main objective is to find out a global preference structure on a set of alternatives, which reflects the evaluation of alternatives on each criterion and the preference among the criteria. As it only takes into account the rankings of alternatives and criteria, it is mainly suited for problems with ordinal data, but it can also be used for problems with both ordinal and cardinal data. The main advantage of this method is that it utilizes only the ordinal ranking of criteria which avoids the occurrence of lengthy discussions among the decision-makers to set priority importance of the criteria, thus speeding up the decision-making process. There are three distinct phases in the procedural application of ORESTE method. In the first phase, different alternatives and criteria are ranked, and a position table (projection matrix) is developed. In the second phase, projections are assigned with global ranks, and in the last phase, mean

Multi-Criteria Decision-Making Methods in Manufacturing Environments: Models and Applications.
Shankar Chakraborty, Prasenjit Chatterjee, & Partha Protim Das (Authors)

ranks are calculated by aggregating the global ranks. The steps for ORESTE method are presented as below [3]:

- **Step 1:** From the decision matrix, a weak order of criteria is constructed, which indicates their relative importance. The complete weak order is a preference structure in which the relative importance of the criteria is defined by considering the indifference (*I*) and preference (*P*) relations as shown below. By constructing weak orders, assigning weights to different criteria can be avoided.

$$C_1 \, P \, C_2 \, I \, C_3 \, P \, C_4 \ldots C_n$$

It denotes that C_1 is the most important and preferred criterion, while C_2 and C_3 are tied as the intermediate important criteria, C_4 is the least important criterion, and so on.

- **Step 2:** In the same way, as defined in Step (1), for each criterion, a weak order of the considered alternatives is also defined. Such a weak order is constructed by ranking the alternatives with respect to each criterion.

C_1: $A_1 \, P \, A_2 \, P \, A_3 \ldots A_m$
C_2: $A_1 \, P \, A_2 \, I \, A_3 \ldots A_m$
C_3: $A_1 \, P \, A_2 \, I \, A_3 \ldots A_m$
...
C_n: $A_1 \, I \, A_2 \, P \, A_3 \ldots A_m$

- **Step 3:** In this step, a procedure is required to assign ranks in terms of numerical values to each of the pre-determined weak orders of the criteria as well as alternatives. Therefore, Besson's ranks are calculated by the preference relations, as obtained in Steps (1) and (2), respectively. The Besson ranking is nothing but to rank the alternatives or criteria such that their sum is equal to the sum of the natural numbers of the alternatives or criteria and ties receive equal ranks. The Besson rank of an alternative A_i with respect to the jth criterion is denoted by $r_j(A_i)$ and the Besson rank of jth criterion is expressed as rC_j. If a weak order of the considered criteria is assumed as C_1 is preferred to C_2, C_2 is preferred to C_3 and C_3 is preferred to C_4, then the corresponding Besson ranks for these four criteria would be $rC_1 = 1$, $rC_2 = 2$, $rC_3 = 3$, and $rC_4 = 4$. If two criteria are both ranked 1, they would both receive a Besson rank of $(1 + 2)/2 = 1.5$ (average rank of the first two places they occupy).
- **Step 4:** Calculate the projection distances. The projection distances correspond to the relative positions of the alternatives with respect

to an arbitrary origin O and are defined by $d(O,A_i)$. Pastijn & Leysen [2] discussed about various types of projections. In this illustrative example, the linear orthogonal projection is employed, as expressed by the following equation:

$$d_j(O,A_i) = (1/2)[rC_j + r_j(A_i)] \tag{11.1}$$

The projection distances are such calculated that if an alternative A_1 is preferred to another alternative A_2 (A_1 P A_2) for j^{th} criterion, then $d_j(A_1) < d_j(A_2)$, i.e., smaller is projection distance, better is the position of the alternative.

- **Step 5:** Calculate the global Besson ranks. For constructing a complete order of the alternatives, the projections are again ranked by means of Besson's mean ranks as their relative positions are important and not their exact values. These ranks are called global ranks. A global Besson rank $r_j(A_i)$ is assigned to all the projection distances from the lowest to the highest ones. Smaller $r_j(A_i)$ indicates better position of a particular alternative.
- **Step 6:** Calculate the mean ranks. For each alternative, a mean rank is computed by the summation of their global Besson ranks over the entire set of criteria using the following expression, which yields a complete ranking order of the alternatives.

$$r(A_i) = \sum_{j=1}^{n} r_j(A_i) \tag{11.2}$$

These mean ranks are simply sorted increasingly to determine the global weak order of the alternatives.

11.1 LITERATURE REVIEW

In the tourism management field, Givescu [4] demonstrated the application of ORESTE method in formulating strategies with respect to tourism product, market development and penetration, organization distribution and advertisement to help managers as well as other stakeholders in effective decision-making. Chatterjee & Chakraborty [5] adopted ORESTE along with PROMETHEE II, COPRAS-G, and OCRA techniques to solve material selection problems, and later compared their relative ranking performance. Chatterjee & Chakraborty [6] employed six preference ranking methods for identification of the most apposite FMS for a given manufacturing organization. The superiority and advantageous features of ORESTE method over the other considered techniques were also validated. The application of ORESTE method was demonstrated by Arjun Raj & Vinodh [7] while solving an agile concept design problem for a rotary switch component. Adali & Işik [8] proposed the application of ORESTE method in the selection and ranking of web design firms, and also proved its efficacy as a structured, rational, and consistent approach while dealing with complex decision-making problems. Chatterjee & Chakraborty [9] presented the application of ORESTE method while solving five advanced manufacturing system selection problems, i.e., (a) an industrial robot; (b) an FMS; (c) a RP process; (d) an appropriate machine for a flexible manufacturing cell; and (e) an NTM process. Its advantageous features, like no requirement of criteria weights or formulation of any preference function were also demonstrated. While identifying the best material for geomembranes in construction, Chingo et al. [10] first determined the corresponding criteria weights using AHP technique and subsequently ranked the candidate alternative materials using ORESTE method. Its performance was also contrasted against other MCDM tools, like COPRAS-G, OCRA, TOPSIS, VIKOR, and PROMETHEE II. Kaya et al. [11] developed a decision support system using ORESTE method to generate all possible machine schedules based on several tangible traditional constraints, and rank those schedules with respect to human-related linguistic variables and ergonomic factors.

11.2 INDUSTRIAL ROBOT SELECTION PROBLEM

Rapid developments in computer engineering and science have been the main reason for the increased application of industrial robots in a variety of automated manufacturing systems. An industrial robot is defined as a multifunctional manipulator designed to move materials, parts, and tools by means of variable programmed motions. Since industrial robots are very costly and have numerous characteristics, their selection needs a thorough assessment and analysis of performance in a number of key areas. To properly evaluate and select a robot for a particular industrial application, several objective and subjective attributes, like precision, repeatability, degrees of freedom, control resolution, maximum tip speed (MTS), memory capacity (MC), load carrying capacity, programming flexibility, man-machine interfacing ability and vendor's service quality are usually taken into consideration. Objective attributes can be numerically defined, e.g., cost and load capacity (LC) of a robot; on the other hand, subjective attributes are qualitative in nature, like programming flexibility, vendor's service quality. Rao [12] considered a robot selection problem for some pick-n-place operations to avoid certain obstacles. In that example, five different robot selection criteria were considered as LC (in kg), repeatability (R) (in mm), MTS (in mm/s), MC (in points/step) and manipulator reach (MR) (in mm). Among these, LC, MTS, MC, and MR are beneficial attributes, whereas, R is the only non-beneficial attribute. Repeatability is the measure of the ability of a robot to return to a programmed position and it is basically calculated in terms of positional deviation from the average of displacements. It is expressed as a deviational error which always needs to be minimized. Table 11.1 represents the performance of seven alternative robots with respect to the five considered criteria. The robot is to be chosen in such a way that it should meet the minimum requirements of LC = 2 kg, repeatability = 0.05 mm, maximum tip speed = 255 mm/s, memory capacity = 250 points/step, manipulator reach = 500 mm, type of drives (actuators) = electrical only and degrees of freedom = 5.

TABLE 11.1 Quantitative Data for the Robot Selection Problem [12]

SL. No.	Robot	LC (C_1)	R (C_2)	MTS (C_3)	MC (C_4)	MR (C_5)
1.	ASEA-IRB 60/2	60	0.4	2,540	500	990
2.	Cincinnati Milacrone T3-726	6.35	0.15	1,016	3,000	1,041
3.	Cybotech V15 Electric Robot	6.8	0.1	1727.2	1,500	1,676
4.	Hitachi America Process Robot	10	0.2	1,000	2,000	965
5.	Unimation PUMA 500/600	2.5	0.1	560	500	915
6.	United States Robots Maker 110	4.5	0.08	1,016	350	508
7.	Yaskawa Electric Motoman L3C	3	0.1	177	1,000	920

Source: Reprinted with permission from Ref. [12]. © 2021 Elsevier.

Rao [12] determined the criteria weights as $w_{LC} = 0.036$, $w_R = 0.192$, $w_{MTS} = 0.326$, $w_{MC} = 0.326$ and $w_{MR} = 0.120$ using AHP method, and these weights are used here for ORESTE method-based analysis. At first, the weak order of the five criteria is determined based on their weight values, indicating their relative importance (preference). The weak order of the criteria is obtained as $C_3\ I\ C_4\ P\ C_2\ P\ C_5\ P\ C_1$ which indicates that MTS (C_3) and memory capacity (C_4) are the most preferred criteria, and LC (C_1) is the least important criterion. Then for each criterion, a weak order of the alternatives is determined, as shown in Table 11.2. Now, based on Step (3) of the procedural steps of ORESTE method, the Besson ranking of all the robot selection criteria and also the Besson ranking of the considered robot alternatives are determined, as given in Tables 11.3 and 11.4, respectively. Then, applying Eqn. (11.1), the corresponding projection distances are computed, as shown in Table 11.5. From this table, the rankings of the projections (global ranks) are obtained and are given in Table 11.6. In the last step of this method, the mean ranks of the alternative robots are estimated from which the final ranking of the candidate alternatives is derived, as shown in Table 11.7. Table 11.7 also exhibits the results obtained by Rao [12] and Chatterjee et al. [13]. From this table, the ranking of the robot alternatives as determined by ORESTE method is observed as 3-2-1-4-6-7-5, which indicates that Cybotech V15 electric robot is the best choice and Unimation PUMA 500/600 is the worst chosen alternative. Using AHP method, Rao [12] derived the ranking of the alternative robots as 3-2-7-1-4-6-5. On the other hand, Chatterjee et al. [13] obtained those rankings as 3-2-7-4-1-6-5 and 3-2-1-7-4-6-5 when VIKOR and ELECTRE methods were respectively applied. It can be noted that for all these four methods, the top two and the last ranked robots exactly match.

TABLE 11.2 Weak Order of Alternative Robots

Criteria	Weak Order of Alternative Robots
LC (C_1)	1 *P* 4 *P* 3 *P* 2 *P* 6 *P* 7 *P* 5
R (C_2)	6 *P* 3 *I* 5 *I* 7 *P* 2 *P* 4 *P* 1
MTS (C_3)	1 *P* 3 *P* 2 *I* 6 *P* 4 *P* 5 *P* 7
MC (C_4)	2 *P* 4 *P* 3 *P* 7 *P* 1 *I* 5 *P* 6
MR (C_5)	3 *P* 2 *P* 1 *P* 4 *P* 7 *P* 5 *P* 6

TABLE 11.3 Besson Ranking of Criteria for the Robot Selection Problem

Criteria	C_1	C_2	C_3	C_4	C_5
Besson Ranking	5	3	1.5	1.5	4

TABLE 11.4 Besson Ranking of Alternative Robots

Robot	C_1	C_2	C_3	C_4	C_5
1	1	7	1	5.5	3
2	4	5	3.5	1	2
3	3	3	2	3	1
4	2	6	5	2	4
5	7	3	6	5.5	6
6	5	1	3.5	7	7
7	6	3	7	4	5

TABLE 11.5 Projection Distances of Alternative Robots

Robot	C_1	C_2	C_3	C_4	C_5
1	3	5	1.25	3.5	3.5
2	4.5	4	2.5	1.25	3
3	4	3	1.75	2.25	2.5
4	3.5	4.5	3.25	1.75	4
5	6	3	3.75	3.5	5
6	5	2	2.5	4.25	5.5
7	5.5	3	4.25	2.75	4.5

TABLE 11.6 Mean Global Besson Ranks of Projections for the Alternative Robots

Robot	C_1	C_2	C_3	C_4	C_5
1	13	31	1.5	18.5	18.5
2	28	23	8	1.5	13
3	23	13	3.5	6	8
4	18.5	28	16	3.5	23
5	35	13	21	18.5	31
6	31	5	8	25.5	33.5
7	33.5	13	25.5	10	28

TABLE 11.7 Mean and Overall Ranks of the Alternative Robots

Robot	1	2	3	4	5	6	7
Mean rank	82.5	73.5	53.5	89	118.5	103	110
ORESTE method	3	2	1	4	7	5	6
AHP method [12]	4	2	1	5	7	6	3
VIKOR method [13]	5	2	1	4	7	6	3
ELECTRE method [13]	3	2	1	5	7	6	4

KEYWORDS

- **load capacity**
- **manipulator reach**
- **maximum tip speed**
- **MCDM method**
- **memory capacity**
- **Organization**
- **Rangement et Synthese de Donnes Relationnelles**

REFERENCES

1. Roubens, M., (1982). Preference relations on actions and criteria in multicriteria decision making. *European Journal of Operational Research, 10*(1), 51–55.
2. Pastijn, H., & Leysen, J., (1989). Constructing an outranking relation with ORESTE. *Mathematical and Computer Modeling, 12*(10, 11), 1255–1268.
3. Teghem, J., Delhaye, C., & Kunsch, P. L., (1989). An interactive decision support system (IDSS) for multicriteria decision aid. *Mathematical and Computer Modeling, 12*(10, 11), 1311–1320.
4. Givescu, O., (2007). The Oreste's method in the multicriteria's decision process for the management of tourism field. *Economia. Seria Management, 10*(1), 37–51.
5. Chatterjee, P., & Chakraborty, S., (2012). Material selection using preferential ranking methods. *Materials and Design, 35*, 384–393.
6. Chatterjee, P., & Chakraborty, S., (2014). Flexible manufacturing system selection using preference ranking methods: A comparative study. *International Journal of Industrial Engineering Computations, 5*, 315–338.
7. Arjun, R. A. S., & Vinodh, S., (2016). A case study on application of ORESTE for agile concept selection. *Journal of Engineering, Design and Technology, 14*(4), 781–801.
8. Adali, E. A., & Işik, A. T., (2017). Ranking web design firms with the ORESTE method. *EGE Academic Review, 17(2), 243–254.*
9. Chatterjee, P., & Chakraborty, S., (2017). Advanced manufacturing systems selection using ORESTE method. *International Journal of Advanced Operations Management, 5*(4), 337–361.
10. Chingo, C., Martínez-Gómez, J., & Narváez, R. A., (2020). Material selection using multi-criteria decision making methods for geomembranes. *International Journal of Mathematics in Operational Research, 16*(1), 24–52.
11. Kaya, B. Y., Adem, A., & Dağdeviren, (2020). A DSS-based novel approach proposition employing decision techniques for system design. *International Journal of Information Technology and Decision Making, 19*(2), 413–445.
12. Bhangale, P. P., Agrawal, V. P., & Saha, S. K., (2004). Attribute based specification, comparison and selection of a robot. *Mechanism and Machine Theory, 39*(12), 1345–1366.
13. Chatterjee, P., Athawale, V. M., & Chakraborty, S., (2010). Selection of industrial robots using compromise ranking and outranking methods. *Robotics and Computer-Integrated Manufacturing, 26*(5), 483–489.

CHAPTER 12

OPERATIONAL COMPETITIVENESS RATING ANALYSIS (OCRA) METHOD

The OCRA method was initially introduced to evaluate the performance measure of a set of production units that consumes resources to create value-added products. The first step towards the application of this method is to determine preference ratings of the non-beneficial or input criteria. The second step is to determine the preference ratings of the output criteria. In the third or final step, the overall preference ratings of the feasible alternatives are calculated considering both the cardinal and ordinal data. This method takes into consideration the decision maker's instinctual preferences in terms of the relative significance of the criteria [1]. The estimated preference ratings of the considered alternatives reflect the preference of the decision maker for the criteria. In addition to this, the main benefit of this method is that it can take care of those MCDM situations where the relative criteria weights are dependent on the alternatives along with different weight distributions being allotted to the criteria for different alternatives, and few of these criteria are not relevant to the candidate alternatives.

The procedural steps involved in OCRA method are discussed as below [2]:

- **Step 1:** Determine the preference ratings of the non-beneficial criteria. In this stage, OCRA method only considers the scores assigned to the alternatives for various input criteria without taking into consideration the scores allotted to the beneficial criteria. The non-beneficial or input criteria having their lower values are mostly preferred. Thus, the aggregate performance of ith alternative taking into consideration all the input criteria is estimated using the equation as follows:

Multi-Criteria Decision-Making Methods in Manufacturing Environments: Models and Applications.
Shankar Chakraborty, Prasenjit Chatterjee, & Partha Protim Das (Authors)

$$\overline{I_i} = \sum_{j=1}^{n} w_j \frac{\max(x_j^m) - x_j^i}{\min(x_j^m)} \quad (i = 1,2,3,\ldots,m; j = 1,2,3,\ldots,n; i \neq m) \qquad (12.1)$$

where; $\overline{I_i}$ is calculated as the measure of ith alternative's relative performance; and x_j^i indicates the performance score of ith alternatives for j^{th} input criterion. If i^{th} alternative is preferred to m^{th} alternative for j^{th} criterion, then $x_j^i < x_j^m$. The term $\frac{\max(x_j^m) - x_j^i}{\min(x_j^m)}$ describes the difference in performance scores for j^{th} criterion, between i^{th} alternative and the alternative having the maximum score among all the candidate options. On the other hand, w_j (relative importance of criterion j) is the calibration constant considered for increasing or decreasing the influence of this difference on the rating $\overline{I_i}$ for j^{th} criterion.

- **Step 2:** Determine the linear preference rating for the input criteria.

$$\overline{\overline{I_i}} = \overline{I_i} - \min(\overline{I_i}) \qquad (12.2)$$

The basic objective of this linear scaling is to allocate a zero rating to the alternative having the least preference. $\overline{\overline{I_i}}$ denotes the aggregate preference rating for i^{th} alternative in respect of the input criteria.

- **Step 3:** Determine the preference ratings for the beneficial criteria. In this step, the aggregate performance rating for i^{th} alternative with respect to all the beneficial or output criteria is computed employing the following equation:

$$\overline{O_i} = \sum_{h=1}^{H} w_h \frac{x_h^i - \min(x_h^m)}{\min(x_h^m)} \qquad (12.3)$$

where; $h = 1, 2, 3, \ldots, H$ signifies the total number of beneficial criteria; and w_h is the corresponding calibration constant (importance weight) of hth output criteria. The higher is the score for an alternative for an output criterion, the stronger is the preference for that alternative.

Thus, it can be stated that $\sum_{j=1}^{n} w_j + \sum_{h=1}^{H} w_h = 1$.

- **Step 4:** Compute the linear preference rating for the output criteria. The linear preference rating for the output criteria can be calculated as follows:

$$\overline{\overline{O_i}} = \overline{O_i} - \min(\overline{O_i}) \qquad (12.4)$$

- **Step 5:** Calculate the overall preference ratings. For each of the candidate alternatives, the overall preference rating is calculated by scaling the sum $(\overline{\overline{I}}_i + \overline{\overline{O}}_i)$ such that the alternative with the least preference receives a rating of zero.

The overall preference rating (P_i) can now be calculated as:

$$P_i = (\overline{\overline{I}}_i + \overline{\overline{O}}_i) - \min(\overline{\overline{I}}_m + \overline{\overline{O}}_m) \tag{12.5}$$

The alternatives can now be ranked in descending order of their overall preference ratings such that the alternative with the highest rating obtains the first rank.

12.1 SURVEY OF THE LITERATURE

Along with different preferential ranking-based MCDM techniques, Chatterjee & Chakraborty [3] also applied OCRA method for identifying the most apposite gear material. Madić et al. [4] explored the application potentiality of OCRA method while solving three real-time NTM process selection problems. It was also claimed that the derived rankings of the alternative NTM processes for the considered case studies would closely match with the observations of the past researchers. Özbek [5] evaluated the efficiency of nine foreign-capital banks in Turkey between 2005 and 2014 using OCRA method, and also compared its ranking performance against MOORA and SAW methods. Considering a hotel selection problem, Işık & Adalı [6] first adopted OCRA method for determining weights of the evaluation criteria and then ranked the candidate hotels using OCRA method. In a laser cutting (LC) process, Madić et al. [7] proposed the application of OCRA method to determine the most suitable settings of laser power, cutting speed, assist gas pressure and focus position for achieving the desired product characteristics. Kundakçı [8] applied OCRA method to rank six competing suppliers in a manufacturing organization. The same problem also solved using efficiency analysis technique with output satisfying (EATWOS) to prove the accuracy of the ranking results derived based on OCRA method. Baloyi & Meyer [9] presented the application of OCRA method for solving a mining method selection problem. Its ranking performance and stability were also contrasted with other popular MCDM tools, like TOPSIS, TODIM, VIKOR, PROMETHEE, OCRA, COPRAS, and SAW. The appropriateness of OCRA method in solving the considered mining method selection problem was also validated using sensitivity analysis. Chingo et al. [10] solved a material selection problem for geomembranes using OCRA method while determining the corresponding criteria weights based on AHP technique. The same problem was also solved employing

COPRAS, TOPSIS, VIKOR, PROMETHEE, and ORESTE methods, and the derived ranking results were compared with that obtained using OCRA method. Görçün [11] presented the hybrid application of entropy and OCRA methods for efficiency estimation of container seaports in Black Sea region. The criteria weights were determined using entropy method, while OCRA was adopted to rank the candidate container seaports.

12.2 FMS SELECTION PROBLEM

Karsak [12] applied a distance-based fuzzy MCDM approach to evaluate the performance of eight mutually exclusive FMS alternatives based on eight criteria, i.e., operating costs (COS), required floor space (RFS), expansion flexibility (EF), work-in-progress (WIP), volume flexibility (VF), product flexibility (PF), quality improvement (QI) and lead time reduction (LTR). The initial decision matrix is provided in Table 12.1. Among the considered criteria, COS and RFS are quantitative attributes. On the other hand, EF, VF, PF, QI, and LTR are expressed using linguistic ratings (poor, fair, good, very good), and WIP is stated using triangular fuzzy numbers. PF can be defined as the ability of the considered FMS to start processing new part/product types rapidly and economically. Volume flexibility expresses the ability of an FMS to operate profitably at varying levels of production volume. On the other hand, EF can be denoted as the ability of an FMS to easily expand its existing capability and capacity. In Table 12.1, the linguistic ratings for criteria EF, VF, PF, QI, and LTR are converted into the corresponding numerical values using a scale as poor = 0.2, fair = 0.5, good = 0.7 and very good = 0.9. Similarly, for WIP criterion, the middle value of the considered triangular fuzzy set is treated as the performance measure of an FMS with respect to this criterion.

Now, this decision-making problem of FMS selection is solved using OCRA method. At first, the aggregated performances of the FMS alternatives considering all the input criteria are estimated. Based on these aggregated performance scores, the linear preference rating for each alternative FMS with respect to all the input criteria is determined. This linear preference rating allocates a zero value to the least favored alternative. Now, the aggregated performance scores for the alternatives with respect to all the output criteria are determined and accordingly, the linear preference rating for the output criteria is evaluated. Lastly, the overall preference rating for each alternative FMS is determined, which indicates the preference preorder of the FMS alternatives for the given problem. The calculation steps of this method are presented in Table 12.2. From this table, it can be easily observed that FMS 3 (A_3) and FMS 7 (A_7) appear as the first and the last alternative choices. On the other hand, using a distance-based fuzzy MCDM approach, Karsak [12] also identified FMS 3 (A_3) as the most favored choice.

TABLE 12.1 Numerical Data for the FMS Selection Problem [12]

FMS	COS ($ Millions)	RFS (m^2)	WIP (Units)	PF	VF	EF	LTR	QI
A_1	3.8	630	42	0.7	0.5	0.7	0.5	0.2
A_2	3.1	620	37	0.7	0.7	0.5	0.9	0.5
A_3	5	425	32	0.9	0.7	0.7	0.7	0.7
A_4	6.4	500	54	0.7	0.5	0.5	0.7	0.9
A_5	3.6	600	44	0.2	0.7	0.7	0.2	0.7
A_6	6.7	780	59	0.9	0.9	0.5	0.5	0.5
A_7	3.4	740	37	0.5	0.2	0.2	0.7	0.9
A_8	3.7	550	36	0.5	0.9	0.9	0.5	0.5

Source: Reprinted with permission from Ref. [12]. © 2021 Taylor and Francis.

TABLE 12.2 Ranking of the FMS Alternatives Using OCRA Method

FMS	$\bar{I}_i$	$\bar{\bar{I}}_i$	$\bar{O}_i$	$\bar{\bar{O}}_i$	P_i	Rank
A_1	0.2458	0.2458	0.9466	0.0696	0.0677	6
A_2	0.3003	0.3003	1.3558	0.4788	0.5314	3
A_3	0.3078	0.3078	1.5881	0.7111	0.7712	1
A_4	0.1280	0.1280	1.3492	0.4722	0.3525	4
A_5	0.2555	0.2555	0.8792	0.0022	0.0100	7
A_6	0	0	1.3515	0.4745	0.2269	5
A_7	0.2477	0.2477	0.8770	0	0	8
A_8	0.3030	0.3030	1.3558	0.4788	0.5341	2

KEYWORDS

- **efficiency analysis technique**
- **expansion flexibility**
- **operational competitiveness rating analysis**
- **required floor space**
- **volume flexibility**
- **work-in-progress**

REFERENCES

1. Parkan, C., & Wu, M. L., (1997). On the equivalence of operational performance measurement and multiple attribute decision making. *International Journal of Production Research, 35*, 2963–2988.
2. Parkan, C., & Wu, M. L., (2000). Comparison of three modern multicriteria decision-making tools. *International Journal of Systems Science, 31*, 497–517.
3. Chatterjee, P., & Chakraborty, S., (2012). Material selection using preferential ranking methods. *Materials and Design, 35*, 384–393.
4. Madić, M., Petković, D., & Radovanović, M., (2015). Selection of non-conventional machining processes using the OCRA method. *Serbian Journal of Management, 10*(1), 61–73.
5. Özbek, A., (2015). Efficiency analysis of foreign-capital banks in Turkey by OCRA and MOORA. *Research Journal of Finance and Accounting, 6*(13), 21–30.
6. Işık, A. T., & Adalı, E. A., (2016). A new integrated decision making approach based on SWARA and OCRA methods for the hotel selection problem. *International Journal of Advanced Operations Management, 8*(2), 140–151.
7. Madić, M., Antucheviciene, J., Radovanović, M., & Petković, D., (2016). Determination of manufacturing process conditions by using MCDM methods: Application in laser cutting. *Engineering Economics, 27*(2), 144–150.
8. Kundakçı, N., (2019). A comparative analyze based on EATWOS and OCRA methods for supplier evaluation. *The Journal of Operations Research, Statistics, Econometrics and Management Information Systems, 7*(1), 104–112.
9. Baloyi, V. D., & Meyer, L. D., (2020). The development of a mining method selection model through a detailed assessment of multi-criteria decision methods. *Results in Engineering, 8,* 100172.
10. Chingo, C., Martínez-Gómez, J., & Narváez, C. R. A., (2020). Material selection using multi-criteria decision making methods for geomembranes. *International Journal of Mathematics in Operational Research, 16*(1), 24–52.
11. Görçün, Ö. F., (2020). Efficiency analysis of Black Sea container seaports: Application of an integrated MCDM approach. *Maritime Policy & Management.* doi: 10.1080/03088839.2020.1783467.
12. Karsak, E. E., (2002). Distance-based fuzzy MCDM approach for evaluating flexible manufacturing system alternatives. *International Journal of Production Research, 40*(13), 3167–3181.

CHAPTER 13

ADDITIVE RATIO ASSESSMENT (ARAS) METHOD

The additive ratio assessment (ARAS) method is based on quantitative measurements and utility theory. In this method, a utility function value determines the relative efficiency of an alternative over the other alternatives. This utility function is directly proportional to the relative effect of the criteria values and weight importance of the considered criteria. The utility value of an alternative is determined by a comparison of variant with the ideally best alternative. The steps of ARAS method are as follows [1, 2]:

- **Step 1:** For beneficial attributes, determine the normalized decision matrix using linear normalization procedure [2]. For non-beneficial attributes, the normalization procedure follows two steps. At first, the reciprocal of each criterion with respect to all the alternatives is taken as shown below:

$$x_{ij}^{*} = \frac{1}{x_{ij}} \left(i = 1,2,\ldots,m; j = 1,2,\ldots,n\right) \tag{13.1}$$

In the second step, the normalized values are calculated.

$$R = [r_{ij}]_{m \times n} = \frac{x_{ij}^{*}}{\sum_{i=1}^{m} x_{ij}^{*}} \tag{13.2}$$

- **Step 2:** Determine the weighted normalized decision matrix, $[y_{ij}]_{m \times n}$, while multiplying the elements of the normalized decision matrix by the corresponding criteria weights.
- **Step 3:** Determine the optimality function (S_i) for i^{th} alternative.

$$S_i = \sum_{j=1}^{n} y_{ij} \tag{13.3}$$

Multi-Criteria Decision-Making Methods in Manufacturing Environments: Models and Applications.
Shankar Chakraborty, Prasenjit Chatterjee, & Partha Protim Das (Authors)

Higher is the S_i value, better is the alternative. The optimality function S_i has a direct and proportional relationship with values in the decision matrix and criteria weights.

- **Step 4:** Calculate the degree of utility (K_i) for each alternative. It is determined by a comparison of the variant with the most efficient alternative (S_0). The equation used for calculating the value of K_i is given as below:

$$K_i = \frac{S_i}{S_0} \tag{13.4}$$

The utility values of the alternatives usually range from 0 to 1. The alternative with the highest utility value is the best choice among the candidate solutions.

13.1 LITERATURE REVIEW

Zavadskas & Turskis [2] first introduced the concept of ARAS method while demonstrating its application for assessment of microclimate in office rooms based on several important criteria, like humidity, temperature, illumination intensity, air flow rate and dew point, influencing human performance. Based on several critical evaluation criteria, Zavadskas & Vilutiene [3] proposed the application of ARAS method for selection of the most suitable foundation installment alternative for safe building construction on aquiferous soil. While solving a gear material selection problem, Chatterjee & Chakraborty [4] compared the rankings of the candidate materials derived based on ARAS method with those obtained using the COPRAS method. Niyazi & Tavakkoli-Moghaddam [5] considered four alternative facility locations, and identified the best location using ARAS method along with TOPSIS and COPRAS techniques. The ranking performance of the considered MCDM tools was finally compared while developing the corresponding REGIME model. Karabašević et al. [6] attempted to establish the role of SWARA and ARAS methods as the effective MCDM tools while identifying the most befitting sales manager (middle management post) in a telecommunication industry with the help of a numerical example. Stanujkic et al. [7] first computed the weights of the considered evaluation criteria using SWARA method and subsequently ranked the competing personnel based on ARAS method.

Adali & Işik [8] endeavored to select the most appropriate air conditioner using ARAS method and the derived ranking was also contrasted with that obtained based on COPRAS technique. Karabašević et al. [9] demonstrated the applicability, usability, and efficacy of ARAS method as an efficient MCDM tool for singling out the best-suited software testing platform. Koçak et al. [10] demonstrated the application of ARAS method using a real-time example while identifying the most suitable sub-contractor for a given task. Considering 17 ratios, broadly including profitability, solvency, stability, managerial efficiency and liquidity as the evaluation criteria, Avinash [11] applied ARAS method for apprizing the financial performance of 24 Indian steel manufacturing organizations. The derived ranking results were also compared with those derived using SAW and TOPSIS methods. Bahrami et al. [12] proposed the hybridized application of BWM and ARAS methods for mapping of copper prospectivity areas in Iran. The BWM was employed for determination of the criteria weights, whereas, the ARAS method was implemented for ranking of the alternatives. Fu [13] considered a catering supplier selection problem, and solved the same while integrating ARAS method with AHP technique and multi-choice goal programming method. Martin & Deepak [14] solved material selection problems for optimal design of engineering components while proposing a new variant of ARAS method and later compared its ranking performance with respect to TOPSIS. Baloyi & Meyer [15] treated the selection of the most suitable mining method as an MCDM problem and solved the same using ARAS method. Its performance with respect to ranking of the candidate alternatives was later validated against other popular MCDM techniques, like TOPSIS, TODIM, VIKOR, gray relational analysis (GRA), PROMETHEE, OCRA, ARAS, COPRAS, SAW, and compromise programming (CP).

Kuvat [16] presented the application of ARAS method as a robust multi-response optimization tool for determining the most favorable combination of nugget steam cooking process parameters for having enhanced product performance. Mostafaeipour et al. [17] explored the application of ARAS method in identifying the most apposite location for construction of geothermal power plant for hydrogen production in Afghanistan. At first, SWARA method was adopted to estimate the weights of nine evaluation criteria and 17 alternative locations were subsequently ranked from the best to the worst for geothermal-based hydrogen production. While machining AM60 work material using a high-speed cutting tool, Paramasivam et al. [18] integrated Taguchi methodology with ARAS method to identify the optimal drilling parameters to achieve minimum values of geometric dimensional deviation, average surface roughness and tool wear. Relative weights were assigned to each of the considered responses based on the experts' opinions. Vijayakumar [19] studied the ranking performance of SAW and ARAS methods while evaluating the performance of some of the major Indian pharmaceutical companies. The considered companies were finally ranked from the best to the worst. In an attempt to solve a mobile network operator selection problem, Bošković et al. [20] first employed CRITIC method to estimate the importance of the considered evaluation criteria, and ARAS method was later adopted to reach at the final selection decision. Based on four evaluation criteria, like quality, price/cost, delivery, and service, Tembhare et al. [21] applied ARAS method to rank five competing vendors in a manufacturing organization. The weights of the considered criteria were estimated using entropy method.

13.2 CNC MACHINE TOOL SELECTION PROBLEM

With the rapid advancement of computational facilities, applications of advanced manufacturing technologies, mainly in the form of industrial robots, CNC machine tools, FMS, automated material handling systems, automated inspection devices, etc., have become indispensable in the modern-day manufacturing environment. Their augmentation would dramatically help in enhancing product quality, fast production and delivery, and increased product variety and flexibility. But, implementation of any of the advanced manufacturing technologies is capital intensive with irreversible investment. Thus, evaluation and selection of an advanced manufacturing technology in a manufacturing organization is a complicated strategic decision-making task with the presence of a set of almost similar performing alternatives and conflicting criteria.

With the continuous demand for highly precise machined components with higher dimensional accuracy and close tolerance, the conventional machining centers are now being replaced by the CNC machine tools. These CNC machine tools have several advantages, like accuracy, reliability, versatility, safety in operation, less/minimum intervention of the operator, no possibility of human error, capability to machine complex shape geometries, uniformity in designs, less requirement of maintenance, etc. On the other hand, they are costly and require highly skilled operators. The performance of a CNC machine tool can be measured with respect to its various technical parameters which often contradict with other. Thus, to identify the most feasible CNC machine tool for performing a specific machining operation, application of a sound mathematical technique in the form of an MCDM tool is mandatory.

Sun [22] considered a CNC lathe selection problem consisting of 21 candidate alternatives for a turning center for small-size shell production. The performance of those alternatives was evaluated based on seven criteria, i.e., capital cost (CC) (in NT$), spindle speed range (SS) (in rpm), tool capacity (TC), rapid traverse rate of X-axis (TX) (in m/min), rapid traverse rate of Z-axis (TZ) (in m/min), maximum machining diameter (MD) (in mm) and maximum machining length (ML) (in mm). Among these criteria, SS, TC, MD, and ML are beneficial attributes, whereas, CC, TX, and TZ are non-beneficial characteristics. Using AHP method, Athawale & Chakraborty [23] determined the corresponding criteria weights as $w_{CC} = 0.1148$, $w_{SS} = 0.1808$, $w_{TC} = 0.1884$, $w_{TX} = 0.1197$, $w_{TZ} = 0.1148$, $w_{MD} = 0.1546$, and $w_{ML} = 0.1268$, which would be later utilized for this ARAS method-based selection of the most suitable CNC machine tool. In a CNC lathe, capable of performing a wide range of turning operations, the primary axes are the Z- and X-axes, and the movements of these two axes, usually controlled by specific CNC commands, lead to the movement of the entire turning center. Rapid traverse rates of these two axes represent the positioning capability of the turning center. The operational speed of the turning center is measured with respect to spindle speed range and it allows to maintain a constant cutting speed regardless of the part diameter. On the other hand, the tool turret is measured by the tool capacity. Fewer tools in the turret are responsible for more time required to change the tools identified for use in a particular program. The machining diameter and length denote the capability of the turning center with respect to job dimension. Sun [22] applied data envelopment analysis to solve this CNC machine tool selection problem, and identified YCM-TC-15, VTURN 16, EX-106, ECOCA SJ20, FEMCO WNCL-30, TOPPER TNL-85A, -85T, and -100AL, and ATECH MT-52L, -75S, and -75L as the most efficient alternative machine tools. The decision matrix consisting of 21 alternative CNC machine tools and seven evaluation criteria, considered for subsequent solution using the ARAS method, is provided in Table 13.1. For this problem, the optimal (most desired) values of the considered criteria are assumed as CC = 10,00,000 NT$, SS = 6,000 rpm, TC = 15, TX = 5 m/min, TZ = 10 m/min, MD = 320 mm, and ML = 1,200 mm. Now, in order to make all the elements of Table 13.1 dimensionless and comparable, the corresponding normalized decision matrix is formulated, as shown in Table 13.2. The elements of this normalized decision matrix are multiplied by the predetermined criteria weights to develop the weighted normalized decision matrix, presented in Table 13.3. Using Eqns. (13.3) and (13.4), the corresponding values of optimality function (S_i) and utility degree (K_i) are finally computed in Table 13.4, and based on K_i

values, a ranking order of all the CNC machine tool alternatives is derived. It can be observed from this ranking list that FEMCO WNCL-20 occupies the top position, followed by VTURN 16. On the other hand, TOPPER TNL-100A is the least preferred choice. The ranking list of all the alternative CNC machine tools is 6-4-21-3-7-5-19-20-10-11-18-14-8-16-1-2-17-9-12-15-13. On the other hand, Athawale & Chakraborty [23] adopted TOPSIS method to solve the same problem and obtained the ranking of the CNC machine tool alternatives as 4-21-3-6-5-7-19-20-10-11-18-1-14-8-2-16-9-12-17-13-15. It can be interestingly noticed that both these rankings closely match with the observations of Sun [22].

TABLE 13.1 Quantitative Information for 21 CNC Machine Tools [22]

SL. No.	CNC Machine Tool	CC	SS	TC	TX	TZ	MD	ML
Optimal (0)	–	10,00,000	6,000	15	5	10	320	1,200
1.	YANG ML-5A	12,00,000	5,590	8	24	24	205	350
2.	YANG ML-25A	15,50,000	3,465	8	20	20	280	520
3.	YCM TC-15	14,00,000	5,950	12	15	20	250	469
4.	VTURN 16	11,00,000	5,940	12	12	15	230	600
5.	FEMCO HL-15	12,00,000	5,940	12	12	16	150	330
6.	FEMCO WNCL-20	15,00,000	3,465	12	6	12	260	420
7.	FEMCO WNCL-30	26,00,000	3,960	12	12	16	300	625
8.	EX-106	13,20,000	4,950	12	24	30	240	340
9.	ECOCA SJ20	11,80,000	4,480	8	24	24	250	330
10.	ECOCA SJ25	15,50,000	3,950	12	15	20	280	460
11.	ECCOA SJ30	16,00,000	3,450	12	15	20	280	460
12.	TOPPER TNL-85A	12,00,000	3,465	8	20	24	264	400
13.	TOPPER TNL-100A	13,50,000	2,970	8	20	24	264	400
14.	TOPPER TNL-100AL	14,00,000	2,970	12	24	30	300	600
15.	TOPPER TNL-85T	13,50,000	3,465	12	30	30	264	350
16.	TOPPER TNL-100T	14,50,000	2,970	12	20	24	300	400
17.	TOPPERTNL-120T	15,20,000	2,475	12	20	24	300	400
18.	ATECH MT-52S	13,76,000	4,752	12	20	24	235	350
19.	ATECH MT-52L	14,40,000	4,752	12	20	24	235	600
20.	ATECH MT-75S	18,24,000	3,790	10	12	20	300	530
21.	ATECH MT-75L	19,20,000	3,790	10	12	20	300	1,030

Source: Reprinted with permission from Ref. [22]. © 2021 Taylor and Francis.

TABLE 13.2 Normalized Decision Matrix for the CNC Machine Tool Selection Problem

SL. No.	CNC Machine Tool	CC	SS	TC	TX	TZ	MD	ML
Weight	–	0.1148	0.1808	0.1884	0.1197	0.1148	0.1546	0.1268
0	–	0.0636	0.0648	0.0617	0.1304	0.0902	0.0551	0.1075
1.	YANG ML-5A	0.053	0.0604	0.0329	0.0272	0.0376	0.0353	0.0314
2.	YANG ML-25A	0.041	0.0374	0.0329	0.0326	0.0451	0.0482	0.0466
3.	YCM TC-15	0.0454	0.0643	0.0494	0.0435	0.0451	0.0431	0.042
4.	VTURN 16	0.0578	0.0642	0.0494	0.0543	0.0602	0.0396	0.0537
5.	FEMCO HL-15	0.053	0.0642	0.0494	0.0543	0.0564	0.0258	0.0296
6.	FEMCO WNCL-20	0.0424	0.0374	0.0494	0.1087	0.0752	0.0448	0.0376
7.	FEMCO WNCL-30	0.0245	0.0428	0.0494	0.0543	0.0564	0.0517	0.056
8.	EX-106	0.0482	0.0535	0.0494	0.0272	0.0301	0.0413	0.0305
9.	ECOCA SJ20	0.0539	0.0484	0.0329	0.0272	0.0376	0.0431	0.0296
10.	ECOCA SJ25	0.041	0.0427	0.0494	0.0435	0.0451	0.0482	0.0412
11.	ECCOA SJ30	0.0397	0.0373	0.0494	0.0435	0.0451	0.0482	0.0412
12.	TOPPER TNL-85A	0.053	0.0374	0.0329	0.0326	0.0376	0.0455	0.0358
13.	TOPPER TNL-100A	0.0471	0.0321	0.0329	0.0326	0.0376	0.0455	0.0358
14.	TOPPER TNL-100AL	0.0454	0.0321	0.0494	0.0272	0.0301	0.0517	0.0537
15.	TOPPER TNL-85T	0.0471	0.0374	0.0494	0.0217	0.0301	0.0455	0.0314
16.	TOPPER TNL-100T	0.0438	0.0321	0.0494	0.0326	0.0376	0.0517	0.0358
17.	TOPPERTNL-120T	0.0418	0.0267	0.0494	0.0326	0.0376	0.0517	0.0358
18.	ATECH MT-52S	0.0462	0.0514	0.0494	0.0326	0.0376	0.0405	0.0314
19.	ATECH MT-52L	0.0442	0.0514	0.0494	0.0326	0.0376	0.0405	0.0537
20.	ATECH MT-75S	0.0349	0.041	0.0412	0.0543	0.0451	0.0517	0.0475
21.	ATECH MT-75L	0.0331	0.041	0.0412	0.0543	0.0451	0.0517	0.0923

TABLE 13.3 Weighted Normalized Decision Matrix for the CNC Machine Tool Selection Problem

SL. No.	CNC Machine Tool	CC	SS	TC	TX	TZ	MD	ML
0.	–	0.0073	0.0117	0.0116	0.0156	0.0104	0.0085	0.0136
1.	YANG ML-5A	0.0061	0.0109	0.0062	0.0033	0.0043	0.0055	0.004
2.	YANG ML-25A	0.0047	0.0068	0.0062	0.0039	0.0052	0.0075	0.0059
3.	YCM TC-15	0.0052	0.0116	0.0093	0.0052	0.0052	0.0067	0.0053
4.	VTURN 16	0.0066	0.0116	0.0093	0.0065	0.0069	0.0061	0.0068
5.	FEMCO HL-15	0.0061	0.0116	0.0093	0.0065	0.0065	0.004	0.0037
6.	FEMCO WNCL-20	0.0049	0.0068	0.0093	0.013	0.0086	0.0069	0.0048
7.	FEMCO WNCL-30	0.0028	0.0077	0.0093	0.0065	0.0065	0.008	0.0071
8.	EX-106	0.0055	0.0097	0.0093	0.0033	0.0035	0.0064	0.0039
9.	ECOCA SJ20	0.0062	0.0088	0.0062	0.0033	0.0043	0.0067	0.0037
10.	ECOCA SJ25	0.0047	0.0077	0.0093	0.0052	0.0052	0.0075	0.0052
11.	ECCOA SJ30	0.0046	0.0067	0.0093	0.0052	0.0052	0.0075	0.0052
12.	TOPPER TNL-85A	0.0061	0.0068	0.0062	0.0039	0.0043	0.007	0.0045
13.	TOPPER TNL-100A	0.0054	0.0058	0.0062	0.0039	0.0043	0.007	0.0045
14.	TOPPER TNL-100AL	0.0052	0.0058	0.0093	0.0033	0.0035	0.008	0.0068
15.	TOPPER TNL-85T	0.0054	0.0068	0.0093	0.0026	0.0035	0.007	0.004
16.	TOPPER TNL-100T	0.005	0.0058	0.0093	0.0039	0.0043	0.008	0.0045
17.	TOPPERTNL-120T	0.0048	0.0048	0.0093	0.0039	0.0043	0.008	0.0045
18.	ATECH MT-52S	0.0053	0.0093	0.0093	0.0039	0.0043	0.0063	0.004
19.	ATECH MT-52L	0.0051	0.0093	0.0093	0.0039	0.0043	0.0063	0.0068
20.	ATECH MT-75S	0.004	0.0074	0.0078	0.0065	0.0052	0.008	0.006
21.	ATECH MT-75L	0.0038	0.0074	0.0078	0.0065	0.0052	0.008	0.0117

TABLE 13.4 Value of Optimality Function and Utility Degree

SL. No.	CNC Machine Tool	S_i	K_i	Rank
0		0.0788	1	
1.	YANG ML-5A	0.0402	0.5104	15
2.	YANG ML-25A	0.0401	0.5094	16
3.	YCM TC-15	0.0485	0.6158	4
4.	VTURN 16	0.0539	0.6842	2
5.	FEMCO HL-15	0.0477	0.6057	6
6.	FEMCO WNCL-20	0.0543	0.689	1
7.	FEMCO WNCL-30	0.0479	0.6083	5
8.	EX-106	0.0415	0.5263	13
9.	ECOCA SJ20	0.0391	0.4965	18
10.	ECOCA SJ25	0.0448	0.5686	9
11.	ECCOA SJ30	0.0437	0.5544	10
12.	TOPPER TNL-85A	0.0388	0.4931	19
13.	TOPPER TNL-100A	0.0372	0.4723	21
14.	TOPPER TNL-100AL	0.0418	0.531	12
15.	TOPPER TNL-85T	0.0385	0.4892	20
16.	TOPPER TNL-100T	0.0409	0.5191	14
17.	TOPPERTNL-120T	0.0397	0.5039	17
18.	ATECH MT-52S	0.0423	0.5375	11
19.	ATECH MT-52L	0.0449	0.5706	7
20.	ATECH MT-75S	0.0449	0.5694	8
21.	ATECH MT-75L	0.0503	0.6389	3

KEYWORDS

- **additive ratio assessment**
- **capital cost**
- **grey relational analysis**
- **maximum machining diameter**
- **maximum machining length**
- **speed range**
- **tool capacity**

REFERENCES

1. Turskis, Z., & Zavadskas, E. K., (2010). A novel method for multiple criteria analysis: Grey additive ratio assessment (ARAS-G) method. *Informatica, 21*, 597–610.
2. Zavadskas, E. K., & Turskis, Z., (2010). A new additive ratio assessment (ARAS) method in multicriteria decision-making. *Technological and Economic Development of Economy, 16*(2), 159–172.
3. Zavadskas, E. K., & Vilutiene, T., (2010). Multiple criteria analysis of foundation instalment alternatives by applying additive ratio assessment (ARAS) method. *Archives of Civil and Mechanical Engineering, 10*(3), 123–141.
4. Chatterjee, P., & Chakraborty, S., (2013). Gear material selection using complex proportional assessment and additive ratio assessment-based approaches: A comparative study. *International Journal of Materials Science and Engineering, 1*(2), 104–111.
5. Niyazi, M., & Tavakkoli-Moghaddam, R., (2014). Solving a facility location problem by three multi-criteria decision making methods. *International Journal of Research in Industrial Engineering, 3*(4), 41–56.
6. Karabašević, D., Stanujkić, D., & Urošević, S., (2015). The MCDM model for personnel selection based on SWARA and ARAS methods. *Management, 77*, 43–52.
7. Stanujkic, D., Djordjevic, B., & Karabasevic, D., (2015). Selection of candidates in the process of recruitment and selection of personnel based on the SWARA and ARAS methods. *Quaestus Multidisciplinary Research Journal, 53*(1), 53–64.
8. Adali, E. A., & Işik, A. T., (2016). Air conditioner selection problem with COPRAS and ARAS methods. *MANAS Journal of Social Studies, 5*(2), 124–138.
9. Karabašević, D. M., Maksimović, M. V., Stanujkić, D. M., Jocić, G. B., & Rajčević, D. P., (2018). Selection of software testing method by using ARAS method. *Standardization Metrology, 18*(5), 724–729.
10. Koçak, S., Kazaz, A., & Ulubeyli, S., (2018). Subcontractor selection with additive ratio assessment method. *Journal of Construction Engineering, Management & Innovation, 1*(1), 18–32.
11. Avinash, V. R., (2019). Financial performance evaluation of selected steel companies in India by using multi criteria decision technique of ARAS, SAW and TOPSIS with DVA based weight determination. *International Journal of Scientific Research and Reviews, 8*(2), 4261–4277.
12. Bahrami, Y., Hassani, H., & Maghsoudi, A., (2019). BWM-ARAS: A new hybrid MCDM method for Cu prospectivity mapping in the Abhar area, NW Iran. *Spatial Statistics, 33*, 100382.
13. Fu, Y. K., (2019). An integrated approach to catering supplier selection using AHP-ARAS-MCGP methodology. *Journal of Air Transport Management, 75*, 164–169.
14. Martin, N., & Deepak, F. X. E., (2019). Application of new additive ratio assessment (NARAS) method in selection of material for optimal design of engineering components. *Materials Today: Proceedings, 11*, 1049–1053.
15. Baloyi, V. D., & Meyer, L. D., (2020). The development of a mining method selection model through a detailed assessment of multi-criteria decision methods. *Results in Engineering, 8*, 100172.
16. Kuvat, Ö., (2020). Multi response robust optimization of nugget steam cooking process parameters. *Journal of the Faculty of Engineering and Architecture of Gazi University, 35*(3), 1171–1185.

17. Mostafaeipour, A., Dehshiri, S. J. H., & Dehshiri, S. S. H., (2020). Ranking locations for producing hydrogen using geothermal energy in Afghanistan. *International Journal of Hydrogen Energy, 45*, 15924–15940.
18. Paramasivam, S. S. S. S., Durai, K., & Natarajan, H., (2020). Taguchi additive ratio assessment (ARAS) method in multi-criteria decision making: A case study for treated drill tools performance. *International Journal of Modern Manufacturing Technologies, 12*(1), 114–124.
19. Vijayakumar, A., (2020). Comparison of multi-criteria decision making methods SAW and ARAS: An application to performance of Indian pharmaceutical companies. *Journal of Economics and Technology Research, 1*(2), 23–46.
20. Bošković, S., Radonjić-Djogatović, V., Ralević, P., Dobrodolac, M., & Jovčić, S., (2021). Selection of mobile network operator using the CRITIC-ARAS method. *International Journal for Traffic and Transport Engineering, 11*(1), 17–29.
21. Tembhare, S. K., Amrendra, & George, J., (2021). Evaluating alternatives using ARAS method integrated with entropy weight method for vendor selection in supply chain management. *Journal of Emerging Technologies and Innovative Research, 8*(2), 1271–1275.
22. Sun, S., (2002). Assessing computer numerical control machines using data envelopment analysis. *International Journal of Production Research, 40*(9), 2011–2039.
23. Athawale, V. M., & Chakraborty, S., (2010). A TOPSIS method-based approach to machine tool selection. In: *Proceedings of the International Conference on Industrial Engineering and Operations Management* (pp. 1–8). Dhaka, Bangladesh.

CHAPTER 14

EVALUATION BASED ON DISTANCE FROM AVERAGE SOLUTION (EDAS) METHOD

The evaluation based on the distance from average solution (EDAS) method is a new distance-based measurement approach, contemporary to other recently developed MCDM methods [1, 2]. It has been observed that TOPSIS, another distance-based MCDM technique, has extensively been applied to solve diverse complex decision-making problems. The TOPSIS method identifies the best solution which is closest to the positive-ideal solution (PIS) and farthest from the negative-ideal solution (NIS). The major limitation of the TOPSIS method is that it measures two target points, but does not project the importance of the distances from those two points. It proposes that the most preferred alternative may not be the nearest to the ideal solution [3]. The mathematical modeling of EDAS method is slightly different from that of TOPSIS as it is based on the distances from the average solution rather than positive- and negative-ideal solutions. The arithmetic mean of the performance values of the candidate alternatives is considered here to evaluate the average solution. The EDAS method performs better than TOPSIS when the rank reversal occurs [4]. It is simple, efficient and has fewer computations giving rise to faster solutions. In this method, the determination of the best alternative mainly depends on the calculation of the positive distance (*PD*) and negative distance (*ND*) from the average solution for each alternative with respect to each criterion. These distances project the differences between the corresponding alternatives and the average solution. Higher *PD* values signify better solutions than the average solution, whereas, higher *ND* values correspond to worse solutions than the average solution. The steps of EDAS method are presented as below:

Multi-Criteria Decision-Making Methods in Manufacturing Environments: Models and Applications.
Shankar Chakraborty, Prasenjit Chatterjee, & Partha Protim Das (Authors)

- **Step 1:** Construct the decision matrix (*X*), having *m* alternatives and *n* criteria:

$$X = [X_{ij}]_{m \times n} = \begin{bmatrix} x_{11} & x_{12} & \dots & x_{1n} \\ x_{21} & x_{22} & \dots & x_{2n} \\ \dots & \dots & \dots & \dots \\ x_{m1} & x_{m2} & \dots & x_{mn} \end{bmatrix} \quad (14.1)$$

- **Step 2:** Determine the average solution (*AVR*) with respect to all the criteria.

$$AVR = [AVR_j]_{1 \times n} \quad (14.2)$$

where; $AVR_j = \frac{\sum_{i=1}^{m} x_{ij}}{m}$ $(j = 1, 2, \dots, n)$

- **Step 3:** Calculate the *PD* from the average and *ND* from the average matrixes depending on the type of the considered criterion (beneficial or non-beneficial).

$$PD = [PD_{ij}]_{m \times n}, \quad (14.3)$$

$$ND = [ND_{ij}]_{m \times n}, \quad (14.4)$$

If j^{th} criterion is beneficial in nature, then:

$$PD_{ij} = \frac{\max(0,(x_{ij} - AVR_j))}{AVR_j}, \quad (14.5)$$

$$ND_{ij} = \frac{\max(0,(AVR_j - x_{ij}))}{AVR_j} \quad (14.6)$$

If j^{th} criterion is non-beneficial, then:

$$PD_{ij} = \frac{\max(0,(AVR_j - x_{ij}))}{AVR_j}, \quad (14.7)$$

$$ND_{ij} = \frac{\max(0, (x_{ij} - AVR_j))}{AVR_j} \quad (14.8)$$

where; PD_{ij} and ND_{ij} respectively denote the positive and negative distances of ith alternative from the average solution with respect to j^{th} criterion.

- **Step 4:** Determine the weighted sums of *PD* (*WSPD*) and *ND* (*WSND*) for all the alternatives.

$$WSPD_i = \sum_{j=1}^{n} w_j PD_{ij} \tag{14.9}$$

$$WSND_i = \sum_{j=1}^{n} w_j ND_{ij} \tag{14.10}$$

where; w_j is the weight of *the* j^{th} criterion.

- **Step 5:** Normalize the values of *WSPD* (*NWSPD*) and *WSND* (*NWSND*) for all the considered alternatives.

$$NWSPD_i = \frac{WSPD_i}{\max_i (WSPD_i)} \tag{14.11}$$

$$NWSND_i = 1 - \frac{WSND_i}{\max_i (WSND_i)} \tag{14.12}$$

- **Step 6:** Calculate the appraisal score (*APS*) for each of the alternatives.

$$APS_i = \frac{1}{2}(NWSPD_i + NWSND_i), \tag{14.13}$$

where; $0 \leq APS_i \leq 1$.

- **Step 7:** Rank the candidate alternatives based on the decreasing values of *APS*. The alternative with the highest *APS* value is the best choice among the available options. If the *PD* value of an alternative is greater than zero, the corresponding *ND* value equals to zero, and when *ND* value is greater than zero, the *PD* value becomes zero.

14.1 LITERATURE SURVEY

Ghorabaee et al. [1] presented the application of EDAS method as an effective MCDM tool for classification of inventory items in stock keeping units. A comparison with other MCDM techniques and a sensitivity analysis study with respect to changing criteria weights proved its effectiveness in classifying inventory items. Using EDAS method, Zavadskas et al. [6] appraised the performance of 21 neighborhoods in Vilnius in the context of a healthy and safe built environment for sustainable development. The corresponding objective and subjective criteria weights were first determined, and the performance of EDAS method was later validated against other MCDM techniques, like COPRAS, SAW, and TOPSIS methods. Chatterjee et al. [2] attempted to integrate the design of experiments (DoE) and EDAS method while developing a hybrid metamodel to solve material selection problems in manufacturing environment. The applicability of the proposed approach was demonstrated through a gear material selection problem. Karabasevic et al. [8] identified selection and evaluation of personnel in an IT industry as an MCDM problem, and subsequently applied EDAS method to solve the same. The weights of the considered criteria were determined using SWARA technique. Mathew & Sahu [9] solved two material handling equipment selection problems, i.e., a conveyor selection problem with six criteria and four alternatives, and an AGV selection problem having six criteria and eight alternatives with the help of EDAS method, and contrasted its ranking performance against MOORA, WASPAS, and CODAS methods. Sahaya Sudha [10] considered an agricultural situation in a locality and employed EDAS method to estimate water requirements for various crops treating duration, water consumption, yield, market price, etc., as the important factors. Yalçin & Uncu [11] presented the application of EDAS method in solving an industrial robot selection problem based on a set of conflicting criteria. Considering seven selected biomass materials and seven criteria, Dhanalakshmi et al. [12] presented simultaneous application of TOPSIS and EDAS methods to evaluate possibilities of utilizing locally available biomass while yielding maximum bio-oil during the pyrolysis process. Das & Chakraborty [7] integrated gray correlation and EDAS methods for simultaneous optimization of three NTM processes, i.e., photochemical machining process, laser-assisted jet electrochemical machining process and abrasive water jet drilling process.

14.2 CUTTING TOOL MATERIAL SELECTION PROBLEM

Selection of the most appropriate cutting tool while machining a given workpiece material is deemed to be a crucial task in the manufacturing environment as it directly affects the achievable machining efficiency and surface finish of the machined component. A wrongly selected cutting tool may adversely affect the machining process, while increasing the product cost and premature tool failure. While selecting a cutting tool for a specific machining operation, it should possess the correct combination of all the desired physical properties. Thus, it is extensively important to select the most appropriate cutting tool material with the desired properties for enhanced machining performance. It has already been established that the performance of cutting tools depends on various mechanical and physicochemical properties of the tool materials, like strength, hardness, fatigue limit, thermal conductivity, corrosion, and wear resistance, and adhesion intensity with the workpiece material. Thus, identification of the most apposite cutting tool for a specific machining operation needs to be treated with utmost importance so as to aid the manufacturing engineers in achieving the end products with the desired dimensional accuracy and tolerance.

Li et al. [5] considered a cutting tool material selection problem while treating YG6X cemented carbide (YG6X), YG8 cemented carbide (YG8), polycrystalline diamond (PCD), cubic boron nitride (CBN) and ceramic as the candidate alternatives. The workpiece material was Al-Si piston alloy and the performance of all the tool materials was evaluated based on seven properties, i.e., density (D) (in g/mm^3), Young's modulus (YM) (in GPa), hardness (H) (in HV), fracture toughness (FT) (in MPa.m$^{1/2}$), thermal stability (TS), thermal conductivity (TC) (in W/m.°C) and thermal expansion coefficient (TEC) (in 10^{-6}/°C). The original decision matrix is exhibited in Table 14.1. Among these physical properties of the cutting tool materials, D, YM, H, FT, TS, and TC are beneficial in nature, whereas, TEC is the sole non-beneficial characteristic.

Now, while solving this cutting tool material selection problem using EDAS method, the corresponding values of the positive and negative distances from the average solution are determined using Eqns. (14.3) and (14.4), respectively, as shown in Tables 14.2 and 14.3. Following the procedural steps of EDAS method, weighted sums of the positive and negative distances from the average, their normalized values and the corresponding appraisal scores are calculated in Table 14.4. In these calculation steps, equal weights are assigned to all the material properties. Based on these appraisal scores, the alternative cutting tool materials are now ranked, PCD occupying the top position in the raking list, followed by YG8. On the other hand, ceramic is opted with the least preference. The same ranking order was also derived by Li et al. [5] while solving this problem applying a compromised weighting technique consisting of subjective and objective judgment methods.

TABLE 14.1 Decision Matrix for the Cutting Tool Material Selection Problem [5]

Tool Material	D	YM	H	FT	TS	TC	TEC
YG6X	14.8	610	1,600	10.8	950	100	5.4
YG8	14.5	626	1,500	11.2	1,000	118	5.9
PCD	3.9	745	4,500	7.8	600	560	4.6
CBN	3.9	630	3,750	5	1,500	70	4.3
Ceramic	4.4	350	2,150	3	1,550	35	7.8

Source: Reprinted with permission from Ref. [5]. © 2021 Springer.

TABLE 14.2 Positive Distance from Average Solution

Tool Material	D	YM	H	FT	TS	TC	TEC
YG6X	0.7831	0.0301	0	0.4286	0	0	0.0357
YG8	0.747	0.0571	0	0.4815	0	0	0
PCD	0	0.258	0.6667	0.0317	0	2.171	0.1786
CBN	0	0.0638	0.3889	0	0.3393	0	0.2321
Ceramic	0	0	0	0	0.3839	0	0

TABLE 14.3 Negative Distance from Average Solution

Tool Material	D	YM	H	FT	TS	TC	TEC
YG6X	0	0	0.4074	0	0.1518	0.4337	0
YG8	0	0	0.4444	0	0.1071	0.3318	0.0536
PCD	0.5301	0	0	0	0.4643	0	0
CBN	0.5301	0	0	0.3386	0	0.6036	0
Ceramic	0.4699	0.409	0.2037	0.6032	0	0.8018	0.3929

TABLE 14.4 Weighted Sums of PD and ND Values, Their Normalized Values and Appraisal Scores

Tool Material	WSPD	WSND	NWSPD	NWSND	APS	Rank
YG6X	1.2775	0.9929	0.3864	0.6553	0.5208	3
YG8	1.2855	0.937	0.3889	0.6747	0.5318	2
PCD	3.306	0.9944	1	0.6548	0.8274	1
CBN	1.0241	1.4724	0.3098	0.4888	0.3993	4
Ceramic	0.3839	2.8804	0.1161	0	0.0581	5

KEYWORDS

- **cubic boron nitride**
- **design of experiments**
- **evaluation based on distance from average solution**
- **negative distance**
- **polycrystalline diamond**
- **positive distance**
- **Young's modulus**

REFERENCES

1. Keshavarz, G. M., Zavadskas, E. K., Olfat, L., & Turskis, Z., (2015). Multi-criteria inventory classification using a new method of evaluation based on distance from average solution (EDAS). *Informatica, 26*(3), 435–451.
2. Chatterjee, P., Banerjee, A., Mondal, S., Boral, S., & Chakraborty, S., (2018). Development of a hybrid meta-model for material selection using design of experiments and EDAS method. *Engineering Transactions, 66*(2), 187–207.
3. Chakraborty, S., & Chatterjee, P., (2013). Selection of materials using multi-criteria decision-making methods with minimum data. *Decision Science Letters, 2*(3), 135–148.
4. Keshavarz-Ghorabaee, M., Amiri, M., Zavadskas, E. K., Turskis, Z., & Antucheviciene, J. A., (2018). Comparative analysis of the rank reversal phenomenon in the EDAS and TOPSIS methods. *Economic Computation & Economic Cybernetics Studies & Research, 52*(3), 121–134.
5. Li, A., Zhao, J., Gong, Z., & Lin, F., (2016). Optimal selection of cutting tool materials based on multi-criteria decision-making methods in machining Al-Si piston alloy. *International Journal of Advanced Manufacturing Technology, 86*, 1055–1062.
6. Zavadskas, E. K., Cavallaro, F., Podvezko, V., Ubarte, I., & Kaklauskas, A., (2017). MCDM assessment of a healthy and safe built environment according to sustainable development principles: A practical neighborhood approach in Vilnius. *Sustainability, 9*, doi: 10.3390/su9050702.
7. Das, P. P., & Chakraborty, S., (2021). Application of grey correlation-based EDAS method for parametric optimization of non-traditional machining processes. *Scientia Iranica.* doi: 10.24200/sci.2020.53943.3499.
8. Karabasevic, D., Zavadskas, E. K., Stanujkic, D., Popovic, G., & Brzakovic, M., (2018). An approach to personnel selection in the IT industry based on the EDAS method. *Transformations in Business & Economics, 17*(2), 54–65.
9. Mathew, M., & Sahu, S., (2018). Comparison of new multi-criteria decision making methods for material handling equipment selection. *Management Science Letters, 8*, 139–150.
10. Sahaya, S. A., (2019). Application of EDAS method on water requirement in agriculture. *International Journal of Engineering Research & Technology, 8*(12), 558–561.
11. Yalçin, N., & Uncu, N., (2019). Applying EDAS as an applicable MCDM method for industrial robot selection. *SIGMA Journal of Engineering and Natural Sciences, 37*(3), 779–796.
12. Dhanalakshmi, C. S., Madhu, P., Karthick, A., Mathew, M., & Vignesh, K. R., (2020). A comprehensive MCDM-based approach using TOPSIS and EDAS as an auxiliary tool for pyrolysis material selection and its application. *Biomass Conversion and Biorefinery.* https://doi.org/10.1007/s13399-020-01009-0.

CHAPTER 15

COMBINATIVE DISTANCE-BASED ASSESSMENT (CODAS) METHOD

In COmbinative Distance-based ASsessment (CODAS) method, the alternatives are evaluated and subsequently ranked based on the computation of their desirability with respect to two measures. Between these two measures, the first one is associated with the Euclidean distance of the alternatives from the negative-ideal solution (NIS) based on the formation of l^2-norm indifference space for criteria. The secondary measure is the Taxicab distance based on the l^1-norm indifference space. Thus, in this method, the alternative having the maximum distance from the NIS would be the most suitable choice. If any two alternatives are not comparable according to the Euclidean distance, the Taxicab distance is adopted as the secondary measure. Although the l^2-norm indifference space is mainly considered in this method, it is always preferred to take into account two types of indifference space for better solution of a decision-making problem. The procedural steps of CODAS method are presented as below [1]:

- **Step 1:** Develop the initial decision matrix (X) consisting of m alternatives and n criteria.

$$X = \begin{bmatrix} x_{11} & x_{12} & \cdots & x_{1n} \\ x_{21} & x_{22} & \cdots & x_{2n} \\ \cdots & \cdots & \cdots & \cdots \\ x_{m1} & x_{m2} & \cdots & x_{mn} \end{bmatrix} \quad (15.1)$$

  where; x_{ij} is the performance of ith alternative ($i = 1, 2, \ldots, m$) with respect to j^{th} criterion ($j = 1, 2, \ldots, n$).
- **Step 2:** Apply linear normalization technique to formulate the corresponding normalized decision matrix. For normalization, the following equations can be employed:

Multi-Criteria Decision-Making Methods in Manufacturing Environments: Models and Applications.
Shankar Chakraborty, Prasenjit Chatterjee, & Partha Protim Das (Authors)

i. **For Beneficial Criterion:**

$$y_{ij} = \frac{x_{ij}}{\max_i x_{ij}} \quad (15.2)$$

ii. **For Non-Beneficial Criterion:**

$$y_{ij} = \frac{\min_i x_{ij}}{x_{ij}} \quad (15.3)$$

- **Step 3:** Calculate the corresponding weighted normalized decision matrix. Using weights of the considered criteria, compute the weighted normalized decision matrix while multiplying elements of the normalized decision matrix by the criteria weights.

$$r_{ij} = w_j \times y_{ij} \quad (15.4)$$

where; w_j is the weight of jth criterion.

- **Step 4:** Determine the NIS (point).

$$ns = [ns_j]_{1\times n} \quad (15.5)$$

$$ns_j = \min_i r_{ij} \quad (15.6)$$

- **Step 5:** Compute the Euclidean and Taxicab distances of the alternatives from the NIS.

$$E_i = \sqrt{\sum_{j=1}^{n}\left(r_{ij} - ns_j\right)^2} \qquad (i = 1, 2, ..., m) \quad (15.7)$$

$$T_i = \sum_{j=1}^{n}\left|r_{ij} - ns_j\right| \quad (15.8)$$

- **Step 6:** Develop the following relative assessment matrix:

$$Ra = [h_{ik}]_{m\times m} \quad (15.9)$$

where; $h_{ik} = (E_i - E_k) + (\psi(E_i - E_k)\times(T_i - T_k))$ (for $k = 1, 2, \ldots, m$), and ψ represents a threshold function to provide equality of the Euclidean distances of two alternatives. Its value can be defined as below:

$$\psi(x) = \begin{cases} 1 & \text{if} \left|x\right| \geq \tau \\ 0 & \text{if} \left|x\right| < \tau \end{cases} \quad (15.10)$$

In the above function, the value of the threshold parameter (τ) needs to be specified by the concerned decision maker. The most preferred value of τ usually ranges between 0.01 and 0.05. In this method, if the difference between the Euclidean distances of two alternatives is less than τ, these two alternatives are also compared using the corresponding Taxicab distance.

- **Step 7:** Compute the assessment score of each alternative.

$$H_i = \sum_{k=1}^{m} h_{ik} \quad (15.11)$$

- **Step 8:** Rank the candidate alternatives based on the decreasing values of their assessment scores (H_i). The alternative having the maximum H_i value would be the best choice.

15.1 REVIEW OF THE LITERATURE

Badi et al. [2] adopted CODAS method for identification of the best location of desalination plant in the north-western coast of Libya in order to minimize the water transportation cost and investigate influences of external criteria on the location selection decision. Bakir & Alptekin [3] considered a dataset consisting of 11 airlines having seven evaluation criteria and ranked their service performance using CODAS method. Tuş & Adalı [4] endeavored to validate the applicability of CODAS method in solving personnel assessment problem in a textile firm. The weights of the corresponding weights were calculated using CRITIC method and the ranking performance of CODAS method was contrasted with that of PSI technique. Laha & Biswas [5] applied CODAS method to compare the relative performance of some selected private and public sector banks in India on a five-year time horizon. The *k*-means clustering algorithm was also adopted for checking the consistency of performance-based ranking with asset management efficiency. Maghsoodi et al. [6] proposed the integrated application of SWARA and CODAS methods to single out the most appropriate cement material type for a real-world conceptual dam construction project in Iran. Considering 13 sub-criteria under four criteria, including technical, economic, environmental, and socio-political aspects, Ali et al. [7] adopted AHP and CODAS methods to explore the most suitable renewable energy technology from a set of three alternatives, i.e., solar-wind hybrid, solar mini-grid, and wind mini-grid energy systems. The AHP method was employed to determine the corresponding criteria weights, while the alternative renewable energy technologies were ranked based on CODAS method. Taking into account six hybrid energy systems and 13 evaluation criteria, and applying CODAS method, Ali et al. [8] identified photovoltaic cell/battery/biogas/diesel generator-based hybrid energy system as the most feasible renewable energy source in the off-grid areas of Bangladesh. Karakuş et al. [9] treated geological structure, land ability class, transportation, groundwater, surface water, residential areas, and existing land use as the important evaluation criteria for identification of the most apposite sanitary landfill site in a certain city. Geographical information system (GIS) was integrated with CODAS method to arrive at the optimal decision. While solving a dry port terminal location selection problem, Tadić et al. [10] first applied AHP technique to estimate relative weights of the considered criteria and later employed CODAS method in gray environment to single out the best location. Adalı & Tuş [11] applied CODAS method along with two other distances based MCDM techniques, i.e., TOPSIS and EDAS for selecting the most suitable site for a hospital based on a set of evaluation criteria. The adoption of CRITIC method identified 'market conditions' as the most important criterion.

15.2 RENEWABLE ENERGY TECHNOLOGY SELECTION PROBLEM

Keeping in mind the requirements of renewable energy as an effective alternative to fossil fuel, Maghsoodi et al. [12] considered a renewable energy technology selection problem in Iran. Six alternative renewable energy sources, i.e., wind energy (wind turbine) (A_1), photovoltaic system (A_2), solar thermal energy (A_3), geothermal energy/power (A_4), hydropower energy (A_5) and biogas and biofuel (A_6) were considered and their performance was compared with respect to 11 evaluation criteria, i.e., investment cost (C_1), operation and maintenance cost (C_2), income generation potential (C_3), energy efficiency (C_4), energy production amount (C_5), degree of sophistication (C_6), reliability (C_7), emission and pollution (C_8), job creation (C_9), long-term risk levels (C_{10}), and international and domestic support (C_{11}). Based on SWARA method, the corresponding weights of all the criteria were computed and MULTIMOORA approach was later implemented to rank the alternatives under consideration. Identification of the related quantitative and qualitative evaluation criteria was performed depending on the opinions of the subject experts. Based on the application of the proposed hybrid approach, alternative A_2 (photovoltaic system) emerged out as the best renewable energy technology, followed by alternatives A_3 (solar thermal energy) and A_1 (wind energy).

Now, the same problem dealing with renewable energy technology selection is solved employing CODAS method. The original decision matrix consisting of six renewable energy technologies and 11 evaluation criteria is presented in Table 15.1. It has already been mentioned that the criteria weights are determined using SWARA method. It is noticed from this table that investment cost (C_1) is the most important criterion, followed by operation and maintenance cost (C_2). On the other hand, degree of sophistication (C_6) is the least preferred evaluation criterion. Among these criteria, C_3, C_4, C_5, C_7, C_9, and C_{11} are beneficial attributes requiring their higher values, while C_1, C_2, C_6, C_8, and C_{10} are non-beneficial attributes. Based on the type of criterion (beneficial or non-beneficial) and applying Eqns. (15.2) and (15.3), the elements of the initial decision matrix are normalized, as shown in Table 15.2. While multiplying the elements of the normalized decision matrix by the respective criteria weights, the related weighted normalized decision matrix is determined, as provided in Table 15.3. In this table, values of the negative-ideal solutions are also identified. Now, based on these negative-ideal solutions, the corresponding Euclidean and Taxicab distances of the alternatives are computed, as also provided in Table 15.3. Considering $\tau = 0.02$ as the threshold value, the relative assessment matrix is developed in Table 15.4 and the assessment scores of the candidate alternatives are estimated. When these assessment scores are arranged according to their descending values, it is noticed that alternative A_2 (photovoltaic system) occupies the top position in the ranking list, followed by alternatives A_3 (solar thermal energy) and A_1 (wind energy). Hydropower energy (A_5) is the least preferred renewable energy technology. It is interestingly observed that the derived rankings of the alternative renewable energy technologies using CODAS method closely match with the observations of Maghsoodi et al. [12] which proves the efficacy of CODAS method in solving decision-making problems.

KEYWORDS

- **combinative distance-based assessment**
- **geographical information system**
- **negative-ideal solution**
- **PSI technique**
- **relative assessment matrix**
- **threshold parameter**

TABLE 15.1 Decision Matrix for the Renewable Energy Technology Selection Problem [12]

Criteria	C_1	C_2	C_3	C_4	C_5	C_6	C_7	C_8	C_9	C_{10}	C_{11}
Weight	0.24	0.165	0.107	0.034	0.066	0.023	0.049	0.028	0.095	0.057	0.136
Type	Min	Min	Max	Max	Max	Min	Max	Min	Max	Min	Max
A_1	3	3	3	3	3	2	3	1	4	3	3
A_2	2	2	3	3	4	2	4	1	2	2	3
A_3	3	2	3	3	3	3	3	1	2	2	3
A_4	3	3	2	3	3	3	3	2	2	3	2
A_5	4	4	4	4	4	3	2	2	3	4	1
A_6	3	3	3	2	3	3	3	2	3	2	2

TABLE 15.2 Normalized Decision Matrix for the Renewable Energy Technology Selection Problem

Alt.	C_1	C_2	C_3	C_4	C_5	C_6	C_7	C_8	C_9	C_{10}	C_{11}
A_1	0.6667	0.6667	0.75	0.75	0.75	1	0.75	1	1	0.6667	1
A_2	1	1	0.75	0.75	1	1	1	1	0.5	1	1
A_3	0.6667	1	0.75	0.75	0.75	0.6667	0.75	1	0.5	1	1
A_4	0.6667	0.6667	0.5	0.75	0.75	0.6667	0.75	0.5	0.5	0.6667	0.6667
A_5	0.5	0.5	1	1	1	0.6667	0.5	0.5	0.75	0.5	0.3333
A_6	0.6667	0.6667	0.75	0.5	0.75	0.6667	0.75	0.5	0.75	1	0.6667

TABLE 15.3 Weighted Normalized Decision Matrix

Alt.	C_1	C_2	C_3	C_4	C_5	C_6	C_7	C_8	C_9	C_{10}	C_{11}	E_i	T_i
A_1	0.16	0.11	0.0803	0.0255	0.0495	0.023	0.0368	0.028	0.095	0.038	0.136	0.1188	0.2843
A_2	0.24	0.165	0.0803	0.0255	0.066	0.023	0.049	0.028	0.0475	0.057	0.136	0.1793	0.4196
A_3	0.16	0.165	0.0803	0.0255	0.0495	0.0153	0.0368	0.028	0.0475	0.057	0.136	0.1363	0.3032
A_4	0.16	0.11	0.0535	0.0255	0.0495	0.0153	0.0368	0.014	0.0475	0.038	0.0907	0.0687	0.1431
A_5	0.12	0.0825	0.107	0.034	0.066	0.0153	0.0245	0.014	0.0713	0.0285	0.0453	0.0631	0.1108
A_6	0.16	0.11	0.0803	0.017	0.0495	0.0153	0.0368	0.014	0.0713	0.057	0.0907	0.0816	0.2041
Negative-ideal solution	0.12	0.0825	0.0535	0.017	0.0495	0.0153	0.0245	0.014	0.0475	0.0285	0.0453	–	

TABLE 15.4 Relative Assessment Matrix and Assessment Scores of the Alternatives

Alt.	A_1	A_2	A_3	A_4	A_5	A_6	H	Rank
A_1	0	–0.1958	–0.0175	0.1913	0.2292	0.1175	0.3248	3
A_2	0.1958	0	0.1594	0.3871	0.425	0.3132	1.4805	1
A_3	0.0175	–0.1594	0	0.2276	0.2656	0.1538	0.505	2
A_4	–0.1913	–0.3871	–0.2276	0	0.0056	–0.0128	–0.8133	5
A_5	–0.2292	–0.425	–0.2656	–0.0056	0	–0.0184	–0.9438	6
A_6	–0.1175	–0.3132	–0.1538	0.0128	0.0184	0	–0.5533	4

REFERENCES

1. Ghorabaee, M. K., Zavadskas, E. K., Turskis, Z., & Antucheviciene, A., (2016). A new combinative distance-based assessment (CODAS) method for multi-criteria decision making. *Economic Computation and Economic Cybernetics Studies and Research, 50*(3), 25–44.
2. Badi, I., Ballem, M., & Shetwan, A., (2018). Site selection of desalination plant in Libya by using combinative distance-based assessment (CODAS) method. *International Journal for Quality Research, 12*(3), 609–624.
3. Bakir, M., & Alptekin, N., (2018). A new approach in service quality assessment: An application on airlines through CODAS method. *Business & Management Studies: An International Journal, 6*(4), 1336–1353.
4. Tuş, A., & Adalı, E. A., (2018). Personnel assessment with CODAS and PSI methods. *The Journal of Operations Research, Statistics, Econometrics and Management Information Systems, 6*(2), 243–256.
5. Laha, S., & Biswas, S., (2019). A hybrid unsupervised learning and multi-criteria decision-making approach for performance evaluation of Indian banks. *Accounting, 5,* 169–184.
6. Maghsoodi, A., Maghsoodi, A. I., Poursoltan, P., Antucheviciene, J., & Turskis, Z., (2019). Dam construction material selection by implementing the integrated SWARA-CODAS approach with target-based attributes. *Archives of Civil and Mechanical Engineering, 19*(4), 1194–1210.
7. Ali, T., Nahian, A. J., & Ma, H., (2020). A hybrid multi-criteria decision-making approach to solve renewable energy technology selection problem for Rohingya refugees in Bangladesh. *Journal of Cleaner Production, 273,* 122967.
8. Ali, T., Ma, H., & Nahian, A. J., (2020). A multi-criteria decision-making approach to determine the optimal hybrid energy system in coastal off-grid areas: A case study of Bangladesh. *Process Integration and Optimization for Sustainability, 4,* 265–277.
9. Karakuş, C. B., Demiroğlu, D., Çoban, A., & Ulutaş, A., (2020). Evaluation of GIS-based multi-criteria decision-making methods for sanitary landfill site selection: The case of Sivas city, Turkey. *Journal of Material Cycles and Waste Management, 22,* 254–272.
10. Tadić, S., Krstić, M., Roso, V., & Brnjac, N., (2020). Dry port terminal location selection by applying the hybrid grey MCDM model. *Sustainability, 12*. doi: 10.3390/su12176983.
11. Adalı, E. A., & Tuş, A., (2021). Hospital site selection with distance-based multi-criteria decision-making methods. *International Journal of Healthcare Management, 14*(2), 534–544.
12. Maghsoodi, A. I., Maghsoodi, A. I., Mosavi, A., Rabczuk, T., & Zavadskas, E. K., (2018). Renewable energy technology selection problem using integrated H-SWARA-MULTIMOORA approach. *Sustainability, 10*. doi: 10.3390/su10124481.

CHAPTER 16

EVALUATION OF MIXED DATA (EVAMIX) METHOD

The EValuation of MIXed data (EVAMIX) method was originally introduced by Voogd in 1983 [1], and later promoted by Martel & Matarazzo [2]. It is based on the determination of the dominance score of an alternative on a criterion-by-criterion basis. The main difference between EVAMIX and other MCDM methods is that it can separately deal with both ordinal (qualitative) and cardinal (quantitative) criteria. Both types of data are initially normalized between 0 and 1 by employing a linear normalization approach. This method estimates the level of pair-wise dominance as the difference in scores received for each pair of alternatives by comparing the higher-performance alternative to the lesser-performance alternative. The weighted sum of the dominance scores is then calculated for each alternative. This is similar to the aggregation procedure involved in the weighted sum method.

The procedural steps involved in EVAMIX method are explained as follows [2, 3]:

- **Step 1:** The ordinal and cardinal data are first separated in the decision matrix.
- **Step 2:** For higher-the-better criterion, the data in the decision matrix is normalized using the following equation:

$$r_{ij} = [x_{ij} - \min(x_{ij})]/[\max(x_{ij}) - \min(x_{ij})]$$
$$(\mathrm{i} = 1, 2, \ldots, m; \mathrm{j} = 1, 2, \ldots, n) \tag{16.1}$$

where; x_{ij} is the performance measure of ith alternative with respect to j^{th} criterion and r_{ij} is the normalized value of x_{ij}.

For lower-the-better criterion, Eqn. (16.2) can be employed:

$$r_{ij} = [\max(x_{ij}) - x_{ij}]/[\max(x_{ij}) - \min(x_{ij})] \tag{16.2}$$

Multi-Criteria Decision-Making Methods in Manufacturing Environments: Models and Applications.
Shankar Chakraborty, Prasenjit Chatterjee, & Partha Protim Das (Authors)

- **Step 3:** Compute the deviations of ith alternative with respect to other alternatives for each ordinal and cardinal criterion. This is estimated by taking the differences in criteria values between pairs of alternatives.
- **Step 4:** Calculate the dominance scores for each alternative pair, (i,i') with respect to all the ordinal and cardinal criteria using the following equation:

$$\alpha_{ii'} = \left[\sum_{j \in O} \left\{ w_j \operatorname{sgn}(r_{ij} - r_{i'j}) \right\}^c \right]^{1/c} \quad (16.3)$$

where;

$$\operatorname{sgn}(r_{ij} - r_{i'j}) = \begin{cases} +1 \text{ if } r_{ij} > r_{i'j} \\ 0 \text{ if } r_{ij} = r_{i'j} \\ -1 \text{ if } r_{ij} < r_{i'j} \end{cases}$$

$$\gamma_{ii'} = \left[\sum_{j \in C} \left\{ w_j \operatorname{sgn}(r_{ij} - r_{i'j}) \right\}^c \right]^{1/c} \quad (16.4)$$

where; c denotes a scaling parameter that can take any arbitrary positive odd number, such as 1, 3, 5, …; O represents the set of ordinal criteria; C symbolizes the set of cardinal criteria; $\alpha_{ii'}$ and $\gamma_{ii'}$ are the dominance scores for the pair of alternatives; (i,i') for ordinal and cardinal criteria, respectively; and w_j is the weight (relative importance) of jth criterion.

- **Step 5:** Calculate the standardized dominance scores. Martel & Matarazzo [2] proposed an additive interval method for the alternative pair, (i,i') to derive the standardized ordinal dominance score $(\delta_{ii'})$ and cardinal dominance score $(d_{ii'})$ as follows:

$$\text{Standardized ordinal dominance score } (\delta_{ii'}) = \frac{(\alpha_{ii'} - \alpha^-)}{(\alpha^+ - \alpha^-)} \quad (16.5)$$

where; α^+ (α^-) is the maximum (minimum) ordinal dominance score for the pair of alternatives, (i,i').

$$\text{Standardized cardinal dominance score } (d_{ii'}) = \frac{(\gamma_{ii'} - \gamma^-)}{(\gamma^+ - \gamma^-)} \quad (16.6)$$

where; γ^+ (γ^-) is the maximum (minimum) cardinal dominance score for the pair of alternatives, (i,i').

- **Step 6:** Determine the overall dominance score. The overall dominance score, $(D_{ii'})$ for each pair of alternatives, (i,i') is calculated to measure the degree by which alternative i dominates alternative i'.

$$D_{ii'} = w_O\delta_{ii'} + w_C d_{ii'} \tag{16.7}$$

where; w_O is the sum of weights for the ordinal criteria $(w_o = \sum_{j\in O} w_j)$; and w_C is the sum of weights for the cardinal criteria $(w_C = \sum_{j\in C} w_j)$.

- **Step 7:** Calculate the appraisal score.

$$\text{Appraisal score } (S_i) = \sum_{i'}\left(\frac{D_{i'i}}{D_{ii'}}\right)^{-1} \tag{16.8}$$

Lastly, the appraisal score (S_i) for i^{th} alternative is calculated to provide the ranking attained by that alternative. An alternative with the highest appraisal score indicates its best performance as compared to others.

16.1 LITERATURE SURVEY

Chatterjee et al. [4] compared the ranking performance of EVAMIX with COPRAS methods while solving two real-time engineering material selection problems. Using both these methods, the competing materials were shortlisted from the most preferable to the least preferable. Coronado et al. [5] proposed a two-step methodology for quantification and waste management analysis of construction and demolition waste in a northern Spanish region. The EVAMIX method and three other MCDM techniques were implemented to search out the most appropriate construction and demolition alternative for waste management. Dosal et al. [6] endeavored to identify the best-suited location for construction and demolition facilities for waste recycling in a Northern Spanish region using EVAMIX method. The derived ranking of the alternative locations was also contrasted with that obtained by SAW, ELECTRE II and REGIME methods. Sensitivity analysis with respect to varying criteria weights was finally conducted to study the robustness of the ranking results. Using three illustrative examples, Chatterjee & Chakraborty [7] demonstrated the application of EVAMIX method in identifying the most apposite NTM processes for different machining applications. Darji & Rao [8] explored the applicability of EVAMIX method in dealing with both ordinal and cardinal data for solving complicated decision-making problems in industrial environment. For the considered demonstrative examples, the criteria weights were estimated using the AHP technique.

Chatterjee et al. [9] adopted EVAMIX and PROMETHEE II methods as two preference dominance-based MCDM techniques while considering an industrial robot selection problem. The ranking performance of those methods was also compared with that of other MCDM tools based on four performance tests. In the sugar industry, Darji & Rao [10] employed EVAMIX method along with other popular intelligent MCDM techniques, like Extended TODIM, ARAS, and OCRA to single out the most appropriate material from a pool of five alternatives. In the healthcare sector, Adalı [11] solved a nurse selection problem based on the simultaneous application of both EVAMIX and TODIM methods, and also compared their relative ranking performance. Işık & Adalı [12] considered an agricultural tractor selection problem and solved the same using EVAMIX method. Its ranking stability and performance were compared with respect to TOPSIS and COPRAS methods. The ranking results were finally aggregated using Borda and Copeland methods.

Chojnacka & Górecka [13] applied EVAMIX method as an effective assessment tool for appraising the relative performance of 11 public benefit organizations in Poland engaged in the area of 'animals, ecology, and heritage protection.' While solving an advanced manufacturing technology selection problem, Nath & Sarkar [14] derived the ranking list of the competing alternatives using EVAMIX method and compared the same with COPRAS-G method. Spina [15] adopted the EVAMIX method as an integrated evaluation tool to classify the adaptive reutilization approaches of unexplored cultural heritage assets while supporting the concerned decision makers in implementing appropriate development strategies in susceptible contexts. Depending on the type of qualitative and quantitative criteria values, Yazır et al. [16] extended the application of EVAMIX method under fuzzy group decision-making environment for selection of ships to be deployed as effective gas carriers.

16.2 NTM PROCESS SELECTION PROBLEM

Selection of the most appropriate NTM process to generate a desired shape feature on a given work material is a time-consuming and challenging task as it requires consideration of numerous conflicting machining quality measures (like maximization of machining efficiency, material removal rate (MRR), and minimization of surface finish, power requirement, etc.), and a diverse range of machining capabilities and quality characteristics of the NTM processes. For efficient use of the inherent capabilities of different NTM processes and for their maximized machining performance, the available machining process parameters need to be carefully selected for a given machining application. Typically, it is considered to be an arduous task requiring human expertise, being affected by several conflicting criteria. Thus, the need for a structured and mathematically sound method arises for appropriate selection of a viable NTM process for a given machining application. It has been observed that the criteria influencing the NTM process selection decision are quantitative as well as qualitative in nature and it is also quite difficult to take into account both these types of attributes in a single decision-making framework. The application of an MCDM method in this direction helps in identifying the most suitable NTM process for a given machining application based on some parametric requirements, like material type, process economy, shape feature and several other process competencies, such as MRR, surface finish, surface damage depth, tolerance, machining medium contamination, efficiency, etc.

In this demonstrative example, surface of revolution feature is to be produced on stainless steel work material. For selecting the most suitable NTM process for generating surface of revolution feature on stainless steel, firstly various important criteria that affect the decision selection of NTM processes are identified. These criteria are tolerance and surface finish (C_1), power requirement (C_2), MRR (C_3), cost (C_4), efficiency (C_5), tooling and fixtures (C_6), tool consumption (C_7), safety (C_8), work material (C_9), and shape feature (C_{10}). Tolerance and surface finish are the two important product quality attributes used to measure the performance of an NTM process for a particular application. Tolerance is measured as the deviation between the maximum and minimum dimensions of a component. Based on the type of application, the allowable variation in dimension is set according to the available standard grades. It also relates to the capability of an NTM process stating how closely the process can achieve the required surface finish on the work material. Power requirement deals with the power rating of the machine/equipment for the selected NTM operation. The MRR is the most important criterion leading to the fact that higher MRR leads to lower machining time, and the effectiveness of an NTM process is usually measured in terms of MRR. Cost is concerned with the initial investment and acquisition cost required to install an NTM process-based machining system for a given application. Sometimes, it may also indicate the cost of machining. Efficiency of an NTM process can be defined as the ratio of the quantity of output energy available for material removal to the input energy. Tooling and fixtures are associated with the cost of tooling and work fixtures that need to be occasionally changed during the NTM operation. Some of the NTM processes require frequent changing of tools and for this reason, tool consumption criterion is taken into consideration to account for the cost related to tool changes. Safety is an important characteristic of an NTM process with respect to emission of toxic gases/fumes, contamination effects, secured environment and less hazardous effects on the operators. Work material mainly relates to the fact that how often a particular NTM process can be used on a given material for machining operation. It basically shows the easiness of an NTM process to machine a given material. Shape feature corresponds to the capability of an NTM process to generate a specific shape feature on a given material to be machined. Among all these criteria, C_1 (in μm), C_2 (in kW) and C_3 (in mm^3/min) are quantitative in nature, having absolute numerical values [17], whereas, C_4, C_5, C_6, C_7, C_8, C_9, and C_{10} have qualitative measures which are measured based on a ranked value judgment scale from 1 to 5 (1 = lowest, 3 = moderate and 5 = highest). In this example, C_3, C_5, C_8, C_9, and C_{10} are beneficial attributes where higher values are desired; whereas, C_1, C_2, C_4, C_6, and C_7 are non-beneficial attributes for which lower values are always preferred. The data for C_1, C_2, and C_3 are cardinal in nature, whereas, the data for other criteria, such as C_4, C_5, etc., are ordinal, so the relatedness is ensured using a five-point scale. Das Chakladar & Chakraborty [18] determined the criteria weights as w_{C1} = 0.0783, w_{C2} = 0.0611, w_{C3} = 0.1535, w_{C4} = 0.1073, w_{C5} = 0.0383, w_{C6} = 0.0271, w_{C7} = 0.0195, w_{C8} = 0.0146, w_{C9} = 0.2766, and w_{C10} = 0.2237 using AHP method. These same criteria weights are used here for further analysis.

While solving this NTM process selection problem, the original decision matrix, as shown in Table 16.1, is developed, and the corresponding ordinal and cardinal data are separated. Applying Eqns. (16.1) or (16.2), the decision matrix is normalized, as given in Table 16.2.

TABLE 16.1 Data for the NTM Process Selection Problem

NTM	C_1	C_2	C_3	C_4	C_5	C_6	C_7	C_8	C_9	C_{10}
USM	0.85	8.5	550	3	5	1	3	2	5	2
WJM	2.125	0.187	0.88	2	5	1	2	4	5	2
AJM	2.125	0.204	0.55	2	5	1	2	4	5	2
ECM	2.55	85	440	6	3	2	1	4	6	5
CHM	2.55	0.34	16.5	4	4	1	1	4	6	2
EDM	2.975	2.295	880	4	5	3	4	4	6	2
WEDM	2.975	2.125	660	4	5	3	4	4	6	2
EBM	2.125	0.17	1.76	5	6	1	1	4	5	2
LBM	1.7	1.19	0.11	4	6	1	1	2	5	2

TABLE 16.2 Normalized Decision Matrix for the NTM Process Selection Problem

NTM	C_1	C_2	C_3	C_4	C_5	C_6	C_7	C_8	C_9	C_{10}
USM	1	0.9018	0.625	0.75	0.6667	1	0.3333	0	0	0
WJM	0.4	0.9998	0.0009	1	0.6667	1	0.6667	1	0	0
AJM	0.4	0.9996	0.0005	1	0.6667	1	0.6667	1	0	0
ECM	0.2	0	0.4999	0	0	0.5	1	1	1	1
CHM	0.2	0.998	0.0186	0.5	0.3333	1	1	1	1	0
EDM	0	0.9749	1	0.5	0.6667	0	0	1	1	0
WEDM	0	0.977	0.75	0.5	0.6667	0	0	1	1	0
EBM	0.4	1	0.0019	0.25	1	1	1	1	0	0
LBM	0.6	0.988	0	0.5	1	1	1	0	0	0

From the normalized decision matrix, the evaluative differences of an NTM alternative for both the ordinal and cardinal criteria with respect to all other alternatives are calculated. Now, the dominance scores of each pair of NTM alternatives for all the ordinal and cardinal criteria are computed using Eqns. (16.3) and (16.4), respectively, and are given in Table 16.3. While calculating these dominance scores, the value of c is taken as 1. Using the additive interval technique, the standardized dominance scores for all the pairs of NTM alternatives are now determined using Eqns. (16.5) and (16.6), respectively for both the ordinal and cardinal criteria, as given in Table 16.4.

TABLE 16.3 Dominance Scores for NTM Process Pairs for NTM Process Selection Problem

NTM Process	$\alpha_{ii'}$	$\gamma_{ii'}$	NTM Process	$\alpha_{ii'}$	$\gamma_{ii'}$	NTM Process	$\alpha_{ii'}$	$\gamma_{ii'}$
(1,2)	–0.1414	0.1707	(4,1)	0.3617	–0.2929	(7,1)	0.1373	0.1363
(1,3)	–0.1414	0.1707	(4,2)	0.3471	0.0141	(7,2)	0.1227	0.0141
(1,4)	–0.3617	0.2929	(4,3)	0.3471	0.0141	(7,3)	0.1227	0.0141
(1,5)	–0.1651	0.1707	(4,5)	0.051	0.0924	(7,4)	–0.1247	0.1363
(1,6)	–0.1373	–0.1363	(4,6)	0.1247	–0.1363	(7,5)	–0.0083	0.0141
(1,7)	–0.1373	–0.1363	(4,7)	0.1247	–0.1363	(7,6)	0	–0.0924
(1,8)	0.0349	0.1707	(4,8)	0.3276	0.0141	(7,8)	0.299	0.0141
(1,9)	0.0495	0.1707	(4,9)	0.3422	0.0141	(7,9)	0.2063	0.0141
(2,1)	0.1414	–0.1707	(5,1)	0.1651	–0.1707	(8,1)	–0.0349	–0.1707
(2,3)	0	0.2146	(5,2)	0.1505	0.0141	(8,2)	–0.0495	0.2146
(2,4)	–0.3471	–0.0141	(5,3)	0.1505	0.0141	(8,3)	–0.0495	0.2146
(2,5)	–0.1505	–0.0141	(5,4)	–0.051	–0.0924	(8,4)	–0.3276	–0.0141
(2,6)	–0.1227	–0.0141	(5,6)	0.0083	–0.0141	(8,5)	–0.3456	–0.0141
(2,7)	–0.1227	–0.0141	(5,7)	0.0083	–0.0141	(8,6)	–0.299	–0.0141

TABLE 16.3 *(Continued)*

NTM Process	$\alpha_{ii'}$	$\gamma_{ii'}$	NTM Process	$\alpha_{ii'}$	$\gamma_{ii'}$	NTM Process	$\alpha_{ii'}$	$\gamma_{ii'}$
(2,8)	0.0495	–0.2146	(5,8)	0.3456	0.0141	(8,7)	–0.299	–0.0141
(2,9)	0.0641	0.1363	(5,9)	0.2529	0.1363	(8,9)	–0.0927	0.1363
(3,1)	0.1414	–0.1707	(6,1)	0.1373	0.1363	(9,1)	–0.0495	–0.1707
(3,2)	0	–0.2146	(6,2)	0.1227	0.0141	(9,2)	–0.0641	–0.1363
(3,4)	–0.3471	–0.0141	(6,3)	0.1227	0.0141	(9,3)	–0.0641	–0.1363
(3,5)	–0.1505	–0.0141	(6,4)	–0.1247	0.1363	(9,4)	–0.3422	–0.0141
(3,6)	–0.1227	–0.0141	(6,5)	–0.0083	0.0141	(9,5)	–0.2529	–0.1363
(3,7)	–0.1227	–0.0141	(6,7)	0	0.0924	(9,6)	–0.2063	–0.0141
(3,8)	0.0495	–0.2146	(6,8)	0.299	0.0141	(9,7)	–0.2063	–0.0141
(3,9)	0.0641	0.1363	(6,9)	0.2063	0.0141	(9,8)	0.0927	–0.1363

TABLE 16.4 Standardized Dominance Scores for NTM Process Pairs for NTM Process Selection Problem

NTM Process	$\delta_{ii'}$	$d_{ii'}$	NTM Process	$\delta_{ii'}$	$d_{ii'}$	NTM Process	$\delta_{ii'}$	$d_{ii'}$
(1,2)	0.3045	0.7914	(4,1)	1	0	(7,1)	0.6898	0.7327
(1,3)	0.3045	0.7914	(4,2)	0.9798	0.5241	(7,2)	0.6696	0.5241
(1,4)	0	1	(4,3)	0.9798	0.5241	(7,3)	0.6696	0.5241
(1,5)	0.2718	0.7914	(4,5)	0.5705	0.6577	(7,4)	0.3276	0.7327
(1,6)	0.3102	0.2673	(4,6)	0.6724	0.2673	(7,5)	0.4885	0.5241
(1,7)	0.3102	0.2673	(4,7)	0.6724	0.2673	(7,6)	0.5	0.3423
(1,8)	0.5482	0.7914	(4,8)	0.9529	0.5241	(7,8)	0.9133	0.5241
(1,9)	0.5684	0.7914	(4,9)	0.973	0.5241	(7,9)	0.7852	0.5241
(2,1)	0.6955	0.2086	(5,1)	0.7282	0.2086	(8,1)	0.4518	0.2086
(2,3)	0.5	0.8663	(5,2)	0.708	0.5241	(8,2)	0.4316	0.8663
(2,4)	0.0202	0.4759	(5,3)	0.708	0.5241	(8,3)	0.4316	0.8663
(2,5)	0.292	0.4759	(5,4)	0.4295	0.3423	(8,4)	0.0471	0.4759
(2,6)	0.3304	0.4759	(5,6)	0.5115	0.4759	(8,5)	0.0223	0.4759
(2,7)	0.3304	0.4759	(5,7)	0.5115	0.4759	(8,6)	0.0867	0.4759
(2,8)	0.5684	0.1337	(5,8)	0.9777	0.5241	(8,7)	0.0867	0.4759
(2,9)	0.5886	0.7327	(5,9)	0.8496	0.7327	(8,9)	0.3719	0.7327
(3,1)	0.6955	0.2086	(6,1)	0.6898	0.7327	(9,1)	0.4316	0.2086
(3,2)	0.5	0.1337	(6,2)	0.6696	0.5241	(9,2)	0.4114	0.2673
(3,4)	0.0202	0.4759	(6,3)	0.6696	0.5241	(9,3)	0.4114	0.2673
(3,5)	0.292	0.4759	(6,4)	0.3276	0.7327	(9,4)	0.027	0.4759
(3,6)	0.3304	0.4759	(6,5)	0.4885	0.5241	(9,5)	0.1504	0.2673
(3,7)	0.3304	0.4759	(6,7)	0.5	0.6577	(9,6)	0.2148	0.4759
(3,8)	0.5684	0.1337	(6,8)	0.9133	0.5241	(9,7)	0.2148	0.4759
(3,9)	0.5886	0.7327	(6,9)	0.7852	0.5241	(9,8)	0.6281	0.2673

The overall dominance score for each pair of NTM alternatives is calculated using Eqn. (16.7) which symbolizes the degree by which an NTM alternative dominates the other. These overall dominance scores for all the pairs of NTM processes are shown in Table 16.5. The appraisal score for each NTM process is finally calculated using Eqn. (16.8) and based on the descending values of this score, the final ranking of NTM alternatives is obtained, as shown in Table 16.6. The best choice for generating surface of revolution feature on stainless steel material is observed to be the ECM process. Chemical machining (CHM) is the second-best process and LBM is the worst chosen process. Das Chakladar & Chakraborty [18] derived the ranking of NTM processes as ECM-EDM-WEDM-USM-EBM-CHM-LBM-WJM-AJM while solving a similar type of problem applying a combined AHP and TOPSIS approach. It is observed that in both the cases, the best choice of NTM process remains the same.

TABLE 16.5 Overall Dominance Scores for NTM Process Selection Problem

NTM Process	$D_{ii'}$	NTM Process	$D_{ii'}$	NTM Process	$D_{ii'}$	NTM Process	$D_{ii'}$
(1,2)	0.4471	(3,4)	0.1537	(5,6)	0.5011	(7,8)	0.7993
(1,3)	0.4471	(3,5)	0.3458	(5,7)	0.5011	(7,9)	0.7087
(1,4)	0.2929	(3,6)	0.373	(5,8)	0.8449	(8,1)	0.3805
(1,5)	0.424	(3,7)	0.373	(5,9)	0.8154	(8,2)	0.5589
(1,6)	0.2976	(3,8)	0.4411	(6,1)	0.7024	(8,3)	0.5589
(1,7)	0.2976	(3,9)	0.6308	(6,2)	0.627	(8,4)	0.1727
(1,8)	0.6195	(4,1)	0.7071	(6,3)	0.627	(8,5)	0.1551
(1,9)	0.6337	(4,2)	0.8463	(6,4)	0.4463	(8,6)	0.2007
(2,1)	0.5529	(4,3)	0.8463	(6,5)	0.4989	(8,7)	0.2007
(2,3)	0.6073	(4,5)	0.5961	(6,7)	0.5462	(8,9)	0.4775
(2,4)	0.1537	(4,6)	0.5537	(6,8)	0.7993	(9,1)	0.3663
(2,5)	0.3458	(4,7)	0.5537	(6,9)	0.7087	(9,2)	0.3692
(2,6)	0.373	(4,8)	0.8273	(7,1)	0.7024	(9,3)	0.3692
(2,7)	0.373	(4,9)	0.8415	(7,2)	0.627	(9,4)	0.1585
(2,8)	0.4411	(5,1)	0.576	(7,3)	0.627	(9,5)	0.1846
(2,9)	0.6308	(5,2)	0.6542	(7,4)	0.4463	(9,6)	0.2913
(3,1)	0.5529	(5,3)	0.6542	(7,5)	0.4989	(9,7)	0.2913
(3,2)	0.3927	(5,4)	0.4039	(7,6)	0.4538	(9,8)	0.5225

TABLE 16.6 Appraisal Scores for NTM Processes

NTM Process	Appraisal Score	Rank
USM	6.9735	6
WJM	7.1808	5
AJM	6.281	7
ECM	27.4862	1
CHM	17.6895	2
EDM	15.1425	3
WEDM	14.7697	4
EBM	4.9572	8
LBM	4.0794	9

KEYWORDS

- **appraisal score**
- **chemical machining**
- **evaluation of mixed data**
- **literature survey**
- **material removal rate**
- **quantitative criteria values**

REFERENCES

1. Voogd, H., (1982). Multicriteria evaluation with mixed qualitative and quantitative data. *Environment and Planning B: Planning and Design, 9*(2), 221–236.
2. Martel, J. M., & Matarazzo, B., (2005). Other outranking approaches. In: Figueira, J., Salvatore, G., & Ehrgott, M., (eds.), *Multiple Criteria Decision Analysis: State of the Art Surveys*. Springer: New York.
3. Hajkowicz, S., & Higgins, A., (2008). A comparison of multiple criteria analysis techniques for water resource management. *European Journal of Operational Research, 184*, 255–265.
4. Chatterjee, P., Athawale, V. M., & Chakraborty, S., (2011). Materials selection using complex proportional assessment and evaluation of mixed data methods. *Materials and Design, 32*, 851–860.
5. Coronado, M., Dosal, E., Coz, A., Viguri, J. R., & Andrés, A., (2011). Estimation of construction and demolition waste (C&DW) generation and multicriteria analysis of

C&DW management alternatives: A case study in Spain. *Waste and Biomass Valorization, 2*, 209–225.
6. Dosal, E., Coronado, M., Muñoz, I., Viguri, J. R., & Andrés, A., (2012). Application of multi-criteria decision-making tool to locate construction and demolition waste (C&DW) recycling facilities in a Northern Spanish region. *Environmental Engineering and Management Journal, 11*(3), 545–556.
7. Chatterjee, P., & Chakraborty, S., (2013). Nontraditional machining processes selection using evaluation of mixed data method. *International Journal of Advanced Manufacturing Technology, 68,* 1613–1626.
8. Darji, V. P., & Rao, R. V., (2013). Application of AHP/EVAMIX method for decision making in the industrial environment. *American Journal of Operations Research, 3,* 542–569.
9. Chatterjee, P., Mondal, S., & Chakraborty, S., (2014). A comparative study of preference dominance-based approaches for selection of industrial robots. *Advances in Production Engineering & Management, 9*(1), 5–20.
10. Darji, V. P., & Rao, R. V., (2014). Intelligent multi-criteria decision making methods for material selection in the sugar industry. *Procedia Materials Science, 5*, 2585–2594.
11. Adalı, E. A., (2016). Personnel selection in health sector with EVAMIX and TODIM methods. *Journal of Operations Research, Statistics, Econometrics and Management Information Systems, 4*(2), 70–83.
12. Işık, A. T., & Adalı, E. A., (2016). A comparative study for the agricultural tractor selection problem. *Decision Science Letters, 5*, 569–580.
13. Chojnacka, E., & Górecka, D., (2016). Evaluating public benefit organizations in Poland with the EVAMIX method for mixed data. *Multiple Criteria Decision Making, 11*, 36–51.
14. Nath, S., & Sarkar, B., (2017). An exploratory analysis for the selection and implementation of advanced manufacturing technology by fuzzy multi-criteria decision-making methods: A comparative study. *Journal of Institution of Engineers (India) Series C, 98*(4), 493–506.
15. Spina, L. D., (2020). Adaptive sustainable reuse for cultural heritage: A multiple criteria decision aiding approach supporting urban development processes. *Sustainability, 12*, 20. doi: 10.3390/su12041363.
16. Yazır, D., Şahin, B., & Yip, T. L., (2021). Selection of new design gas carriers by using fuzzy EVAMIX method. *The Asian Journal of Shipping and Logistics, 37*, 91–104.
17. Pandey, P. C., & Shan, H. S., (1981). *Modern Machining Processes*. Tata McGraw-Hill Publishing Com. Ltd., New Delhi.
18. Das, C. N., & Chakraborty, S., (2008). A combined TOPSIS-AHP method based approach for non-traditional machining processes selection. *Proceedings of Institution of Mechanical Engineers Journal of Engineering Manufacture, 222*, 1613–1623.

CHAPTER 17

PREFERENCE SELECTION INDEX (PSI) METHOD

It has been noticed that in almost all the MCDM techniques, there is a requirement to determine weights or relative importance of the considered criteria as the final ranking of the alternatives greatly depends of these criteria weights. But, in the PSI method, proposed by Maniya & Bhatt [1], there is no need to estimate criteria weights, because the overall preference value of the considered criteria is computed using the concepts of statistics. This method seems to be quite attractive when there is a conflict in deciding the relative importance between the participating criteria. While calculating the overall preference values, the candidate alternatives are ranked based on their preference selection indexes (PSIs). The procedural steps of PSI method are highlighted as below:

- **Step 1: Identify the Objective:** For a given decision-making problem, identify the feasible alternatives and the corresponding set of evaluation criteria.
- **Step 2: Develop the Decision Matrix:** Considering the performance scores of all the alternatives with respect to different criteria, the corresponding decision matrix, $X = [x_{ij}]_{m \times n}$ is developed, where m is the number of alternatives, n is the number of criteria and x_{ij} is the performance of ith alternative against j^{th} criterion.
- **Step 3: Normalization of the Decision Matrix:** In order to make all the elements of the developed decision matrix dimensionless and comparable, the matrix is normalized using the following equations:
  i. For Beneficial Criterion:

$$r_{ij} = \frac{x_{ij}}{x_j^{\max}} \tag{17.1}$$

Multi-Criteria Decision-Making Methods in Manufacturing Environments: Models and Applications.
Shankar Chakraborty, Prasenjit Chatterjee, & Partha Protim Das (Authors)

ii. For Non-Beneficial Criterion:

$$r_{ij} = \frac{x_j^{min}}{x_{ij}} \tag{17.2}$$

where; r_{ij} is the normalized value of x_{ij}.

- **Step 4: Computation of the Preference Variation Value:** The preference variation value (PV_j) for each criterion is now estimated using the sample variance concept based on the following expression:

$$PV_j = \sum_{i=1}^{m} (r_{ij} - \bar{r}_j)^2 \quad (j = 1, 2, \ldots, n) \tag{17.3}$$

where; $\bar{r}_j$ is the mean of the normalized values of criterion j.

- **Step 5: Determination of the Overall Preference Value:** The overall preference value (Ψ_j) for each criterion is calculated in this step. For this purpose, the deviation (Φ_j) in the preference variation value (PV_j) for each criterion is computed based on the following equation:

$$\Phi_j = 1 - PV_j \tag{17.4}$$

Now, the overall preference value (Ψ_j) is calculated using the following expression:

$$\Psi_j = \frac{\Phi_j}{\sum_{j=1}^{n} \Phi_j} \tag{17.5}$$

It is worthwhile to mention here that the total overall preference value for all the considered criteria should always be one.

- **Step 6: Derive the Preference Selection Index (PSI):** The PSI (I_i) for i^{th} alternative is determined based on the following expression:

$$I_i = \sum_{j=1}^{n} (r_{ij} \times \Phi_j) \tag{17.6}$$

- **Step 7: Ranking of the Alternatives:** Based on the descending values of the calculated PSI, all the candidate alternatives are finally ranked from the best to the worst. Thus, the most preferred alternative should have the highest PSI.

17.1 REVIEW OF THE LITERATURE

Maniya & Bhatt [1] proposed the novel application of the PSI method while solving three different material selection problems. Its ranking performance was also validated against GTMA and TOPSIS methods. It was also claimed that the PSI method would be a more logical and appropriate approach in solving complex decision-making problems without the involvement of criteria weights in its calculation steps. Using PSI method, Maniya & Bhatt [2] solved an FMS selection problem without taking into account the relative importance of the considered criteria. Its simplicity, solution accuracy and appropriateness in solving MCDM problems were also validated. Sawant et al. [3] proposed the application of PSI method to determine the corresponding selection indexes of the alternative AGVs to be deployed in a manufacturing environment. Akyuza & Akab [4] demonstrated the application of PSI method to calculate the corresponding performance indexes for a glass manufacturing industry in order to enhance its production volume. Based on illustrative examples, Attri & Grover [5] presented the application potentiality of PSI method while solving various decision-making problems confronted during design stage of production system life cycle. Diyaley et al. [6] determined the optimal settings of pulse-on time, pulse-off time, servo voltage and wire tension based on the application of PSI method for a WEDM process while machining EN31 tool steel material. Jian & Ying [7] adopted the PSI method to identify the most suitable machine in a flexible manufacturing cell, and also proved its appropriateness and feasibility as an effective MCDM tool. Using Taguchi's L_{27} orthogonal array as the experimental design plan, and roughness of the cut surface, heat affected zone (HAZ), kerf width and MRR as the responses, Madić et al. [8] employed PSI method to determine the optimal settings of different input parameters while performing CO_2 laser cutting (LC) operation on stainless steel. Applying the PSI method, Petković et al. [9] solved two real-time case studies involving machinability of materials and selection of the most suitable cutting fluid for a given machining application.

Jain [10] adopted both MOORA and PSI methods for evaluation and ranking of different performance factors of FMSs based on six variables influencing three elements of FMS performance, i.e., productivity, flexibility, and quality. Productivity was identified as the primary factor in the manufacturing system. Patel et al. [11] applied PSI method to identify the optimal parametric settings of fused deposition modeling (FDM) process using polylactic acid material in order to achieve the desired values of tensile strength, tensile module, compressive strength, compressive module, and surface roughness. Sahir et al. [12] developed a decision support system using the PSI method to identify the most appropriate location for selling of used laptops based on several criteria, like rental price, area, visitors, distance, etc. Considering speed, feed, and depth of cut as the inputs, and MRR and surface roughness as the responses, Vara Prasad et al. [13] applied the PSI method to determine the optimal combination of different input parameters for a turning operation. Yadav et al. [14] presented the application of a hybrid TOPSIS-PSI approach for selection of materials for marine applications. Based on the results from two real-time examples, it was concluded that 9 wt.% and 6 wt.% reinforced hybrid aluminum nanocomposites would be the most appropriate materials for marine applications having the optimal combination of all the physical, mechanical, and corrosion properties. Based on PSI method, Prabhu et al. [15] evaluated the performance of 10 alternative 3D printers, i.e., Lulzbot TAZ 5, Ultimaker 2, Zortrax, Wanhao duplicator 4, Airwolf 3D AW3D HD2X, Flash forge creator X, Makerbot replicator original, Delta non-turbo WASP 2040, Artifex duo 2 and UP plus 2 based on six criteria, like maximum print volume, speed of operation, minimum thickness, extruder capacity, printer cost and filament material cost. Wanhao duplicator 4 emerged out as the best-suited 3D printer for manufacturing of spoilers for automobile applications.

17.2 MILLING MACHINE SELECTION PROBLEM

In any manufacturing organization, selection of the most appropriate machine/equipment is a very important and complicated task due to the involvement of a large set of available alternatives and conflicting criteria. A wrong machine selection decision may adversely influence the overall performance and productivity of the manufacturing organization. The outputs of a manufacturing system with respect to production rate, cost and quality of the products, etc., largely depend on the type of the machine selected. A properly selected machine/equipment can enhance the production process, provide effective utilization of resources, increase quality, and improve system flexibility. Thus, the problem of machine selection in a manufacturing organization cannot be overlooked.

It has been observed that due to the presence of a set of competing alternatives, selection of a new machine/equipment is a time-consuming and difficult task, requiring the involvement of advanced knowledge of the human experts. So, machine selection problem may be a hard task for the engineers and managers, and also for the equipment manufacturers or vendors, to solve. For an effective evaluation process, the decision maker may need a large amount of machine-related data to be analyzed and many factors to be considered. Dağdeviren [16] developed a decision-making problem dealing with selection of the most appropriate milling machine for a manufacturing organization based on six evaluation criteria, i.e., price of the milling machine (in \$), weight (in kg), power (in W), spindle speed (in rpm), diameter (in mm) and stroke length (in mm). The original decision matrix consisting of five candidate milling machines and six criteria is exhibited in Table 17.1. Amongst the considered criteria, price and weight are the non-beneficial attributes, whereas, power, spindle speed, diameter, and stroke are the beneficial characteristics. This milling machine selection problem is solved here while applying PSI method. Now, depending on the type of the criterion, the elements of Table 17.1 are normalized using Eqns. (17.1) or (17.2), as shown in Table 17.2. Table 17.3 presents the preference variation values for this milling machine selection problem, computed using Eqn. (17.3). Finally, in order to rank the candidate alternatives, their overall preference values and corresponding PSIs are calculated, as provided in Table 17.4. From this table, it can be noticed that MM-5 is the best-chosen milling machine for the considered application, followed by MM-4. On the other hand, MM-3 is the least preferred choice. Dağdeviren [16] applied PROMETHEE to observe that MM-5, MM-2, MM-4, and MM-1 alternatives would be preferred to MM-3 alternative, and MM-4 alternative would be preferred to MM-1 alternative. The MM-5 alternative was also selected as the best milling machine and the derived ranking order was MM-5 → MM-4 → MM-2 → MM-1 → MM-3.

TABLE 17.1 Decision Matrix for the Milling Machine Selection Problem [16]

Milling Machine	Price	Weight	Power	Spindle	Diameter	Stroke
MM-1	936	4.8	1,300	24,000	12.7	58
MM-2	1,265	6	2,000	21,000	12.7	65
MM-3	680	3.5	900	24,000	8	50
MM-4	650	5.2	1,600	22,000	12	62
MM-5	580	3.5	1,050	25,000	12	62

Source: Reprinted with permission from Ref. [16]. © 2021 Springer.

TABLE 17.2 Normalized Decision Matrix

Milling Machine	Price	Weight	Power	Spindle	Diameter	Stroke
MM-1	0.6197	0.7292	0.65	0.96	1	0.8923
MM-2	0.4585	0.5833	1	0.84	1	1
MM-3	0.8529	1	0.45	0.96	0.6299	0.7692
MM-4	0.8923	0.6731	0.8	0.88	0.9449	0.9538
MM-5	1	1	0.525	1	0.9449	0.9538

TABLE 17.3 Computed Preference Variation Values

Milling Machine	Price	Weight	Power	Spindle	Diameter	Stroke
MM-1	0.021	0.0046	0.0012	0.001	0.0092	0.0005
MM-2	0.0937	0.0457	0.0992	0.0077	0.0092	0.0074
MM-3	0.0078	0.0412	0.0552	0.001	0.0751	0.0209
MM-4	0.0163	0.0154	0.0132	0.0023	0.0017	0.0016
MM-5	0.0554	0.0412	0.0256	0.0052	0.0017	0.0016

TABLE 17.4 Overall Preference Values and Preference Selection Indexes

Milling Machine	Price	Weight	Power	Spindle	Diameter	Stroke	PSI	Rank
MM-1	0.0939	0.1168	0.0985	0.1774	0.1699	0.1624	0.8189	4
MM-2	0.0695	0.0935	0.1515	0.1553	0.1699	0.1821	0.8216	3
MM-3	0.1293	0.1602	0.0682	0.1774	0.107	0.14	0.7821	5
MM-4	0.1352	0.1078	0.1212	0.1626	0.1605	0.1737	0.8611	2
MM-5	0.1515	0.1602	0.0795	0.1848	0.1605	0.1737	0.9103	1

KEYWORDS

- **fused deposition modelling**
- **heat affected zone**
- **manufacturing organization**
- **normalized decision matrix**
- **preference selection index**
- **review of the literature**

REFERENCES

1. Maniya, K., & Bhatt, M. G., (2010). A selection of material using a novel type decision-making method: Preference selection index method. *Materials and Design, 31*, 1785–1789.
2. Maniya, K. D., & Bhatt, M. G., (2011). The selection of flexible manufacturing system using preference selection index method. *International Journal of Industrial and Systems Engineering, 9*(3), 330–349.
3. Sawant, V. B., Mohite, S. S., & Patil, R., (2011). A decision-making methodology for automated guided vehicle selection problem using a preference selection index

method. In: Shah, K., Lakshmi, G. V. R., & Phirke, A., (eds.), *Technology Systems and Management. Communications in Computer and Information Science* (Vol. 145, pp. 176–181). Springer, Heidelberg.
4. Akyuza, G., & Akab, S., (2015). An alternative approach for manufacturing performance measurement: Preference selection index (PSI) method. *Business and Economics Research Journal, 6*(1), 63–77.
5. Attri, R., & Grover, S., (2015). Application of preference selection index method for decision making over the design stage of production system life cycle. *Journal of King Saud University - Engineering Sciences, 27*, 207–216.
6. Diyaley, S., Shilal, P., Shivakoti, I., Ghadai, R. K., & Kalita, K., (2017). PSI and TOPSIS based selection of process parameters in WEDM. *Periodica Polytechnica Mechanical Engineering, 61(4)*, 255–260.
7. Jian, S., & Ying, S., (2017). Preference selection index method for machine selection in a flexible manufacturing cell. *MATEC Web of Conferences, 139*, 00167.
8. Madić, M., Antucheviciene, J., Radovanović, M., & Petković, D., (2017). Determination of laser cutting process conditions using the preference selection index method. *Optics & Laser Technology, 89*, 214–220.
9. Petković, D., Madić, M., Radovanović, M., & Gečevska, V., (2017). Application of the performance selection index method for solving machining MCDM problems. *FACTA Universitatis Series: Mechanical Engineering, 15*(1), 97–106.
10. Jain, V., (2018). Application of combined MADM methods as MOORA and PSI for ranking of FMS performance factors. *Benchmarking: An International Journal, 25*(6), 1903–1920.
11. Patel, P. B., Patel, J. D., & Maniya, K. D., (2018). Application of PSI methods to select FDM process parameter for polylactic acid. *Materials Today: Proceedings, 5*, 4022–4028.
12. Sahir, S. H., Afriani, J., Ginting, G., Fachri, B., Siregar, D., Simbolon, R., Lindawati, L., et al., (2018). The preference selection index method in determining the location of used laptop marketing. *International Journal of Engineering & Technology, 7*(3, 4), 260–263.
13. Vara, P. R., Maheswara, R. C., & Naga, R. B., (2018). Application of preference selection index (PSI) method for the optimization of turning process parameters. *International Journal of Modern Trends in Engineering and Research, 5*(5), 140–144.
14. Yadav, S., Pathak, V. K., & Gangwar, S., (2019). A novel hybrid TOPSIS-PSI approach for material selection in marine applications. *Sadhana, 44*, 58.
15. Prabhu, S. R., Ilangkumaran, M., & Mohanraj, T., (2020). 3D printing of automobile spoilers using MCDM techniques. *Materials Testing, 62*(11), 1121–1125.
16. Dağdeviren, M., (2008). Decision making in equipment selection: An integrated approach with AHP and PROMETHEE. *Journal of Intelligent Manufacturing, 19*, 397–406.

CHAPTER 18

TOMADA DE DECISAO INTERATIVA MULTICRITERIO (TODIM) METHOD

The TODIM is a discrete MCDM method based on the concept of prospect theory [1], which was awarded the Nobel Prize for Economics in 2002. It has been observed that most of the existing MCDM methods basically work on the framework that a solution always corresponds to the maximum of any global measure, e.g., the maximum value possible for a utility function in the field of multi-attribute utility theory (MAUT). Oppositely, the foundation of the TODIM method is based on the principle that a global measurement value is calculated by employing the paradigm of probability expectation theory. It is thus based on a description, empirically proved, of how efficiently people are able to formulate decisions taking into consideration the risk present. The framework of value function involved in this method is similar to that of the gain/loss function of prospect theory. The implementation of the TODIM method is concerned with defining a global multi-attribute value function. This function is developed partially through defining the mathematical descriptions representing the gain/loss function of prospect theory. Finally, an aggregated multi-attribute value for all the alternatives is measured considering gains and losses over all the considered criteria.

The multi-attribute value (additive difference) function involved in the TODIM method mainly deals with the projection of differences observed between a pair of alternatives with respect to a criterion. This method employs paired comparisons among the evaluation criteria and at the same time, eliminates infrequent discrepancies that arise from such comparisons. Thus, the main idea of the TODIM method lies in measuring the dominance

Multi-Criteria Decision-Making Methods in Manufacturing Environments: Models and Applications.
Shankar Chakraborty, Prasenjit Chatterjee, & Partha Protim Das (Authors)

degree of each alternative over the remaining ones using the prospect value function. Simply, it calculates the partial and overall dominance degrees of each alternative over the other alternatives and finally, generates a global ranking for the list of alternatives. Further, it can also take into account different verbal scale value judgments by employing fuzzy scale value judgments along with the use of interdependence relationships between the alternatives. It is also a non-compensatory method where tradeoffs do not occur [2].

Let there is a decision-making problem consisting of a set of m alternatives to be evaluated and subsequently ranked based on n quantitative/qualitative criteria, and among those criteria, there is one criterion to be treated as the reference one. For each of the criteria c, the influence of ith alternative to the objective corresponding to the criterion is subsequently estimated. The TODIM method is quite suitable for dealing with both quantitative and qualitative criteria. The verbal scales defining qualitative criteria can be converted to cardinal data, and both types of the data are subsequently normalized. The normalized decision matrix thus contains values between 0 and 1. The measure of supremacy of one alternative with respect to another is determined for each pair of alternatives. It is calculated as the sum over all the criteria considering both relative gain/loss values for the alternatives. The components in this sum may be either losses, gains, or zeros, and depend on the performance characteristics of each alternative for every criterion. After determination of the criteria weights and their normalization, the partial and final matrixes of dominance are developed. In this stage, a criterion r is to be identified as the reference criterion based on which the calculations are performed in accordance to the relative priority assigned for each criterion. Typically, a criterion with the highest weight (importance) is considered as the reference criterion. Hence, w_{cr} can be calculated as the ratio of the weight of criterion c with respect to the weight of the reference criterion r. Use of w_{cr} enables all the paired differences between the performance measurements to be directed towards the reference criterion. Based on the prospect theory, the dominance of alternative A_i over alternative A_j is now measured. Finally, the global measures obtained help in assigning complete ranking to all the alternatives.

The procedural steps of TODIM method are presented as below [3, 4]:

- **Step 1:** In the first step of the TODIM method, the corresponding decision matrix X needs to be developed. In this matrix, there are m and n numbers of alternatives and evaluation criteria, respectively.

$$X = [x_{ic}]_{m\times n} = \begin{bmatrix} x_{11} & x_{12} & \cdots & x_{1n} \\ x_{21} & x_{22} & \cdots & x_{2n} \\ \cdots & \cdots & \cdots & \cdots \\ x_{m1} & x_{m2} & \cdots & x_{mn} \end{bmatrix} \quad (i=1,2,...,m; c=1,2,...,n) \tag{18.1}$$

where; x_{ij} denotes the performance of ith alternative with respect to j^{th} criterion.

- **Step 2:** In order to make this decision matrix dimensionless and all of its elements comparable, it is subsequently normalized. For beneficial criterion, Eqn. (18.2) can be adopted. On the other hand, for non-beneficial criterion, Eqn. (18.3) can be utilized.

$$P_{ic} = \frac{x_{ic}}{\sum_{i=1}^{m} x_{ic}} \tag{18.2}$$

$$P_{ic} = \frac{1/x_{ic}}{\sum_{i=1}^{m} 1/x_{ic}} \tag{18.3}$$

where; P_{ic} is the normalized value of x_{ic}.

- **Step 3:** Using AHP or entropy method, the priority weights (w_c) of all the considered criteria are determined. The relative weight (w_{cr}) of criterion C_c ($c = 1, 2, \ldots, n$) with respect to the reference criterion C_r is computed using Eqn. (18.4).

$$w_{cr} = w_c/w_r \tag{18.4}$$

where; w_r is the weight of the reference criterion. Usually, the reference criterion is selected as that criterion having the maximum importance. The use of w_{cr} allows all pairs of differences between the performance measurements to be converted into identical dimension with respect to the reference criterion.

- **Step 4:** Next, the dominance degree of alternative A_i over alternative A_j is calculated applying the following equation:

$$\delta(A_i, A_j) = \sum_{c=1}^{n} \phi_c(A_i, A_j) \qquad \forall(i, j) \tag{18.5}$$

In the above equation, the dominance degree of alternative A_i over alternative A_j, i.e., $\phi_c(A_i,A_j)$, with respect to criterion C_c ($c = 1, 2, \ldots, n$), is evaluated using Eqn. (18.6).

$$\phi_c(A_i, A_j) = \begin{cases} \sqrt{\dfrac{w_{cr}(P_{ic} - P_{jc})}{\sum_{c=1}^{n} w_{cr}}} & \text{if } (P_{ic} - P_{jc})>0 \\ 0 & \text{if } (P_{ic} - P_{jc})=0 \\ \dfrac{-1}{\theta}\sqrt{\dfrac{(\sum_{c=1}^{n} w_{cr})(P_{ic} - P_{jc})}{w_{cr}}} & \text{if } (P_{ic} - P_{jc})<0 \end{cases} \tag{18.6}$$

where; P_{ic} and P_{jc}, respectively represent the normalized performances of the alternatives A_i and A_j with respect to c, and θ denotes the attenuation factor of losses. The expression $\phi_c(A_i,A_j)$ symbolizes the contribution of criterion c to function $\delta(A_i,A_j)$, while comparing i^{th} alternative with j^{th} alternative. When $(P_{ic} - P_{jc}) > 0$, it represents a gain for function $\delta(A_i,A_j)$; when $(P_{ic} - P_{jc}) = 0$, a zero value is assigned to $\phi_c(A_i,A_j)$; and when $(P_{ic} - P_{jc}) < 0$, it denotes a loss of ith alternative over j^{th} alternative. The expression for $\phi_c(A_i,A_j)$ actually allows an alteration of data to the value function of prospect theory; hence, clarifying aversion and propensity to risk. This developed function has the shape of an '*S*.' Above the horizontal axis, there is a concave curve signifying gains, and underneath the horizontal axis, a convex curve symbolizes losses. Different values of θ may result in different shapes of the prospect theoretical value function in the negative quadrant.

➢ **Step 5:** The overall dominance degree of alternative A_i (ζ_i) is determined applying the following expression:

$$\zeta_i = \frac{\sum \delta(A_i, A_j) - \min\sum \delta(A_i, A_j)}{\max\sum \delta(A_i, A_j) - \min\sum \delta(A_i, A_j)} \tag{18.7}$$

➢ **Step 6:** The alternatives are finally ranked based on the descending values of their dominance scores and the alternative having the maximum dominance score becomes obviously the best choice.

18.1 LITERATURE SURVEY

Based on an illustrative example, Rangel et al. [5] contrasted the relative ranking performance of ELECTRE IV and TODIM methods as the two effective tools of decision-making. Moshkovich et al. [6] adopted the TODIM method for assessing real estate properties in a certain city.

Rangel et al. [7] demonstrated the application of the TODIM method for evaluation of different types of access available to the Broadband Internet facility for a small organization. Kazancoglu & Burmaoglu [8] solved an ERP software selection problem of a steel forming and hot dip-galvanizing firm. With the help of two illustrative examples, Lourenzutti & Krohling [9] explored the application potentiality of the TODIM method in processing heterogeneous data in a more systematic way while partitioning it into sample modules to be subsequently dealt straightforwardly. The effectiveness and practicability of the adopted approach was later validated. Adalı [10] proposed the application of the TODIM method in solving a nurse selection problem of a private hospital and compared the derived ranking results with those obtained from EVAMIX approach. Sen et al. [11] considered two real-time industrial robot selection problems and solved them using the TODIM method while identifying the best performing robots for the given applications. Zindani et al. [12] explored the applicability of the TODIM method in effectively identifying the most apposite materials of an engine flywheel and a metallic gear. A sensitivity analysis was further performed to check the consistency of the rankings of the candidate materials based on the attenuation factor of the losses. Later, finite element analysis was also performed to understand the performance of the considered material alternatives under specific loading conditions. Using 10 qualitative as well as quantitative evaluation criteria, Chakraborty & Chakraborty [13] ranked 14 under-construction housing projects in the city of Kolkata in India while applying the TODIM method. Baloyi & Meyer [14] applied TODIM method to solve a mining method selection problem, and contrasted its ranking performance with other popular MCDM techniques, like TOPSIS, VIKOR, GRA, PROMETHEE, OCRA, ARAS, COPRAS, SAW, and CP with the help of sensitivity analysis, Spearman's rank correlation and Kendall's coefficient. Dehaghi & Khoshfetrat [15] proposed a new variant of the TODIM method and presented its application in ranking water reuse alternatives. The stability of the derived ranking results and sensitivity of the proposed approach were also validated. Sahaya et al. [16] adopted the TODIM method as an efficient MCDM tool in selecting the best system in housing and feeding of goats at a mixed farming agrofarm in India. The derived results were later compared with those obtained using TOPSIS. Zhou et al. [17] considered a location selection problem for photovoltaic charging stations in Beijing. With the help of GIS, seven alternative locations were first identified based on traffic flow and road distribution, and TODIM method was later employed to rank those alternatives.

18.2 AUTOMATED GUIDED VEHICLE (AGV) SELECTION PROBLEM

In an attempt to solve an AGV selection problem, Maniya & Bhatt [18] applied modified GRA technique while considering controllability (CL), accuracy (A), range (RA), reliability (RL) and flexibility (F) as the beneficial criteria, and cost (CO) as the non-beneficial criterion. The corresponding criteria weights were estimated using AHP method as $w_{CL} = 0.346$, $w_A = 0.168$, $w_{CO} = 0.058$, $w_{RA} = 0.073$, $w_{RL} = 0.063$ and $w_F = 0.293$. The related decision matrix for the AGV selection problem thus consists of eight alternatives and six evaluation criteria, as shown in Table 18.1. In this matrix, the performances of all the candidate AGVs are expressed in linguistic terms, which are subsequently converted into quantitative values based on the five-point conversion scale as proposed by Rao [19]. Now, using Eqns. (18.2) and (18.3), and taking the type of criterion under consideration, the original decision matrix of Table 18.1 is normalized, as exhibited in Table 18.2.

As mentioned earlier, Maniya & Bhatt [18] adopted the AHP method to determine weights of the considered criteria. In this problem, 'Controllability' having the maximum weight is treated as the reference criterion and the relative weights of other criteria are estimated, as depicted in Table 18.3. In Table 18.4, based on Eqns. (18.5) and (18.6), the dominance degrees of alternative A_1 over the other alternatives with respect to each considered criterion are evaluated. Finally, the overall dominance degree of alternative A_i (ζ_i) is estimated using Eqn. (18.7), as shown in Table 18.5. Based on the descending values of the calculated dominance scores, the candidate AGV alternatives are ranked and the alternative with the maximum dominance score becomes the best choice. From Table 18.5, the rankings of the candidate AGVs can be derived as $A_4 \rightarrow A_5 \rightarrow A_1 \rightarrow A_2 \rightarrow A_8 \rightarrow A_6 \rightarrow A_7 \rightarrow A_3$. Thus, AGV A_4 is the most preferred alternative, whereas, AGV A_3 is the worst choice. While applying a modified GRA technique, Maniya & Bhatt [18] determined the rankings of the AGV alternatives as $A_5 \rightarrow A_6 \rightarrow A_4 \rightarrow A_1 \rightarrow A_8 \rightarrow A_2 \rightarrow A_7 \rightarrow A_3$. On the other hand, Mathew & Sahu [20] adopted five different MCDM methods in the form of CODAS, EDAS, WASPAS, MOORA, and TOPSIS methods to solve the same AGV selection problem. It is quite interesting to notice that the application of all those MCDM techniques provides almost similar ranking lists of the AGV alternatives as compared to the TODIM method. It thus proves the efficacy of the TODIM method in solving complex MCDM problems with varying alternatives and conflicting criteria.

TABLE 18.1 Decision Matrix for the AGV Selection Problem [18]

Criteria / Alternative	CL	A	CO	RA	RL	F
A_1	High (0.895)	Average (0.495)	Above average (0.695)	Average (0.495)	High (0.895)	Below average (0.295)
A_2	Low (0.115)	High (0.895)	High (0.895)	High (0.895)	Average (0.495)	Average (0.495)
A_3	Low (0.115)	Low (0.115)	High (0.895)	Low (0.115)	Above average (0.695)	High (0.895)
A_4	Below average (0.295)	High (0.895)	Low (0.115)	Average (0.495)	Average (0.495)	High (0.895)
A_5	High (0.895)	Average (0.495)	Low (0.115)	Above average (0.695)	Below average (0.295)	Average (0.495)
A_6	Average (0.495)	Average (0.495)	High (0.895)	Low (0.115)	Above average (0.695)	Above average (0.695)
A_7	Low (0.115)	Below average (0.295)	High (0.895)	Low (0.115)	High (0.895)	High (0.895)
A_8	Low (0.115)	Average (0.495)	Above average (0.695)	Average (0.495)	Average (0.495)	Above average (0.695)

Source: Reprinted with permission from Ref. [18]. © 2021 Taylor and Francis.

TABLE 18.2 Normalized Decision Matrix for the AGV Selection Problem

Alternative	CL	A	CO	RA	RL	F
A_1	0.2944	0.1184	0.0582	0.1447	0.1804	0.055
A_2	0.0378	0.2141	0.0452	0.2617	0.0998	0.0924
A_3	0.0378	0.0275	0.0452	0.0336	0.1401	0.167
A_4	0.097	0.2141	0.3515	0.1447	0.0998	0.167
A_5	0.2944	0.1184	0.3515	0.2032	0.0595	0.0924
A_6	0.1628	0.1184	0.0452	0.0336	0.1401	0.1297
A_7	0.0378	0.0706	0.0452	0.0336	0.1804	0.167
A_8	0.0378	0.1184	0.0582	0.1447	0.0998	0.1297

TABLE 18.3 Relative Weights of the Considered Criteria

Criteria	CL	A	CO	RA	RL	F	Total
w_c	0.346	0.168	0.058	0.073	0.063	0.293	1.001
Relative weight (w_{cr})	1	0.4855	0.1688	0.2110	0.1821	0.8468	2.8942

TABLE 18.4 Dominance Degrees of Alternative A_1 Over Others with Respect to Each Criterion

Alternative	CL	A	CO	RA	RL	F	$\delta(A_i,A_j)$
A_2	0.2977	–0.7553	0.0275	–1.2667	0.0712	–0.3571	–1.9825
A_3	0.2977	0.1235	0.0275	0.09	0.0504	–0.6185	–0.0294
A_4	0.2611	–0.7553	–2.2428	0	0.0712	–0.6185	–3.2842
A_5	0	0	–2.2428	–0.8957	0.0872	–0.3571	–3.4083
A_6	0.2132	0	0.0275	0.09	0.0504	–0.505	–0.1239
A_7	0.2977	0.0896	0.0275	0.09	0	–0.6185	–0.1137
A_8	0.2977	0	0	0	0.0712	–0.505	–0.1361

TABLE 18.5 Overall Dominance Degrees of the AGV Alternatives

Alternative	$\sum_{c=1}^{n} \delta(A_i, A_j)$	ξ_i	Rank
A_1	–9.078	0.7425	3
A_2	–12.6389	0.5299	4
A_3	–21.5186	0	8
A_4	–4.7627	1	1
A_5	–8.7663	0.7611	2
A_6	–16.1688	0.3193	6
A_7	–18.0788	0.2053	7
A_8	–14.8339	0.3989	5

KEYWORDS

- **dominance degrees**
- **flexibility**
- **framework**
- **multi-attribute utility theory**
- **reliability**
- **Tomada de Decisao Interativa Multicriterio**

REFERENCES

1. Gomes, L. F. A. M., & Lima, M. M. P. P., (1992). TODIM: Basics and application to multicriteria ranking of projects with environmental impacts. *Foundations of Computing and Decision Sciences, 16*(4), 113–127.
2. Gomes, L. F. A. M., & Rangel, L. A. D., (2009). An application of the TODIM method to the multicriteria rental evaluation of residential properties. *European Journal of Operational Research, 193*(1), 204–211.
3. Gomes, L. F. A. M., Rangel, L. A. D., & Maranhão, F. J. C., (2009). Multicriteria analysis of natural gas destination in Brazil: An application of the TODIM method. *Mathematical and Computer Modeling, 50*(1, 2), 92–100.
4. Uysal, F., & Tosun, Ö., (2014). Multicriteria analysis of the residential properties in Antalya using the TODIM method. *Procedia-Social and Behavioral Sciences, 109*, 322–326.
5. Rangel, L. A. D., Gomes, L. F. A. M., & Moreira, R. A., (2009). Decision theory with multiple criteria: An application of ELECTRE IV and TODIM to SEBRAE/RJ. *Operational Research, 29*(3), 577–590.
6. Moshkovich, H. M., Gomes, L. F. A. M., & Mechitov, A. I., (2011). An integrated multicriteria decision-making approach to real estate evaluation: Case of the TODIM method. *Operational Research, 31*(1), 3–20.
7. Rangel, L. A. D., Gomes, L. F. A. M., & Cardoso, F. P., (2011). An application of the TODIM method to the evaluation of broadband internet plans. *Operational Research, 31*(2), 235–249.
8. Kazancoglu, Y., & Burmaoglu, S., (2013). ERP software selection with MCDM: Application of TODIM method. *International Journal of Business Information Systems, 13*(4), 435–452.
9. Lourenzutti, R., & Krohling, R. A., (2015). TODIM based method to process heterogeneous information. *Procedia Computer Science, 55*, 318–327.
10. Adalı, E. A., (2016). Personnel selection in health sector with EVAMIX and TODIM methods. *Journal of Operations Research, Statistics, Econometrics and Management Information Systems, 4*(2), 70–83.

11. Sen, D. K., Datta, S., & Mahapatra, S. S., (2016). Application of TODIM (TOmada de decisión ineratíva multicritero) for industrial robot selection. *Benchmarking: An International Journal, 23*(7), 1818–1833.
12. Zindani, D., Maity, S. R., Bhowmik, S., & Chakraborty, S., (2017). A material selection approach using the TODIM (TOmada de decisao interativa multicriterio) method and its analysis. *International Journal of Materials Research, 108*, 345–354.
13. Chakraborty, S., & Chakraborty, A., (2018). Application of TODIM (TOmada de decisao interativa multicriterio) method for under construction housing project selection in Kolkata. *Journal of Project Management, 3,* 207–216.
14. Baloyi, V. D., & Meyer, L. D., (2020). The development of a mining method selection model through a detailed assessment of multi-criteria decision methods. *Results in Engineering, 8*, 100172.
15. Dehaghi, B. F., & Khoshfetrat, A., (2020). Modified TODIM method for water reuse application: A case study from Iran. *Desalination and Water Treatment, 195*, 186–200.
16. Sahaya, S. A., Gomes, L. F. A. M., & Vijayalakshmi, K. R., (2020). Assessment of MCDM problems by TODIM using aggregated weights. *Neutrosophic Sets and Systems, 35*(1), 78–98.
17. Zhou, J., Wu, Y., Wu, C., He, F., Zhang, B., & Liu, F., (2020). A geographical information system based multi-criteria decision-making approach for location analysis and evaluation of urban photovoltaic charging station: A case study in Beijing. *Energy Conversion and Management, 205*, 112340.
18. Maniya, K. D., & Bhatt, M. G., (2011). A multi-attribute selection of automated guided vehicle using the AHP/M-GRA technique. *International Journal of Production Research, 49*(20), 6107–6124.
19. Rao, R. V., (2007). *Decision Making in the Manufacturing Environment: Using Graph Theory and Fuzzy Multiple Attribute Decision Making Methods* (pp. 187–195). Springer.
20. Mathew, M., & Sahu, S., (2018). Comparison of new multi-criteria decision making methods for material handling equipment selection. *Management Science Letters, 8*, 139–150.

CHAPTER 19

RANGE OF VALUE (ROV) METHOD

The ROV method, developed by Yakowitz et al. [1], is a simple MCDM approach with uncomplicated mathematical formulations, and is applied for ranking of the competing alternatives from the best to the worst based on a set of conflicting criteria. Its application involves the following procedural steps [2, 3]:

- **Step 1: Development of the Initial Decision Matrix:**

$$X = [x_{ij}]_{m\times n} = \begin{bmatrix} x_{11} & x_{12} & \dots & x_{1n} \\ x_{21} & x_{22} & \dots & x_{2n} \\ \dots & \dots & \dots & \dots \\ x_{m1} & x_{m2} & \dots & x_{mn} \end{bmatrix} \quad (19.1)$$

  where; x_{ij} is the performance of ith alternative ($i = 1, 2, \dots, m$) for j^{th} criterion ($j = 1, 2, \dots, n$).

- **Step 2: Normalize the Decision Matrix:** Depending on the type of the criterion under consideration, the following equations can be adopted to convert all the entities of the decision matrix into dimensionless elements.

  i. **For Beneficial Criterion:**

$$r_{ij} = [x_{ij} - \min(x_{ij})]/[\max(x_{ij}) - \min(x_{ij})] \quad (\mathrm{i} = 1, 2, \dots, m; \mathrm{j} = 1, 2, \dots, n) \quad (19.2)$$

  ii. **For Non-Beneficial (Cost) Criterion:**

$$r_{ij} = [\max(x_{ij}) - x_{ij}]/[\max(x_{ij}) - \min(x_{ij})] \quad (19.3)$$

- **Step 3: Determine the Best and the Worst Utility Functions:** In this step, the best and the worst utility functions of each alternative are separately computed taking into consideration both the beneficial and cost criteria. Utility functions for beneficial criteria (u_i^+) and cost criteria (u_i^-) are calculated based on the following equations:

Multi-Criteria Decision-Making Methods in Manufacturing Environments: Models and Applications.
Shankar Chakraborty, Prasenjit Chatterjee, & Partha Protim Das (Authors)

$$\text{Maximize: } u_i^+ = \sum_{j=1}^{n} w_j \times r_{ij} \quad (19.4)$$

$$\text{Minimize: } u_i^- = \sum_{j=1}^{n} w_j \times r_{ij} \quad (19.5)$$

where; w_j is the weight of jth criterion.

To rank the alternatives, the average utility function (u_i) is estimated. The best alternative should have the maximum value of u_i.

$$u_i = \frac{u_i^+ + u_i^-}{2} \quad (19.6)$$

19.1 LITERATURE REVIEW

Madić et al. [2] applied ROV method to solve three real-time cutting fluid selection problems. The efficacy of the adopted method was also proved while comparing the derived results against those obtained by the past researchers. While employing ROV method, Madić & Radovanović [3] attempted to identify the most suitable NTM process for generating a specific shape feature on a given material. The relative importance of the considered evaluation criteria was determined using CRITIC method. It was claimed that ROV method could offer an objective and systematic approach for determination of the complete ranking of competitive NTM processes. In a real-time manufacturing environment, Jha et al. [4] solved two supplier selection problems using ROV method.

19.2 SAILING BOAT MAST MATERIAL SELECTION PROBLEM

In order to demonstrate the application of ROV method in MCDM, the problem of selection of the most apposite material for a sailing boat mast in the form of a hollow cylinder of length 1,000 mm [5] is considered here. Based on the application requirement, it is always desired to choose a material having high yield strength, elastic modulus, fracture toughness (FT) and corrosion resistance, and low specific density properties. The initial decision matrix consisting of 15 candidate materials evaluated based on specific strength (in MPa), specific modulus (in GPa), corrosion resistance and cost is exhibited in Table 19.1. In this matrix, corrosion resistance and cost are measured in relative scales. For corrosion resistance, a 1–5 scale (excellent = 5, very good = 4, good = 3, fair = 2 and poor = 1) is taken into account. Similarly, another 1–5 scale (very inexpensive = 5, inexpensive = 4, moderate cost

= 3, expensive = 2 and very expensive = 1) is adopted for cost criterion. For this material selection problem for sailing boat mast, the weights for specific strength, specific modulus, corrosion resistance and cost are determined as 0.333, 0.333, 0.125, and 0.208, respectively using the entropy method. Now, depending on the type of the criterion (beneficial or cost), the decision matrix of Table 19.1 is first normalized while employing Eqns. (19.2) and (19.3). This normalized decision matrix is provided in Table 19.2. The elements of this normalized decision matrix are now multiplied by the respective criteria weights to formulate the corresponding weighted normalized decision matrix, as shown in Table 19.3. From this weighted normalized decision matrix, the best and the worst utility functions of each candidate material are separately computed taking into account both the beneficial and cost criteria, as exhibited in Table 19.4. In the final step, the average utility function (u_i) for each material alternative is calculated. It can be observed from Table 19.4 that among the considered materials, material M14 emerges out as the most suitable sailing boat mast material, followed by material M15. On the other hand, M1 is the worst chosen material. Employing a simplified fuzzy logic approach, Khabbaz et al. [5] also identified M14 and M1 as the best and the worst materials respectively for the considered problem. It proves the effectiveness of ROV method in deriving almost accurate ranking results while solving this sailing boat mast material selection problem.

TABLE 19.1 Properties of Candidate Materials for a Sailing Boat Mast [5]

Material	Specific Strength	Specific Modulus	Corrosion Resistance	Cost
M1	35.9	26.9	1	5
M2	51.3	26.9	1	5
M3	42.3	27.2	1	5
M4	194.9	27.2	4	3
M5	25.6	25.1	4	3
M6	57.1	28.1	4	3
M7	71.4	28.1	4	3
M8	101.9	25.8	3	4
M9	141.9	26.1	3	4
M10	148.2	25.8	3	4
M11	180.4	25.9	3	4
M12	208.7	27.6	5	1
M13	604.8	28	4	2
M14	416.2	66.5	4	1
M15	637.7	27.5	4	1

Source: Reprinted with permission from Ref. [5].

TABLE 19.2 Normalized Matrix for the Sailing Boat Mast Material Selection Problem

Material	Specific Strength	Specific Modulus	Corrosion Resistance	Cost
M1	0.0168	0.0435	0	0
M2	0.0420	0.0435	0	0
M3	0.0273	0.0507	0	0
M4	0.2766	0.0507	0.7500	0.5
M5	0	0	0.7500	0.5
M6	0.0515	0.0725	0.7500	0.5
M7	0.0748	0.0725	0.7500	0.5
M8	0.1247	0.0169	0.5000	0.25
M9	0.1900	0.0242	0.5000	0.25
M10	0.2003	0.0169	0.5000	0.25
M11	0.2529	0.0193	0.5000	0.25
M12	0.2991	0.0604	1.0000	1
M13	0.9463	0.0700	0.7500	0.75
M14	0.6381	1.0000	0.7500	1
M15	1.0000	0.0580	0.7500	1

TABLE 19.3 Weighted Normalized Matrix for Sailing Boat Mast Material Selection Problem

Material	Specific Strength	Specific Modulus	Corrosion Resistance	Cost
M1	0.0056	0.0145	0	0
M2	0.0140	0.0145	0	0
M3	0.0091	0.0169	0	0
M4	0.0921	0.0169	0.0938	0.104
M5	0	0	0.0938	0.104
M6	0.0171	0.0241	0.0938	0.104
M7	0.0249	0.0241	0.0938	0.104
M8	0.0415	0.0056	0.0625	0.052
M9	0.0633	0.0080	0.0625	0.052
M10	0.0667	0.0056	0.0625	0.052
M11	0.0842	0.0064	0.0625	0.052
M12	0.0996	0.0201	0.1250	0.208
M13	0.3151	0.0233	0.0938	0.156
M14	0.2125	0.3330	0.0938	0.208
M15	0.3330	0.0193	0.0938	0.208

TABLE 19.4 Ranks of Sailing Boat Mast Materials Using ROV Method

Material	u_i^+	u_i^-	u_i	Rank
M1	0.0201	0	0.0100	15
M2	0.0285	0	0.0142	13
M3	0.0260	0	0.0130	14
M4	0.2027	0.104	0.1534	5
M5	0.0938	0.104	0.0989	9
M6	0.1350	0.104	0.1195	7
M7	0.1428	0.104	0.1234	6
M8	0.1096	0.052	0.0808	12
M9	0.1338	0.052	0.0929	11
M10	0.1348	0.052	0.0934	10
M11	0.1532	0.052	0.1026	8
M12	0.2447	0.208	0.2264	4
M13	0.4322	0.156	0.2941	3
M14	0.6392	0.208	0.4236	1
M15	0.4461	0.208	0.3270	2

KEYWORDS

- **conflicting criteria**
- **corrosion resistance**
- **elastic modulus**
- **literature review**
- **normalized matrix**
- **range of value**

REFERENCES

1. Yakowitz, D. S., Lane, L. J., & Szidarovszky, F., (1993). Multi-attribute decision making: Dominance with respect to an importance order of the attributes. *Applied Mathematics and Computation, 54*(2, 3), 167–181.
2. Madić, M., Radovanović, M., & Manić, M., (2016). Application of the ROV method for the selection of cutting fluids. *Decision Science Letters, 5*(2), 245–254.

3. Madić, M., & Radovanović, M., (2015). Ranking of some most commonly used non-traditional machining processes using ROV and CRITIC methods. *UPB Scientific Bulletin Series D, 77*(2), 193–204.
4. Jha, G. K., Chatterjee, P., Chatterjee, R., & Chakraborty, S., (2013). Suppliers selection in manufacturing environment using range of value method. *I-Manager's Journal on Mechanical Engineering, 3*(3), 15–22.
5. Khabbaz, R. S., Dehghan-Manshadi, B., Abedian, A., & Mahmudi, R., (2009). A simplified fuzzy logic approach for materials selection in mechanical engineering design. *Materials and Design, 30*, 687–697.

CHAPTER 20

SUPERIORITY AND INFERIORITY RANKING (SIR) METHOD

The SIR method [1] is a generalization of the illustration of superiority and inferiority scores introduced by Rebai [2, 3]. Superiority and inferiority scores are estimated while measuring the difference in the criteria (response) values to calculate the degree of preference for alternative x over alternative x'. For instance, the scores for ordinal values can be estimated for two alternatives x and x' related to a criterion c (for maximization problem) while employing the following preference framework $\{P, I\}$:

$$xPx' \ (x \textit{ is preferred to } x') \quad \textit{iff} \quad c(x) > c(x')$$
$$xIx' \ (x \textit{ is indifferent to } x') \quad \textit{iff} \quad c(x) = c(x')$$

where; $c(x)$ and $c(x')$ represent values of the alternatives x and x', respectively with respect to criterion c. Assessment of cardinal data is often more problematic than ordinal data, as it may be affected by inaccuracy, feeble, and vagueness in the estimation of criteria values [1]. A number of techniques have already been presented to overcome such complications, either by defining constraints [4] or by taking into account dissimilarities between criteria values using generalized preference structures [5]. The SIR method takes into account dissimilarities between criteria values with the assistance of different generalized criteria. In order to classify the characteristics of functions representing the stated criteria, these generalized criteria are utilized. The step-by-step procedure for the application of SIR method, when all the criteria need to be maximized, is demonstrated here-in-under.

- **Step 1: Evaluate the Intensity of Preference:** Consider $x_1, x_2, \ldots, x_m$ as a set of m alternatives, $c_1, c_2, \ldots, c_n$ as a set of n cardinal criteria, and $c_j(x_i)$ is the observed value for i^{th} alternative with respect to j^{th} criterion. Accordingly, the subsequent evaluation matrix can be obtained.

Multi-Criteria Decision-Making Methods in Manufacturing Environments: Models and Applications.
Shankar Chakraborty, Prasenjit Chatterjee, & Partha Protim Das (Authors)

$$C = \begin{bmatrix} c_1(x_1) & \cdots & c_j(x_1) & \cdots & c_n(x_1) \\ \cdots & \cdots & \cdots & \cdots & \cdots \\ c_1(x_i) & \cdots & c_j(x_i) & \cdots & c_n(x_i) \\ \cdots & \cdots & \cdots & \cdots & \cdots \\ c_1(x_m) & \cdots & c_j(x_m) & \cdots & c_n(x_m) \end{bmatrix}$$

In this step, the intensity of preference of one criterion value over the other is estimated by comparing the values for each criterion. The preference intensity for alternative x over alternative x' can be calculated by the difference between their criteria values.

$$P(x,x') = f(d) = f(c(x) - c(x')) \tag{20.1}$$

For a maximization type of criterion, $P(x,x') = f(d)$, whereas, $P(x,x') = -f(d)$ for a minimization type of criterion.

Brans et al. [5] proposed six different generalized preference structures, as already presented in Figure 9.1. Each function defines the preference intensity by pair-wise comparison among the criteria values for two alternatives. They help in classifying the function characteristics, representing various attitudes to the preference structure along with the preference intensity. They are mainly adopted to represent small changes in the criteria values at a recognizable scale in terms of preference function values. The suitable preference structure for comparison of the criteria values is to be defined by the decision maker for a specified decision-making problem taking into account the attitude and preference intensity.

➢ **Step 2: Estimate the Corresponding Superiority and Inferiority Indexes:** For alternative x_i, the superiority index $S_j(x_i)$ and inferiority index $I_j(x_i)$ are calculated with respect to j^{th} criterion employing Eqns. (20.2) and (20.3), respectively.

$$S_j(x_i) = \sum_{k=1}^{m} P(x_i, x_k) = \sum_{k=1}^{m} f_j(c_j(x_i) - c_j(x_k)) \tag{20.2}$$

$$I_j(x_i) = \sum_{k=1}^{m} P(x_k, x_i) = \sum_{k=1}^{m} f_j(c_j(x_k) - c_j(x_i)) \tag{20.3}$$

These calculated superiority and inferiority indexes are now organized to formulate S-matrix (superiority matrix) and I-matrix (inferiority matrix), respectively. The S-matrix and I-matrix furnish meaningful representation of the original evaluation matrix C. The S-matrix denotes the intensity of superiority that each alternative has on each criterion, while the intensity of inferiority is represented in the form of *I*-matrix.

$$S = \begin{bmatrix} S_1(x_1) & S_2(x_1) & \cdots & S_n(x_1) \\ S_1(x_2) & S_2(x_2) & \cdots & S_n(x_2) \\ \cdots & \cdots & \cdots & \cdots \\ S_1(x_m) & S_2(x_m) & \cdots & S_n(x_m) \end{bmatrix}$$

$$I = \begin{bmatrix} I_1(x_1) & I_2(x_1) & \cdots & I_n(x_1) \\ I_1(x_2) & I_2(x_2) & \cdots & I_n(x_2) \\ \cdots & \cdots & \cdots & \cdots \\ I_1(x_m) & I_2(x_m) & \cdots & I_n(x_m) \end{bmatrix}$$

- **Step 3: Estimate the Related Flows:** After developing the corresponding *S*-matrix and *I*-matrix, various standard MCDM assemblage techniques can be applied to combine the superiority and inferiority indexes to produce two types of global preference index, i.e., superiority flow (*S*-flow) ($\phi^{>}(.)$) and inferiority flow (*I*-flow) ($\phi^{<}(.)$). The *S*-flow denotes the degree by which an alternative outperforms all other alternatives. On the other hand, *I*-flow indicates the degree by which an alternative is outperformed by the remaining alternatives. Simple aggregation techniques, like SAW, TOPSIS, etc., are generally applied in obtaining the relevant *S*-flows and *I*-flows. Aggregation using the SAW method is one of the simplest MCDM methods being applied as a yardstick to evaluate solutions obtained employing other MCDM techniques. On the other hand, TOPSIS method, having a strong mathematical background, is computationally more complex and involves a rational approach in solving different MCDM problems. The aggregation procedures applying SAW and TOPSIS methods for estimating *S*-flow and *I*-flow are explained as follows:

  i. **SIR-SAW Method:** It is a prominent aggregation approach, accepted for its simplicity and comprehensiveness. The superiority and inferiority flows are calculated applying Eqns. (20.4) and (20.5), respectively.

$$\phi^{>}(x_i) = \sum_{j=1}^{n} w_j S_j(x_i) \tag{20.4}$$

$$\phi^{<}(x_i) = \sum_{j=1}^{n} w_j I_j(x_i) \tag{20.5}$$

  where; the weight of jth criterion (response) is represented by w_j.

ii. **SIR-TOPSIS Method:** The TOPSIS method can also be adopted for aggregation purposes. Using the positive-ideal solution (PIS), x_S^+ and negative-ideal solution (NIS), x_S^-, the superiority flow can be calculated using Eqn. (20.6).

$$\phi^{>}(x_i) = \frac{S^-(x_i)}{(S^-(x_i) + S^+(x_i))} \tag{20.6}$$

where;

$$S^+(x_i) = \left\{ \sum_{j=1}^{n} \left| w_j (S_j(x_i) - S_j^+) \right|^{\lambda} \right\}^{1/\lambda}; \qquad 0 \le \lambda \le \infty$$

$$S^-(x_i) = \left\{ \sum_{j=1}^{n} \left| w_j (S_j(x_i) - S_j^-) \right|^{\lambda} \right\}^{1/\lambda}; \qquad 0 \le \lambda \le \infty$$

$$x_S^+ = \left(\max_i S_1(x_i), \ldots, \max_i S_n(x_i) \right) = (S_1^+, \ldots, S_n^+)$$

$$x_S^- = \left(\min_i S_1(x_i), \ldots, \min_i S_n(x_i) \right) = (S_1^-, \ldots, S_n^-)$$

The aggregation procedure for estimation of superiority and inferiority flows using TOPSIS method mainly depends on varying λ (a non-negative constant) values, i.e., $\lambda = 1$, 2, and 10 representing respectively city block distance, Euclidean distance, and a relatively large distance. Similarly, based on the positive ideal solution, x_I^+ and negative ideal solution, x_I^-, the inferiority flow can be estimated based on Eqn. (20.7).

$$\phi^{<}(x_i) = \frac{I^+(x_i)}{(I^+(x_i) + I^-(x_i))} \tag{20.7}$$

where;

$$I^+(x_i) = \left\{ \sum_{j=1}^{n} \left| w_j (I_j(x_i) - I_j^+) \right|^{\lambda} \right\}^{1/\lambda}; \qquad 0 \le \lambda \le \infty$$

$$I^-(x_i) = \left\{ \sum_{j=1}^{n} \left| w_j (I_j(x_i) - I_j^-) \right|^{\lambda} \right\}^{1/\lambda}; \qquad 0 \le \lambda \le \infty$$

$$x_I^+ = \left(\min_i I_1(x_i), \ldots, \min_i I_n(x_i) \right) = (I_1^+, \ldots, I_n^+)$$

$$x_I^- = \left(\max_i I_1(x_i), \ldots, \max_i I_n(x_i) \right) = (I_1^-, \ldots, I_n^-)$$

The corresponding net flow (n-flow) ($\phi_n(.)$) and relative flow (r-flow) ($\phi_r(.)$) are then computed using the following equations:

$$\phi_n(x_i) = \phi^>(x_i) - \phi^<(x_i) \tag{20.8}$$

$$\phi_r(x_i) = \frac{\phi^>(x_i)}{\phi^>(x_i) + \phi^<(x_i)} \tag{20.9}$$

- **Step 4: Develop the Complete and Partial Rankings**

i. Complete Ranking: Based on *S*-, *I*-, *n*- and *r*-flows, four complete rankings, i.e., *S*-ranking ($\Re_>$), *I*-ranking ($\Re_<$), *n*-ranking ($\Re_n$) and *r*-ranking ($\Re_r$) can be derived, as shown in Figure 20.1. The *S*-ranking ($\Re_> = \{P_>, I_>\}$) can be obtained with respect to the descending order of *S*-flows ($\phi^>(x_i)$), with the largest value receiving the top rank.

$$x_i P_> x_k \quad iff \quad \phi^>(x_i) > \phi^>(x_k)$$
$$x_i I_> x_k \quad iff \quad \phi^>(x_i) = \phi^>(x_k)$$

On the other hand, the *I*-ranking ($\Re_< = \{P_<, I_<\}$) is developed depending on the ascending values of *I*-flows ($\phi^<(x_i)$ with the smallest one having the first rank.

$$x_i P_< x_k \quad iff \quad \phi^<(x_i) < \phi^<(x_k)$$
$$x_i I_< x_k \quad iff \quad \phi^<(x_i) = \phi^<(x_k)$$

Similarly, the *n*-ranking and *r*-ranking can also be obtained with respect to the decreasing sequence of *n*-flows and *r*-flows, respectively.

ii. Partial Ranking: It considers both the *S*-ranking ($\Re_<$) and *I*-ranking($\Re_<$) to develop a combined partial ranking structure, as exhibited in Figure 20.2, and is denoted using the following expression:

$$\Re = \{P, I, R\} = \Re_> \cap \Re_<$$

For partial ranking, the intersection principle, introduced by Brans et al. [5]; and Roy et al. [6], can be adopted in comparing any two alternatives a and a'.

The preference relation, P can be represented as:

$$xPx' \; iff \; (xP_>x' \; and \; xP_<x') \; or \; (xP_>x' \; and \; xI_<x') \; or \; (xI_>x' \; and \; xP_<x')$$

The indifference relation, I is expressed as:

$$xIx' \; iff \; xI_>x' \; and \; xI_<x'$$

The incomparability relation, R is denoted as:

$$xRx' \; iff \; (xP_>x' \; and \; x'P_<x) \; or \; (x'P_>x \; and \; xP_<x')$$

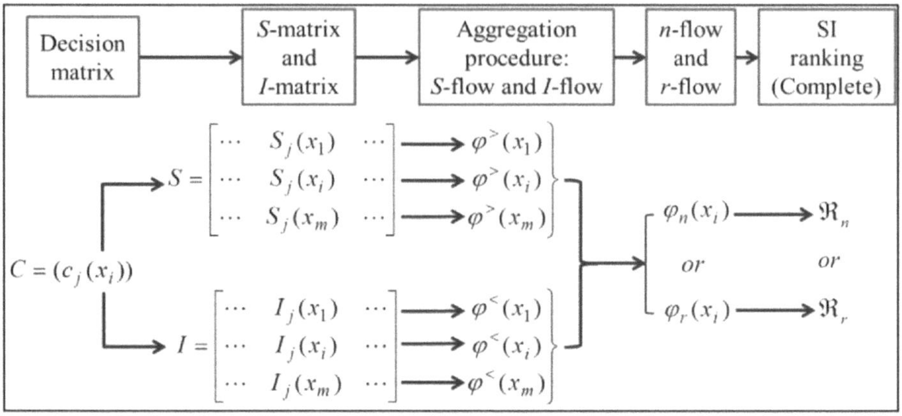

FIGURE 20.1 Complete ranking procedure for the SIR method.

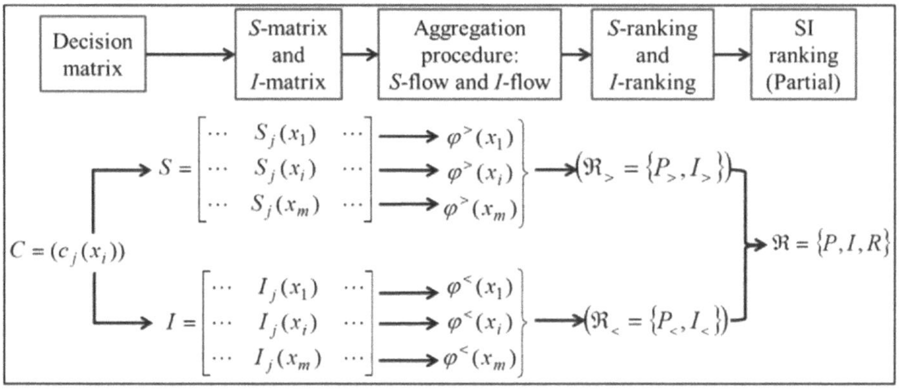

FIGURE 20.2 Partial ranking procedure for the SIR method.

20.1 SURVEY OF THE LITERATURE

Tam et al. [7] demonstrated the application of SIR method in solving a construction pump selection problem with 10 alternatives and nine emulation criteria. It was also concluded that the SIR method would be quite capable in dealing with qualitative, cardinal and/or ordinal data, and would allow imprecise information while incorporating indifference and preference thresholds for each criterion. Marzouk [8] solved a contractor selection problem using SIR method while considering SAW and TOPSIS as the two aggregation procedures to generate the corresponding superiority and inferiority flows. Memarzade et al. [9] integrated SIR and VIKOR methods to propose a new MCDM tool for solving a contractor selection problem. It was also claimed that the proposed method would be more reliable than the other MCDM tools in solving decision-making problems.

Marzouk et al. [10] first applied the AHP method to determine the corresponding criteria weights for an equipment procurement decision problem and then employed the SIR method to identify the most suitable equipment to be installed inside a facility. Vazifehdan & Darestani [11] integrated QFD and SIR methods to develop a comprehensive model to evaluate and rank the outsourcing green logistics, while singling out the main factors influencing the outsourcing selection process. While employing the SIR method, Das & Chakraborty [12] solved two parametric optimization problems for an Nd:YAG and a pulsed CO_2 (carbon dioxide) laser cutting (LC) processes. It was concluded that the derived optimal parametric intermixes would lead to enhanced cutting performance of the said processes. Considering peak current, pulse duration, dielectric level and flushing pressure as the input parameters, and process time, relative tool wear ratio, process energy, concentration of aerosol and dielectric consumption as the responses of an EDM process, Das & Chakraborty [13] first developed the corresponding causal relationship between the process outputs using DEMATEL method. The optimal parametric combination of the said process was later identified using SIR method for better machining performance.

20.2 PLANT LAYOUT DESIGN SELECTION PROBLEM

It has already been observed that the plant or facility layout design problem is one of the most complicated and widely encountered problems in manufacturing domain. In any manufacturing organization, processing of any job is executed in a plant or facility. In a plant or facility layout, various process elements, like machines, work centers, departments, storage places, etc., are effectively arranged for manufacturing of products or delivery of services so as to enhance the overall efficiency of operations. In a plant layout selection problem, the effectiveness of various alternative layout designs is evaluated based on a set of both qualitative and quantitative criteria.

Rao & Singh [14] considered a plant layout design selection problem for a chemical packaging industry and solved the same using a weighted Euclidean distance-based approach. In that problem, the performance of four alternative layout designs was assessed with respect to five evaluation criteria, i.e., interaction with existing facility distance (IEFD) (in m), area available for each assembly group (AAG) (in m^2), material quantity flow

(MQF) (in kg/hr.), accessibility for firefighting (AFF) (in %) and comfort of crew (COC). Among these criteria, COC is expressed in a relative scale. The developed decision matrix consisting of four alternative layout designs and five evaluation criteria is exhibited in Table 20.1. Using entropy method, Rao & Singh [14] determined the criteria weights as $w_{IEFD} = 0.2961$, $w_{AAG} = 0.0664$, $w_{MQF} = 0.0605$, $w_{AFF} = 0.082$, and $w_{COC} = 0.495$. These criteria weights are employed here while solving the plant layout design problem using SIR method.

TABLE 20.1 Decision Matrix for the Plant Layout Design for a Chemical Packaging Industry [14]

Alternative	IEFD	AAG	MQF	AFF	COC
P1	102	3,000	200	94	0.1364
P2	84	1,800	140	82	0.6667
P3	123	2,200	230	56	0.5
P4	224	2,500	180	98	0.3333

Source: Reprinted from Ref. [14]. Open access.

While solving this problem applying SIR method, based on the available criterion function of Figure 9.1, the type of the generalized criterion for each of the evaluation criteria along with its relevant parameters is selected, as shown in Table 20.2. It can be noted from this table that AAG, MQF, AFF, and COC are the beneficial attributes, whereas, IEFD is the sole non-beneficial attribute. Here, two separate scenarios are considered, where scenario 1 (W_1) assumes equal weights to all the criteria and scenario 2 (W_2) employs criteria weights determined using entropy method. Amongst these criteria, COC has the maximum importance, followed by IEFD. Now, adopting Eqns. (20.2) and (20.3), the corresponding S-matrix and I-matrix are subsequently developed, as exhibited in Table 20.3.

TABLE 20.2 Characteristics of the Considered Criteria

Criteria	IEFD	AAG	MQF	AFF	COC
Preference	Min	Max	Max	Max	Max
Type of criterion	Type 4	Type 3	Type 5	Type 1	Type 2
Parameter	$q = 0, p = 100$	$p = 500$	$q = 0, p = 50$		$q = 0$
Weight (W_1)	0.2	0.2	0.2	0.2	0.2
Weight (W_2)	0.2961	0.0664	0.0605	0.082	0.495

TABLE 20.3 *S*-Matrix and *I*-Matrix

Alternative	*S*-Matrix					*I*-Matrix				
	IEFD	AAG	MQF	AFF	COC	IEFD	AAG	MQF	AFF	COC
P1	1.5	3	1.4	2	0	0.5	0	0.6	1	3
P2	2	0	0	1	3	0	2.8	2.8	2	0
P3	1	0.8	2.6	0	2	1	1.6	0	3	1
P4	0	1.6	0.8	3	1	3	1	1.4	0	2

1. **SIR-SAW:** The *S*-flows and *I*-flows are computed for both the scenarios (W_1 and W_2), based on Eqns. (20.4) and (20.5), respectively. Now, applying Eqns. (20.8) and (20.9), the *n*-flow and *r*-flow values for each alternative plant layout design are also calculated. These values of *S*-, *I*-, *n*-, and *r*-flows for both the scenarios are provided in Table 20.4. The partial ranking relationships for the considered weighting scenarios along with the partial ranking diagrams are exhibited in Table 20.5 and Figure 20.3. It can be noticed from Figure 20.3 that for weight set W_1 (all the criteria have the equal importance), the ranking of the alternative plant layouts is achieved as P1 → P3 → P4 → P2. On the other hand, for the same problem when weights are assigned to the considered criteria using entropy method (weight set W_2), the achieved ranking of the layouts is P2 → P3 → P1 ~ P4. In this case, the performances of alternative layouts P1 and P4 are incomparable.

TABLE 20.4 SIR Flows Using SAW Procedure

Alternative	*S*–Flow		*I*–Flow		*N*–Flow		*r*–Flow	
	W_1	W_2	W_1	W_2	W_1	W_2	W_1	W_2
P1	1.58	0.8921	1.02	1.7514	0.56	–0.8593	0.6077	0.3375
P2	1.2	2.1592	1.52	0.5193	–0.32	1.6399	0.4412	0.8061
P3	1.28	1.4965	1.32	1.1433	–0.04	0.3532	0.4923	0.5669
P4	1.28	0.8956	1.48	2.0294	–0.2	–1.1338	0.4638	0.3062

TABLE 20.5 Partial Ranking Relationships between Alternative Layouts for SIR-SAW Method

For Weight Set (W_1)	$P_1 > P_2$	$P_3 > P_2$	$P_4 > P_2$
	$P_1 > P_3$	$P_3 > P_4$	–
	$P_1 > P_4$	–	–
For Weight Set (W_2)	$P_2 > P_1$	$P_3 > P_1$	$P_1 \sim P_4$
	$P_2 > P_3$	$P_3 > P_4$	–
	$P_2 > P_4$	–	–

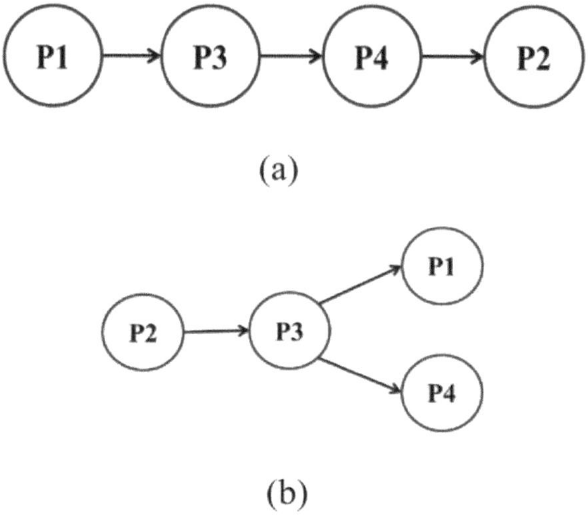

FIGURE 20.3 Partial ranking diagrams using SIR-SAW procedure. (a) Partial ranking diagram (W_1); and (b) partial ranking diagram (W_2).

2. **SIR-TOPSIS:** The *S*-, *I*-, *n*-, and *r*-flows are again calculated using TOPSIS method for both the scenarios in Table 20.6, while taking λ = 1, 2, and 10. The corresponding partial ranking diagrams are portrayed in Figure 20.4. The partial ranking relationships between the alternative plant layouts for two different weighting scenarios and varying λ values are exhibited in Table 20.7. It can be revealed from Figure 20.4 that at weight scenario W_1 (equal weights assigned to all the criteria), alternative plant layout P1 occupies the top position in the ranking list for λ = 1, 2, and 10. On the other hand, layout design P2 is the worst preferred choice for all the values of λ at weight scenario W_1. On the contrary, for weight scenario W_2 (criteria weights allocated using entropy method), alternative P2 emerges out as the most suitable layout for λ = 1 and 2. But for λ = 10, alternative P4 becomes the best facility layout design, followed by P1. Rao & Singh [14] solved the same problem using a weighted Euclidean distance-based approach and determined the rankings of the candidate layout designs as P1 → P4 → P3 → P2, P2 → P3 → P1 → P4 and P2 → P3 → P1 → P4 while considering subjective, objective, and integrated weights of the evaluation criteria. The ranking results based on the SIR method-based analyzes thus closely corroborate with the observations of Rao & Singh [14].

TABLE 20.6 SIR Flows Using SAW-TOPSIS Method

	Alternative	*S*–Flow		*I*–Flow		*n*–Flow		*r*–Flow	
		W_1	W_2	W_1	W_2	W_1	W_2	W_1	W_2
$\lambda = 1$	P1	0.5809	0.3329	0.3493	0.5888	0.2316	–0.2559	0.6245	0.3612
	P2	0.4412	0.8058	0.5205	0.1746	–0.0794	0.6312	0.4587	0.8219
	P3	0.4706	0.5585	0.4521	0.3844	0.0185	0.1741	0.51	0.5923
	P4	0.4706	0.3342	0.5068	0.6822	–0.0363	–0.348	0.4815	0.3288
$\lambda = 2$	P1	0.5482	0.2581	0.4048	0.6537	0.1434	–0.3956	0.5752	0.2831
	P2	0.457	0.8412	0.5044	0.1477	–0.0474	0.6935	0.4754	0.8506
	P3	0.4694	0.6191	0.4697	0.3524	–0.0003	0.2667	0.4998	0.6373
	P4	0.5	0.3271	0.5061	0.7	–0.0061	–0.373	0.497	0.3184
$\lambda = 10$	P1	0.5005	0.2302	0.508	0.6673	–0.0076	–0.4371	0.4963	0.2565
	P2	0.4947	0.8795	0.4832	0.1164	0.0115	0.763	0.5059	0.8831
	P3	0.4649	0.6665	0.5156	0.3334	–0.0507	0.3332	0.4742	0.6666
	P4	0.5791	0.3332	0.5003	0.6731	0.0788	–0.3399	0.5365	0.3311

TABLE 20.7 Partial Ranking Relationships between Alternative Layouts for SIR-TOPSIS Method

$\lambda = 1$	**For Weight Set (W_1)**	P1 > P2	P 3> P2	P4 > P2
		P1 > P3	P3 > P4	–
		P1 > P4	–	–
	For Weight Set (W_2)	P2 > P1	P3 > P1	P1 ~ P4
		P2 > P3	P3 > P4	–
		P2 > P4	–	–
$\lambda = 2$	**For Weight Set (W_1)**	P1 > P2	P3 > P2	P4 ~ P2
		P1 > P3	P3 ~ P4	–
		P1 > P4	–	–
	For Weight Set (W_2)	P2 > P1	P3 > P1	P1 ~ P4
		P2 > P3	P3 > P4	–
		P2 > P4	–	–
$\lambda = 10$	**For Weight Set (W_1)**	P1 > P2	P3 > P2	P4 > P2
		P1 > P3	P3 ~ P4	–
		P1 > P4	–	–
	For Weight Set (W_2)	P4 > P1	P1 ~ P2	P2 > P3
		P4 ~ P2	P1 > P3	–
		P4 > P3	–	–

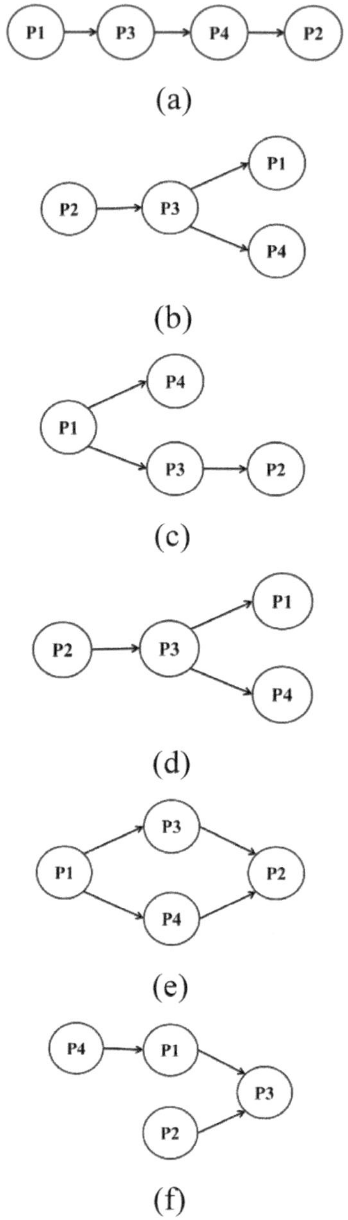

FIGURE 20.4 Partial ranking diagrams using SIR-TOPSIS procedure. (a) Partial ranking diagram (W_1) ($\lambda = 1$); (b) partial ranking diagram (W_2) ($\lambda = 1$); (c) partial ranking diagram (W_1) ($\lambda = 2$); (d) partial ranking diagram (W_2) ($\lambda = 2$); (e) partial ranking diagram (W_1) ($\lambda = 10$); and (f) partial ranking diagram (W_2) ($\lambda = 10$).

KEYWORDS

- **assembly group**
- **comfort of crew**
- **firefighting accessibility**
- **material quantity flow**
- **partial ranking**
- **superiority and inferiority ranking**

REFERENCES

1. Xu, X., (2001). The SIR method: A superiority and inferiority ranking method for multiple criteria decision making. *European Journal of Operational Research, 131*, 587–602.
2. Rebai, A., (1993). BBTOPSIS: A bag-based technique for order preference by similarity to ideal solution. *Fuzzy Sets and Systems, 60*, 143–162.
3. Rebai, A., (1994). Canonical fuzzy bags and bags fuzzy measures as a basis for MADM with mixed non-cardinal data. *European Journal of Operational Research, 78*, 34–48.
4. Roy, B., (2013). *Multicriteria Methodology for Decision Aiding* (Vol. 12). In: Springer Science & Business Media.
5. Brans, J. P., Vincke, P., & Mareschal, B., (1986). How to select and how to rank projects: The PROMETHEE method. *European Journal of Operational Research, 24*, 228–238.
6. Roy, B., Slowinski, R., & Treichel, W., (1992). Multicriteria programming of water supply systems for rural areas. *Journal of the American Water Resources Association, 28*, 13–31.
7. Tam, C. M., Tong, T. K. L., & Wong, Y. W., (2004). Selection of concrete pump using the superiority and inferiority ranking method. *Journal of Construction Engineering and Management, 130*(6), 827–834.
8. Marzouk, M., (2008). A superiority and inferiority ranking model for contractor selection. *Construction Innovation, 8*(4), 250–268.
9. Memarzade, M., Alvandi, M., & Mavi, R. K., (2011). Developing a new MADM method by integrating SIR and VIKOR methods. *American Journal of Scientific Research, 28*, 71–86.
10. Marzouk, M., El Shinnawy, N., & El-Said, O. M. M., (2013). Measuring sensitivity of procurement decisions using superiority and inferiority ranking. *International Journal of Information Technology & Decision Making, 12*(3), 395–423.
11. Vazifehdan, M. N., & Darestani, S. A., (2019). Green logistics outsourcing employing multi criteria decision making and quality function deployment in the petrochemical industry. *Asian Journal of Shipping and Logistics, 35*(4), 243–254.

12. Das, P. P., & Chakraborty, S., (2020). Application of superiority and inferiority multi-criteria ranking method for parametric optimization of laser cutting processes. *Process Integration and Optimization for Sustainability, 4*(4), 481–495.
13. Das, P. P., & Chakraborty, S., (2020). Parametric analysis of a green electrical discharge machining process using DEMATEL and SIR methods. *OPSEARCH, 57*(2), 513–540.
14. Rao, R. V., & Singh, D., (2012). Weighted Euclidean distance-based approach as a multiple attribute decision making method for plant or facility layout design selection. *International Journal of Industrial Engineering Computations, 3,* 365–382.

CHAPTER 21

MEASURING ATTRACTIVENESS BY A CATEGORICAL-BASED EVALUATION TECHNIQUE (MACBETH)

The measuring attractiveness by a categorical-based evaluation technique (MACBETH) is a multi-criteria decision analysis approach motivated by the multi-attribute value theory. The main idea of MACBETH method is to develop an interval scale from the preference information. The decision-makers are asked to provide judgments on the difference of attractiveness between two stimuli at a time from a set of semantic scales having seven categories, i.e., 'no,' 'very weak,' 'weak,' 'moderate,' 'strong,' 'very strong' and 'extreme,' arranged in ascending order of their importance [1]. The significance of these seven semantic scales is represented in Table 21.1. The performance measures are usually qualitative judgments which are further quantified proportionately on a 0–100 scale. The criteria weights can be determined applying MACBETH method based on pair-wise comparison questioning mode [2]. It aids the decision makers to rank the alternatives based on aggregated measurement of relative weighted attractiveness of performance of alternatives with respect to several decision criteria. The MACBETH method is supported by M-MACBETH software (http://www.m-macbeth.com/en/downloads.html), developed with the intention to ease out the overall computation procedure.

In the first step, a set of decision criteria is defined and input to the software in the form of a value tree. For all the criteria, the performance levels are decided, with identification of two reference levels (good and neutral) for each criterion. In order to solve the inter-criteria commensurability problem, it is sufficient to determine, for all the interval scales, two common reference levels, i.e., good (upper reference level) and neutral

Multi-Criteria Decision-Making Methods in Manufacturing Environments: Models and Applications.
Shankar Chakraborty, Prasenjit Chatterjee, & Partha Protim Das (Authors)

(lower reference level) with the corresponding performance values of 100 and 0, respectively. Further, to quantify the qualitative performance levels or convert quantitative performance levels into proportionate MACBETH scale, the performance levels are arranged in a matrix form from left to right according to their importance and the desired MACBETH judgment (range of judgments) is identified representing the difference of attractiveness. Consistent judgments lead to the generation of the converted MACBETH scale. However, the M-MACBETH software can identify inconsistencies, if any, and suggest necessary remedial actions to eliminate them. Further, the alternatives and their performance are entered into the model. The same procedure of preference function is adopted in MACBETH approach to obtain the criteria weights. Finally, the alternatives are ranked based on an overall score, derived from the preference judgments of the decision makers using an additive aggregation function.

TABLE 21.1 Significance of MACBETH Semantic Scales

Semantic Scale	Equivalent Numerical Scale	Significance
Null	0	Indifference between alternatives
Very weak	1	An alternative is very weakly attractive over another
Weak	2	An alternative is weakly attractive over another
Moderate	3	An alternative is moderately attractive over another
Strong	4	An alternative is strongly attractive over another
Very strong	5	An alternative is very strongly attractive over another
Extreme	6	An alternative is extremely attractive over another

In order to obtain quantified MACBETH scores of qualitative performance levels, the following procedure is adopted [3].

Let k be a criterion having L_i ($i = 1, 2, \ldots, n$) performance levels for which the cardinal MACBETH score is to be generated. The performance levels for this criterion are arranged in a matrix form according to the descending order of their importance from left to right and top to bottom, as shown in Table 21.2. In order to understand the procedure, an example with four performance levels, i.e., L_1, L_2, L_3, and L_4 is considered here such that L_2 is the 'good' level, while L_4 is the 'neutral' level. Let the arrangement of preference levels according to their reducing importance is L_2-L_3-L_1-L_4. Therefore, if $v(L_1)$, $v(L_2)$, $v(L_3)$ and $v(L_4)$ are the quantified MACBETH scores for levels L_1, L_2, L_3 and L_4, respectively, then $v(L_2) = 100$, $v(L_4) = 0$ and $v(L_2) > v(L_3)$

> $v(L_1) > v(L_4)$. Furthermore, the strengths of performance of the levels over each other are compared. Here, it can be noted that for n reference levels, maximum $n(n-1)/2$ number of comparisons are possible, but $(n-1)$ number of comparisons, as presented parallel to the diagonal in Table 21.2, are sufficient for conversion of the scale. Comparison of strengths of performance can be performed using one or a range of seven semantic scales. If the decision maker does not provide any strength of performance, it is noted by positive or P in the corresponding cell of the matrix. If the decision maker prefers performance of L_1 over L_2 with a strength:

$$h \in \{0, 1, \ldots, 6\}, \text{ i.e., } L_1 \overset{h}{\succ} L_2$$

Then,

$$v(L_1) - v(L_2) = h\alpha \tag{21.1}$$

where; α is a coefficient necessary to meet the condition that $v(L_1)$ and $v(L_2) \in [0,100]$. The quantified MACBETH scores are obtained by solving the related equations for all the performance levels. Let the decision maker decides the strengths of performance, as expressed in Table 21.2, and when all the strengths of performance levels are provided, the matrix of judgments is ready for quantification of the data.

TABLE 21.2 Strengths of Performance Levels for k^{th} Criterion

Performance Level	L_2	L_3	L_1	L_4
L_2 (Good)	No	Strong	P	P
L_3	–	No	Moderate	P
L_1	–	–	No	Very weak
L_4 (Neutral)	–	–	–	No

The quantitative measures for a seven-point semantic scale are as follows: ‘extreme’ = 6, ‘very strong’ = 5, ‘strong’ = 4, ‘moderate’ = 3, ‘weak’ = 2, ‘very weak’ = 1 and ‘no’ = 0. Therefore, from the judgments provided in Table 21.1, the following system of equations can be extracted:

$$v(L_2) - v(L_3) = 4\alpha \tag{21.2}$$

$$v(L_3) - v(L_1) = 3\alpha \tag{21.3}$$

$$v(L_1) - v(L_4) = \alpha \tag{21.4}$$

On solving Eqns. (21.2)–(21.4), the solutions are obtained as $\alpha = 12.5$, $v(L_1) = 12.5$ and $v(L_3) = 50$. The quantification of performance levels for all the remaining criteria as well as the corresponding criteria weights can be obtained adopting the same procedure. Convergence of the quantitative performance levels into proportionate MACBETH scores can be performed employing the following approach.

Even though MACBETH is well-known for dealing with qualitative performance scores, it can analyze the quantitative performance scores too. In this approach, quantitative performance levels are also converted into proportionate MACBETH scale with two reference levels, i.e., 'good' and 'neutral.' In fact, conversion of qualitative or quantitative data on a 0–100 MACBETH scale can be treated as normalization of data, as it serves the same purpose of normalization, generally used in other decision-making techniques. To transform quantitative performance levels into MACBETH scores, the selected quantitative performance levels are arranged in descending order of their importance from top to bottom and left to right. The desired MACBETH judgments (range of judgments) are selected showing the difference of attractiveness of one performance level over the other. Consistency of judgments is checked and necessary changes are made, if required. The MACBETH scale is generated for consistent judgments by M-MACBETH software. The conversion from quantitative performance levels is based on the following formulas.

Let x_{ij} be the performance measure of ith alternative on j^{th} criterion, ($i = 1, 2, \ldots, n; j = 1, 2, \ldots, m$) and $v(x_{ij})$ be the converted numerical MACBETH scale for x_{ij}. Let (x_j^-) is the lower reference level and (x_j^+) is upper reference level for j^{th} criterion. In the MACBETH approach, the MACBETH numerical scale is anchored on two predefined reference levels, and is assigned the scores of 0 and 100 to the lower and upper reference levels, respectively. Therefore, $v(x_j^-) = 0$ and $v(x_j^+) = 100$. The MACBETH score for any intermediate performance measure x_{ij} can be obtained as follows:

i. **For Beneficial Criterion:**

$$v(x_{ij}) = v(x_j^-) + \frac{(x_{ij} - x_j^-)}{(x_j^+ - x_j^-)}[v(x_j^+) - v(x_j^-)] \tag{21.5}$$

ii. **For Non-Beneficial Criterion:**

$$v(x_{ij}) = v(x_j^-) + \frac{(x_j^- - x_{ij})}{(x_j^- - x_j^+)}[v(x_j^+) - v(x_j^-)] \tag{21.6}$$

Furthermore, the converted MACBETH scores for all the performance measures are multiplied by the respective criteria weights and are added together to find out the overall attractiveness scores for the alternatives. The final overall score is obtained using the following additive value model.

$$V(X_i) = \sum_{j=1}^{m} w_j(v_j) \tag{21.7}$$

with:

$$\sum_{j=1}^{m} w_j = 1, w_j \succ 0 \text{ and } \begin{cases} v_j(x_j^+) = 100 \\ v_j(x_j^-) = 0 \end{cases} \tag{21.8}$$

Finally, the candidate alternatives are ranked based on $V(X_i)$ values in descending order of preference.

21.1 LITERATURE REVIEW

Bana e Costa & Vansnick [4] presented the application of MACBETH method as an effective decision support tool while solving two real-time problems. Bana e Costa et al. [5] extensively reviewed the applications of MACBETH method in diverse domains in order to prove its potentiality in solving complex decision-making problems with subjective assessment of criteria. Clivillé et al. [6] employed both ELECTRE III and MACBETH methods for dealing with tactical decision-making problems in SMEs in order to increase their annual turnover. Employing MACBETH method, Ertay et al. [7] considered four criteria along with 15 sub-criteria for evaluating the renewable energy alternatives in Turkey from a pool of four energy sources, i.e., solar, wind, hydropower, and geothermal. Considering different qualitative performance measures, Karande & Chakraborty [8] applied MACBETH method for solving two real-time supplier selection problems. Its ability to provide a numerical judgment scale for ranking the alternative suppliers and selecting the best one was also validated. Employing MACBETH method, Karande & Chakraborty [9] solved two FMS selection problems in a manufacturing environment while deploying semantic scales for pair-wise comparison of performance of the considered alternatives and determining the criteria weights.

Carnero & Gómez [10] integrated Markov chains with MACBETH method for more rational application of corrective and preventive maintenance policies in four dialysis subsystems of a healthcare unit while taking into account cost, quality of care and impact of care as the predominant evaluation criteria. Demesouka et al. [11] proposed a synergic combination of geographic information system (GIS) and MACBETH methods for landfill suitability analysis based on several conflicting objectives and decision maker's preferences. Its performance was also compared with other well-known MCDM techniques. Kundakcı [12] proposed a combined application of MACBETH and MULTIMOORA techniques for selection of the most apposite automobile in a marble company. In a textile industry, Kundakcı & Işık [13] first applied the MACBETH method for estimating the weights of various qualitative evaluation criteria and later employed the COPRAS method for ranking of the candidate air compressors. Komchornrit [14] integrated confirmatory factor analysis (CFA), MACBETH and PROMETHEE methods for identification of a suitable dry port location. The CFA was employed for establishing the importance of the considered criteria and their interrelationships, MACBETH was utilized to estimate the corresponding criteria weights and PROMETHEE was finally applied for ranking of the candidate dry port location alternatives from the best to the worst. Using MACBETH method, Tosun [15] solved a technology selection and evaluation problem in a manufacturing environment.

Taking into account acceptable product rate, major fault, return rate and delivery performance as the quantitative criteria, and production flexibility, capacity management, reliability, communication, and ecological awareness as the qualitative criteria, Akyüz et al. [16] applied MACBETH method as an effective decision-making tool to appraise the performance of some shortlisted suppliers in a manufacturing organization. Carnero & Gómez [17] developed a multi-criteria decision-making (MCDM) model employing MACBETH method to optimize the external supply of medical oxygen systems, and streamline the supply conditions of nitrous oxide and medicinal air in a healthcare unit. Kundakcı [18] attempted to select the most appropriate steam boiler for a textile dyehouse while proposing a combined application of MACBETH and EDAS methods. The relative importance of the considered subjective criteria was estimated using the MACBETH method, whereas, the competing boiler alternatives were ranked based on EDAS method. Özdağoğlu et al. [19] emphasized on the application of MACBETH method for evaluating importance levels of different criteria influencing the selection of the best ginning machine. Ferreira & Santos [20] conducted an exhaustive bibliometric review on the applications of MACBETH method while focusing on the scope of its future development.

21.2 SUPPLIER SELECTION PROBLEM

Supply chains are becoming an integral part of modern manufacturing organizations. Supply chains help the manufacturing organizations to increase their profit margins by means of effective procurement through a network of best-chosen suppliers. It has been estimated that for each dollar an organization earns on the sale of a product, it spends about 50–60% on goods and services [21]. Efficient supply chains not only help to increase the profit margins by means of enhanced sales volume but also result in savings through reduced procurement costs. Bringing down the procurement costs can have a dramatic effect on the bottom line – a 5% cut can translate into a 30% jump in profits [22, 23]. The success of supply chains mainly depends on a strong team of competent and efficient suppliers, capable of on-time delivery of goods or services of desired quality with consistent reliability at competitive rates. Enhanced industrialization, globalization of business and on-going trend for outsourcing of non-core activities have resulted in the availability of huge supplier selection options for the manufacturing organizations. Therefore, selection of the most suitable supplier from a pool of available choices has now become extremely important and should be dealt with utmost care.

The supplier selection process starts with the identification of the probable suppliers for the required goods or services, and ends with executing supply contracts with the suppliers identified through an evaluation process. Historically, evaluation of suppliers was mostly performed through a single criterion approach, i.e., based on the lowest invoice cost. Ongoing research in the supplier selection domain has proved that this approach not only ignores other sources of indirect supplier costs associated with late delivery, poor quality, etc., but also does not take into account the effects of other important criteria, like service, flexibility, reliability, etc. Therefore, the supplier selection problems are now being treated and solved based on multiple criteria approaches. Additionally, it has been noticed that during the past few years, the supplier selection problem has become multi-dimensional and multi-functional, having increased complexity with respect to effective decision-making. Therefore, choosing the right method for supplier selection has become very important to reduce the associated purchase risk.

Li et al. [24] considered a supplier selection problem with six competing alternatives evaluated based on four criteria, i.e., product quality, service quality, delivery time and price. The performance of all the six alternative suppliers is expressed in the form of ordinal (qualitative) data. Product quality, service quality and delivery time are beneficial criteria, and on the other hand, price is a non-beneficial criterion. These four evaluation criteria are first entered into M-MACBETH software to develop the corresponding value tree, as shown in Figure 21.1. In this value tree, the 'benefit criteria' node consists of all the three beneficial criteria, while the 'cost criteria' node contains only the non-beneficial criterion.

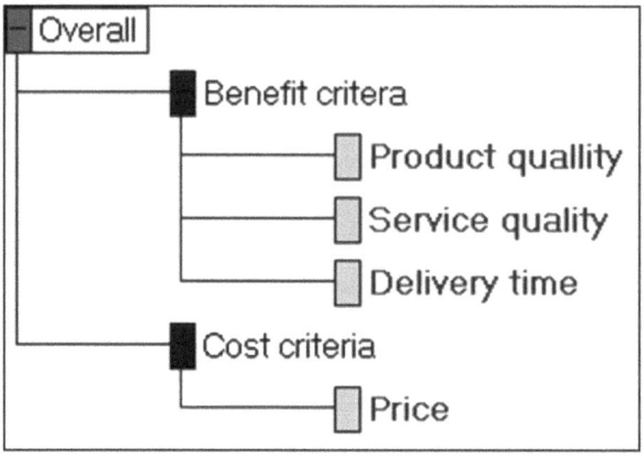

FIGURE 21.1 MACBETH value tree for supplier selection problem.

Li et al. [24] proposed a degree of seven ordinal performance levels to map the performance of the alternatives with respect to four evaluation criteria. For beneficial criteria, the proposed performance levels are arranged in descending order of their attractiveness as 'very good' (VG), 'good' (G), 'medium good' (MG), 'fair' (F), 'medium poor' (MP), 'poor' (P) and 'very poor' (VP). For non-beneficial criteria, the preference order is just reversed. For entering the performance levels into M-MACBETH software, the basis of comparison is selected as 'qualitative performance levels' because the performance of the alternatives with respect to all the considered criteria is measured based on an ordinal scale. For all the criteria, the above-mentioned seven performance levels are entered into M-MACBETH software according to the descending order of their attractiveness. For beneficial criteria, the most attractive performance level, i.e., VG is selected as upper reference level and the least attractive performance level, i.e., VP is chosen as lower reference level; while for non-beneficial criteria, VP is selected as upper reference level and VG as lower reference level. Figure 21.2 shows the selected performance levels for 'product quality' criterion for this example.

The major contribution of the decision maker relies on mapping the difference of attractiveness between the ordinal performance measures for two alternatives at a time. The differences of attractiveness of performance levels from seven semantic scales for 'product quality' and 'price' criteria are respectively shown in Figures 21.3 and 21.4. For example, it is observed in Figure 21.3 that the performance level VG is very weakly attractive to G in case of 'product quality' criterion. Consistency of the judgments is checked and it is revealed that the entered judgments are entirely consistent. Furthermore, based on the provided differences of attractiveness, M-MACBETH software converts the ordinal performance levels into proportionate cardinal MACBETH scales using appropriate linear programming (LP) models. The converted MACBETH scales of the performance levels for 'product quality' criterion is obtained by extracting a system of equations based on Eqn. (21.1), as shown below.

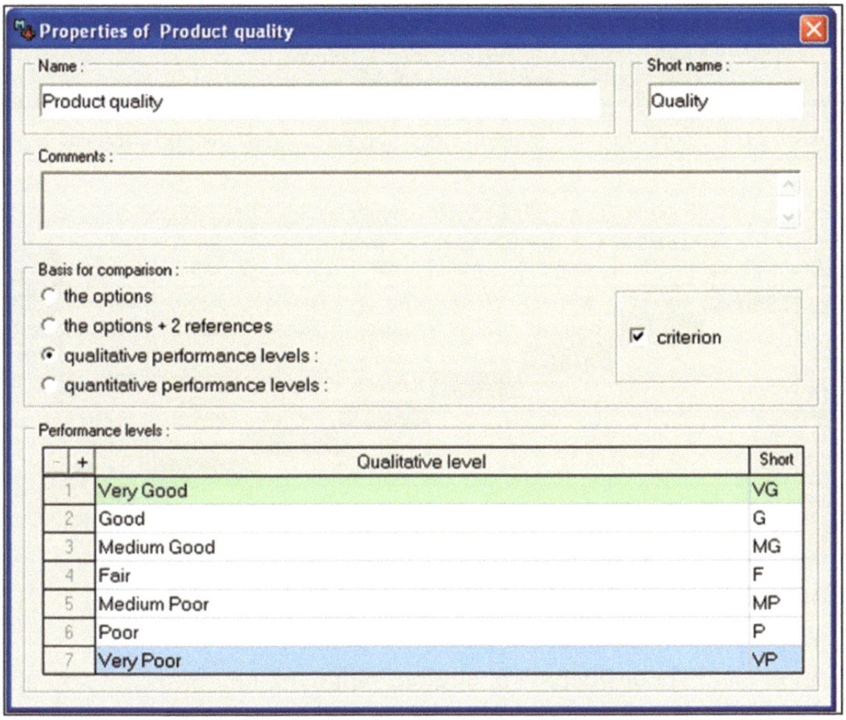

FIGURE 21.2 Performance levels for 'product quality' criterion.

v(VG) – v(G) = α, v(G) – v(MG) = α, v(MG) – v(F) = α, v(F) – v(MP) = α, v(MP) – v(P) = α and v(P) – v(VP) = α. Here, the performance level VG is selected as upper reference level and therefore, v(VG) = 100. On the other hand, the performance level VP is chosen as the lower reference level and hence, v(VP) = 0. Solving this system of equations leads to the results as α = 16.67, v(P) = 16.67, v(MP) = 33.33, v(F) = 50, v(MG) = 66.67 and v(G) = 83.33. The quantification of ordinal performance levels for all the remaining criteria is performed applying the same methodology.

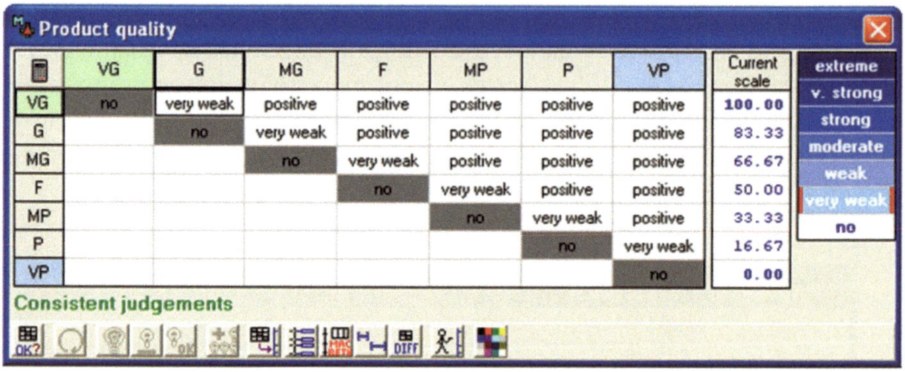

FIGURE 21.3 Comparison of performance levels for 'product quality' criterion.

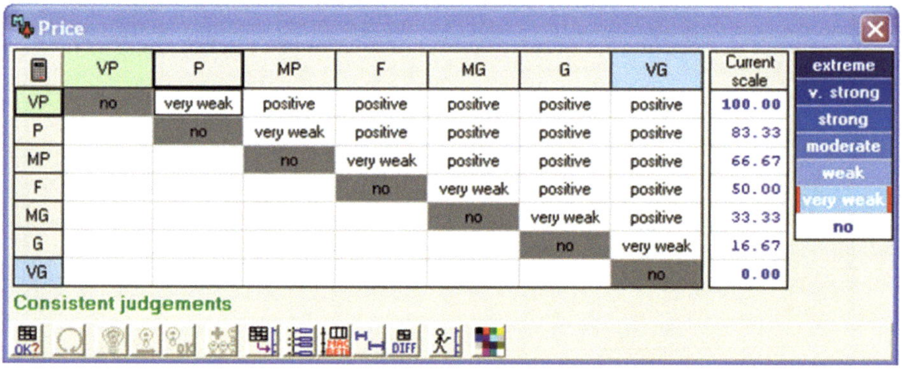

FIGURE 21.4 Comparison of performance levels for 'price' criterion.

Li et al. [24] considered six eligible suppliers, i.e., S_1, S_2, ..., S_6, among which the best one needs to be selected for a given task. A team of four decision makers was formed to evaluate the performance of the alternative suppliers and also to determine the relative criteria weights. It is observed that the mapping of ordinal performance of the suppliers with respect to different criteria is also subjective. Therefore, a result based on a consensus reached by a group of decision makers is more credible than the one based on the views of an individual. In this example, the average of mapping scores suggested by the four decision makers is considered as the performance of a candidate alternative supplier with respect to a particular criterion, as shown in Table 21.3.

Thereafter, the alternative suppliers are defined and their performance with respect to all the criteria is entered into M-MACBETH software, as shown in Figure 21.5. The MACBETH is a method having the capability to calculate the criteria weights of its own. In MACBETH, the methodology adopted for the estimation of the criteria weights is the same as that for quantification of the ordinal data. Based on the decision maker's opinion, the considered criteria are arranged in descending order of their importance from top to bottom and left to right. Further, the attractiveness of one criterion over another is judged using seven classes of semantic scale and the judgments are checked for consistency. The M-MACBETH software determines the corresponding criteria weights and also provides the ranges within which those weights can be adjusted. In this example, as pre-defined criteria weights are used, those same weights are adjusted on the weighing scale. Figure 21.6 shows the MACBETH weighting judgments and criteria weights for this supplier selection problem.

TABLE 21.3 Performance of the Alternative Suppliers

Alternative	Product Quality (Quality)	Service Quality (Service)	Delivery Time (Time)	Price
S_1	G	G	G	G
S_2	MG	G	G	G
S_3	MG	F	MG	VG
S_4	MG	MP	G	G
S_5	MG	MP	MG	MG
S_6	MG	MP	F	VG

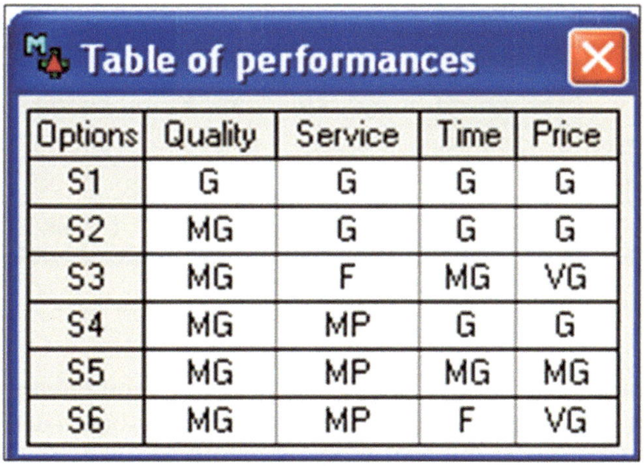
Table of performances

Options	Quality	Service	Time	Price
S1	G	G	G	G
S2	MG	G	G	G
S3	MG	F	MG	VG
S4	MG	MP	G	G
S5	MG	MP	MG	MG
S6	MG	MP	F	VG

FIGURE 21.5 MACBETH table of performances for supplier selection problem.

Weighting (Overall)

	[Service]	[Quality]	[Time]	[Price]	[all lower]	Current scale
[Service]	no	very weak	vweak-weak	weak	positive	30.93
[Quality]		no	vweak-weak	weak-mod	positive	27.84
[Time]			no	weak-mod	positive	22.68
[Price]				no	positive	18.55
[all lower]					no	0.00

extreme
v. strong
strong
moderate
weak
very weak
no

Consistent judgements

FIGURE 21.6 MACBETH weighting judgments for supplier selection problem.

The M-MACBETH software provides the overall attractiveness scores for all the alternative suppliers, as calculated using Eqn. (21.7). Those quantified cardinal scores of performances of the considered suppliers with respect to four criteria and the corresponding overall scores are given in Figure 21.7.

Table of scores

Options	Overall	Quality	Service	Time	Price
S1	70.96	83.33	83.33	83.33	16.67
S2	66.33	66.67	83.33	83.33	16.67
S3	49.15	66.67	50.00	66.67	0.00
S4	50.86	66.67	33.33	83.33	16.67
S5	50.17	66.67	33.33	66.67	33.33
S6	40.21	66.67	33.33	50.00	0.00
[all upper]	100.00	100.00	100.00	100.00	100.00
[all lower]	0.00	0.00	0.00	0.00	0.00
Weights :		0.2784	0.3093	0.2268	0.1855

FIGURE 21.7 MACBETH table of scores for supplier selection problem.

The alternative suppliers are now ranked based on their overall scores, as shown in Table 21.4. Li et al. [24] adopted a gray-based decision-making approach for solving this supplier selection problem and obtained a ranking of the alternative suppliers as S_1-S_2-S_4-S_5-S_3-S_6, as shown in Table 21.5.

On the other hand, for the same problem, MACBETH method provides a ranking of the alternative suppliers as S_1-S_2-S_4-S_5-S_3-S_6. It is observed that the rankings obtained using both these methods perfectly match.

TABLE 21.4 Overall Scores and Ranking for Supplier Selection Problem

Alternative	Product Quality	Service Quality	Delivery Time	Price	Overall Score	Rank
S_1	83.33	83.33	83.33	16.67	70.96	1
S_2	66.67	83.33	83.33	16.67	66.33	2
S_3	66.67	50.00	66.67	0	49.15	5
S_4	66.67	33.33	83.33	16.67	50.86	3
S_5	66.67	33.33	66.67	33.33	50.17	4
S_6	66.67	33.33	50.00	0	40.21	6

TABLE 21.5 Comparison of Results for Supplier Selection Problem

Alternative	Li et al. [24]	MACBETH
S_1	1	1
S_2	2	2
S_3	5	5
S_4	3	3
S_5	4	4
S_6	6	6

Sensitivity analysis shows the stability of the derived results, and in this analysis, the criteria weights or model parameters are changed in a constrained way to check the effects of such changes on the obtained ranking results. Here, the weight of one criterion is consistently varied, adjusting the difference equally over the remaining criteria weights. The M-MACBETH software provides the facility to perform sensitivity analysis with respect to all the evaluation criteria. In this example, as 'service quality' has the maximum weight, the effect of variation in weight of 'service quality' criterion on the ranking of the alternative suppliers is exhibited in Figure 21.8. It is observed that S_1 and S_2 remain as the first and second choices, while S_6 is the last choice for any variation in the weight of 'service quality' criterion. The next important criterion is 'product quality' and it is noticed that variation in the weight for 'service quality' criterion has no effect on the ranking of the alternative suppliers. For 'delivery time' criterion, the rankings of the first, second, and last alternative suppliers remain unaffected due to the change in their weight. In the case of 'price' criterion, S_1 remains as the first choice up to a weight of 0.638, thereafter, S_5 becomes the first choice pushing S_1 to be the second choice. The last choice of supplier (S_6) remains unaffected due to this criterion weight change. It is thus observed that the results obtained using MACBETH method are quite consistent and acceptable, almost unaffected due to the changes in the values of the considered criteria weights.

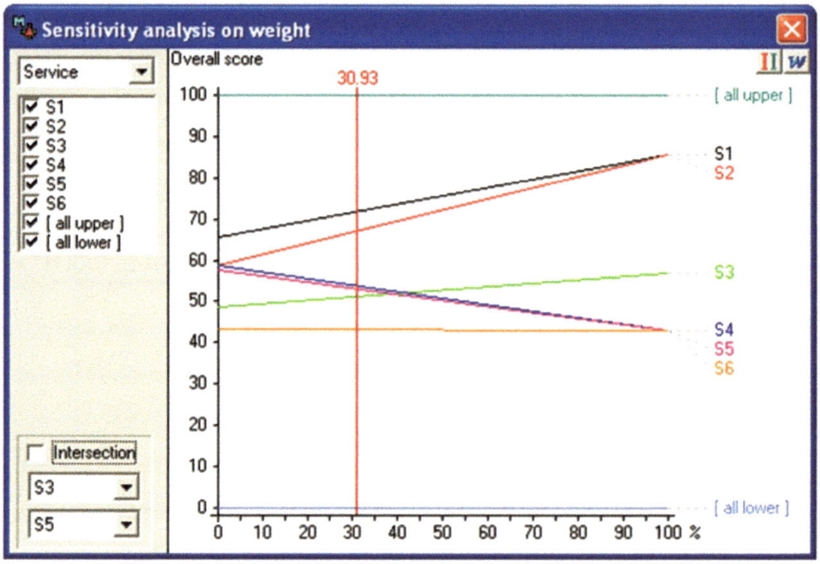

FIGURE 21.8 Sensitivity analysis with respect to 'service quality' criterion.

KEYWORDS

- **categorical-based evaluation technique**
- **confirmatory factor analysis**
- **geographic information system**
- **MACBETH method**
- **service quality**
- **supplier selection problem**

REFERENCES

1. Bana, E. C. C. A., Corrêa, E. C., De Corte, J. M., & Vansnick, J. C., (2002). Facilitating bid evaluation in public call for tenders: A socio-technical approach. *Omega, 30*(3), 227–242.
2. Bana, E. C. C. A., De Corte, J. M., & Vansnick, J. C., (2012). Macbeth. *International Journal of Information Technology & Decision Making, 11*(2), 359–387.
3. Fakhfakh, N., Verjus, H., Pourraz, F., & Moreaux, P., (2011). Measuring the satisfaction degree of quality attributes requirements for services orchestrations. In: *Proceedings of, 4th International Conference on Communication Theory, Reliability, and Quality of Service* (pp. 89–94).
4. Bana, E. C. C. A., & Vansnick, J. C., (1997). Applications of the MACBETH approach in the framework of an additive aggregation model. *Journal of Multi-criteria Decision Analysis, 6*, 107–114.
5. Bana, E. C. C. A., De Corte, J. M., & Vansnick, J. C., (2004). On the mathematical foundations of MACBETH. *Working Paper LSEOR*, 1–38.
6. Clivillé, V., Berrah, L., & Mauris, G., (2013). Deploying the ELECTRE III and MACBETH multi-criteria ranking methods for SMEs tactical. performance improvements. *Journal of Modeling in Management, 8*(3), 348–370.
7. Ertay, T., Kahraman, C., & Kaya, İ., (2013). Evaluation of renewable energy alternatives using MACBETH and fuzzy multi-criteria methods: A case of Turkey. *Technological and Economic Development of Economy, 19*(1), 38–62.
8. Karande, P., & Chakraborty, S., (2013). Using MACBETH method for supplier selection in manufacturing environment. *International Journal of Industrial Engineering Computations, 4*, 259–272.
9. Karande, P., & Chakraborty, S., (2013). Evaluation and selection of flexible manufacturing systems using MACBETH method. *International Journal of Services and Operations Management, 16*(1), 123–144.
10. Carnero, M. C., & Gómez, A., (2016). A multi-criteria decision making approach applied to improving maintenance policies in healthcare organizations. *BMC Medical Informatics and Decision Making, 16*(47), 1–22.
11. Demesouka, O. E., Vavatsikos, A. P., & Anagnostopoulos, K. P., (2016). Using MACBETH multi-criteria technique for GIS-based landfill suitability analysis. *Journal of Environ. Engineering, 142*(10), 04016042.
12. Kundakcı, N., (2016). Combined multi-criteria decision making approach based on MACBETH and multi-MOORA methods. *The Journal of Operations Research, Statistics, Econometrics and Management Information Systems, 4*(1), 17–26.
13. Kundakcı, N., & Işık, A. T., (2016). Integration of MACBETH and COPRAS methods to select air compressor for a textile company. *Decision Science Letters, 5*, 381–394.
14. Komchornrit, K., (2017). The selection of dry port location by a hybrid CFA-MACBETH-PROMETHEE method: A case study of Southern Thailand. *The Asian Journal of Shipping and Logistics, 33*(3), 141–153.
15. Tosun, O., (2017). Using Macbeth method for technology selection in production environment. *American Journal of Data Mining and Knowledge Discovery, 2*(1), 37–41.
16. Akyüz, G., Tosun, Ö., & Aka, S., (2018). Multi criteria decision-making approach for evaluation of supplier performance with MACBETH method. *International Journal of Information and Decision Sciences, 10*(3), 249–262.

17. Carnero, M. C., & Gómez, A., (2019). Optimization of decision making in the supply of medicinal gases used in health care. *Sustainability, 11*. doi: 10.3390/su11102952.
18. Kundakcı, N., (2019). An integrated method using MACBETH and EDAS methods for evaluating steam boiler alternatives. *Journal of Multi-Criteria Decision Analysis, 26*, 27–34.
19. Özdağoğlu, A., Yılmaz, K., & Keleş, M., (2020). Evaluation of machine selection criteria with MACBETH method in a ginnery factory. *Journal of Süleyman Demirel University Institute of Social Sciences, 2*(37), 26–37.
20. Ferreira, F. A. F., & Santos, S. P., (2021). Two decades on the MACBETH approach: A bibliometric analysis. *Annals of Operations Research, 296*, 901–925.
21. Ferreira, L., & Borenstein, D., (2012). A fuzzy-Bayesian model for supplier selection. *Expert Systems with Applications, 39*(9), 7834–7844.
22. De Boer, L., Labro, E., & Morlacchi, P., (2001). A review of methods supporting supplier selection. *European Journal of Purchasing & Supply Management, 7*(2), 75–89.
23. Degraeve, Z., & Roofhooft, F., (2001). A smarter way to buy. *Harvard Business Review, 79*, 22–23.
24. Li, G. D., Yamaguchi, D., & Nagai, M., (2007). A grey-based decision-making approach to the supplier selection problem. *Mathematical and Computer Modeling, 46*(3, 4), 573–581.

CHAPTER 22

MODIFIED SIMILARITY INDEX METHOD

The modified similarity index method is an extension of Deng's similarity measure concept [1]. This method is a customized form of TOPSIS, which employs the concept of ideal solution for comparing two alternatives in such a way that the most preferred alternative should have the highest degree of similarity to the positive-ideal solution (PIS) and lowest similarity to the negative-ideal solution (NIS). Safari et al. [2] emphasized on determining the conflicting index between two alternatives to show the degree of conflict between them. The concept of variable's gradient is employed here to calculate the conflicting index as it can explain conflicts between different evaluation criteria in MCDM problems very well. If A_i and A_j represent two alternatives in a given MCDM problem, they can be considered as two vectors, and the angle between A_i and A_j in the m-dimensional real space measures the conflict between them. It can be interpreted from Figure 22.1 that there is no conflict between A_i and A_j when $\theta_{ij} = 0$, and the conflict occurs when $\theta_{ij} \neq 0$, i.e., $\theta_{ij} \in (0, \pi/2)$ i.e.,. When $\theta_{ij} = 0$, the inclination of both the alternatives A_i and A_j is simultaneously increasing in the same direction and hence, there exists no conflict between them.

The degree of conflict between alternatives A_i and A_j is determined using the following equation:

$$\cos\theta_{ij} = \frac{\sum_{k=1}^{m} x_{ik} x_{jk}}{\left(\sum_{k=1}^{m} x_{ik}\right)^{0.5} \left(\sum_{k=1}^{m} x_{jk}\right)^{0.5}} \tag{22.1}$$

where; θ_{ij} is the angle between the gradients of the alternatives; and $(x_{i1}, x_{i2}, \ldots, x_{im})$ and $(x_{j1}, x_{j2}, \ldots, x_{jm})$ are the gradients of two alternatives A_i and A_j, respectively.

Multi-Criteria Decision-Making Methods in Manufacturing Environments: Models and Applications.
Shankar Chakraborty, Prasenjit Chatterjee, & Partha Protim Das (Authors)

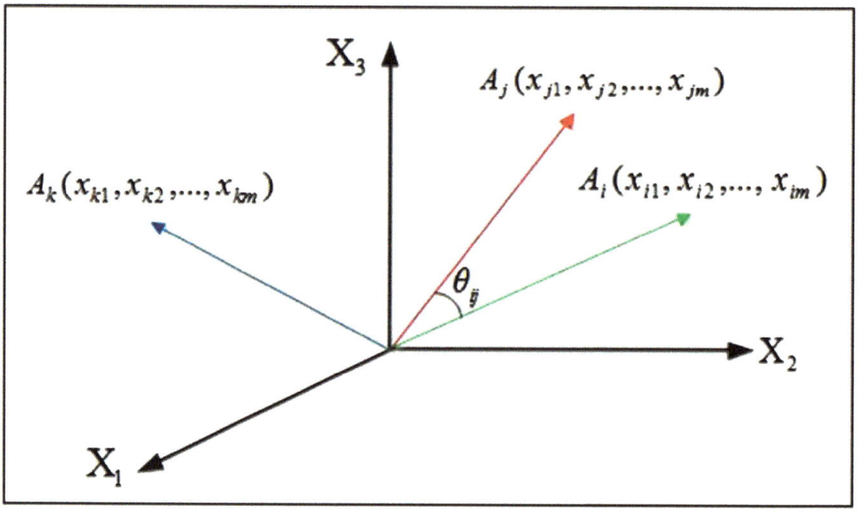

FIGURE 22.1 Degree of conflict between alternatives by gradients.

The conflict index is represented by $\theta_{ij} = 0$ and it depicts that the corresponding gradient vectors lie in the same direction of improvement. In the same manner, the conflict index characterized by $\theta_{ij} = \pi/2$ denotes that their gradient vectors have a vertical relationship with each other. The procedural steps for the modified similarity index approach are enlisted as below:

➢ **Step 1:** This step is characterized by establishing the decision matrix consisting of a set of alternatives A_i ($i = 1, 2, \ldots, n$) to be evaluated against a set of criteria C_j ($j = 1, 2, \ldots, m$).

$$X = \begin{bmatrix} x_{11} & x_{12} & \cdots & x_{1m} \\ x_{21} & x_{22} & \cdots & x_{2m} \\ \vdots & \vdots & \vdots & \vdots \\ x_{n1} & x_{n2} & \cdots & x_{nm} \end{bmatrix} \tag{22.2}$$

➢ **Step 2:** Determining the weighting vector is the next step. Let w_j represents the relative importance of criterion C_j with respect to the overall objective of the MCDM problem. The AHP or entropy method can be employed here for estimating the priority weights of the considered criteria.

$$W = (w_1, w_2, \ldots, w_m) \tag{22.3}$$

➢ **Step 3:** The ranking actually begins with normalizing the decision matrix through the vector (Euclidean) normalization technique to ensure that all the values in the matrix are dimensionless and can be compared with each other.

$$x'_{ij} = \frac{x_{ij}}{\left(\sum_{j=1}^{m} x_{ij}^2\right)^{0.5}} \tag{22.4}$$

Thus, the normalized matrix can be obtained as follows:

$$X' = \begin{bmatrix} x'_{11} & x'_{12} & \cdots & x'_{1m} \\ x'_{21} & x'_{22} & \cdots & x'_{2m} \\ \vdots & \vdots & \vdots & \vdots \\ x'_{n1} & x'_{n2} & \cdots & x'_{nm} \end{bmatrix} \tag{22.5}$$

- **Step 4:** In this step, the weighted performance matrix reflecting the performance of each alternative with respect to the evaluation criterion is obtained while multiplying the normalized decision matrix by the weight vector. This matrix is called the performance matrix.

$$X'_w = \begin{bmatrix} w_1 x'_{11} & w_2 x'_{12} & \cdots & w_m x'_{1m} \\ w_1 x'_{21} & w_2 x'_{22} & \cdots & w_m x'_{2m} \\ \vdots & \vdots & \vdots & \vdots \\ w_1 x'_{n1} & w_2 x'_{n2} & \cdots & w_m x'_{nm} \end{bmatrix} = \begin{bmatrix} y_{11} & y_{12} & \cdots & y_{1m} \\ y_{21} & y_{22} & \cdots & y_{2m} \\ \vdots & \vdots & \vdots & \vdots \\ y_{n1} & y_{n2} & \cdots & y_{nm} \end{bmatrix} \tag{22.6}$$

- **Step 5:** Determining the PIS and NIS is based on the concept that they consist of the best (or worst) criteria values attainable from all the alternatives if each criterion takes monotonically increasing or decreasing values. This notion has been widely used in various multi-criteria analysis models for solving real-time decision-making problems [3]. The PIS and NIS can be obtained while employing the following equations:

$$\begin{cases} A_j^+ = \max\ y_{ij} \\ A_j^- = \min\ y_{ij} \end{cases} \quad i = 1, 2, \ldots, n;\ j = 1, 2, \ldots, m \tag{22.7}$$

- **Step 6:** The next step consists of calculating the degree of conflict between each alternative A_i and the PIS (A_j^+) and NIS (A_j^-) in the m-dimensional real space. The angle between A_i and A_j^+ (A_j^-) in the m-dimensional real space, which is shown by θ_j^+ (θ_j^-), is a good measure of conflict between the vectors. Figure 22.2 shows the degree of conflict between the above-mentioned vectors. The degree of conflict between the alternatives is calculated using the following equations:

$$\begin{cases} \cos \grave{e}_i^+ = \dfrac{\sum_{j=1}^{m} y_{ij} A_j^+}{(\sum_{j=1}^{m} y_{ij}^2)^{0.5} \, (\sum_{j=1}^{m} A_j^{+2})^{0.5}} \\ \cos \grave{e}_i^- = \dfrac{\sum_{j=1}^{m} y_{ij} A_j^-}{(\sum_{j=1}^{m} y_{ij}^2)^{0.5} \, (\sum_{j=1}^{m} A_j^{-2})^{0.5}} \end{cases} \quad i = 1, 2, \ldots, n \qquad (22.8)$$

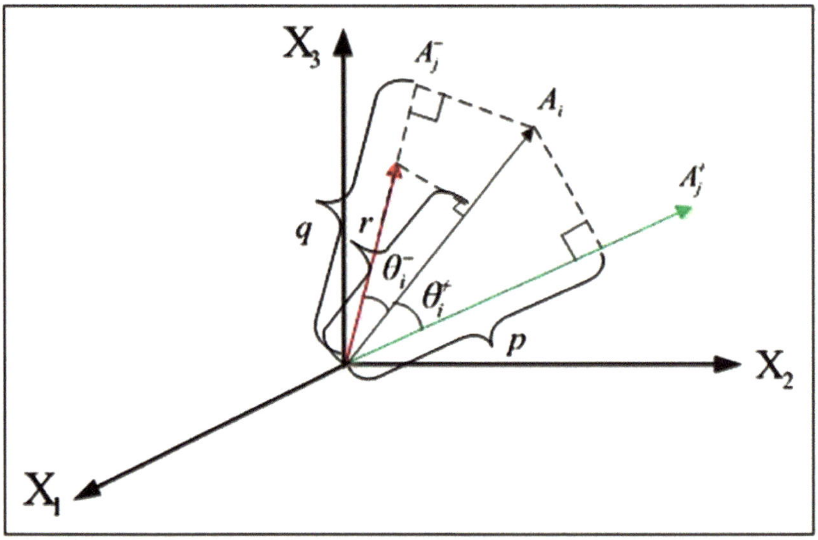

FIGURE 22.2 Degree of conflict between alternatives and A_j^+ (A_j^-).

- **Step 7:** Once the degree of conflict is calculated, determination of the degree of similarity between each pair of alternatives, and the PIS and NIS is the subsequent step. The degrees of similarity of the alternative A_i to A_j^+ and alternative A_i to A_j^- are denoted through S_j^+ and S_j^- respectively. Before computing the similarity index, the following point regarding Deng's similarity method needs to be explained. It is evident from Figure 22.2 that p is the projection of alternative vector A_i on the PIS vector A_j^+. Since the PIS has the largest value among all the alternatives, the A_i vector is equal or shorter than it. As a result, values for S_j^+ range between 0 and 1. Similarly, q is the projection of alternative vector A_i on the NIS vector A_j^-. But in this case, as the NIS has the smallest value among all the alternatives, q is equal or longer than A_j^-, thus causing the problem in Deng's method. In order to resolve this problem, another vector is needed. As shown in Figure 22.2, r is the projection of the NIS vector A_j^- on the alternative vector A_i which is always shorter than the alternative vector [4]. Therefore, the following equations are used to calculate the similarity index.

$$S_i^+ = \frac{\cos \grave{e}_i^+ \times (\Sigma_{j=1}^m y_{ij}^2)^{0.5}}{|A_j^+|}$$
$$S_i^- = \frac{|A_j^-|}{\cos \grave{e}_i^- \times (\Sigma_{j=1}^m y_{ij}^2)^{0.5}} \quad (22.9)$$

- **Step 8:** Calculating the overall performance score (P_i) for each alternative with respect to the considered criteria is the last step of this modified similarity index method. The higher is the performance index, the more preferred is the alternative.

$$P_i = \frac{S_i^+}{S_i^+ + S_i^-} \quad (22.10)$$

In this method, since the angle between alternative A_i and NIS (PIS) is not clear when it is equal to PIS (NIS), the value for the overall performance score is not equal to one (zero). Finally, the ranks for all the alternatives are obtained in decreasing order of their performance scores.

22.1 LITERATURE SURVEY

With the help of an illustrative example, Safari et al. [2] presented the application of modified similarity index-based MCDM approach which had been based on the notion that the most preferred alternative should have the highest degree of similarity to the PIS and the lowest degree of similarity to the NIS. Safari & Ebrahimi [4] employed the modified similarity index method for ranking of 187 countries based on multi-criteria human development index. Suresh et al. [5] studied the influences of types of the insert and cutting fluid, cutting speed, feed, and depth of cut on surface roughness, MRR, tool interface temperature, specific energy and flank wear while turning hardened AISI D3 tool steel material. Deng's similarity index method was later adopted to determine the optimal parametric combination of the said process and the derived results were contrasted with those obtained using WASPAS method. Using two real-time examples, Prasad & Chakraborty [6] demonstrated the applicability of the modified similarity index method in identifying the most suited cutting fluids during specific machining operations. Mallick et al. [7] applied a modified similarity index method for the classification of multiple inventory items of a conveyor belt manufacturing unit into A, B, and C categories. The derived results were also compared with those obtained using TOPSIS and AHP methodologies. Taşabat [8] proposed a new MCDM technique (DSC TOPSIS) based on distance, similarity, and correlation for deriving rational solutions of various decision-making problems. In the proposed method, Euclidean was used as the distance measure, cosine was employed as the similarity measure and Pearson correlation was utilized as the relation measure. Its efficacy was finally validated while comparing human development indexes of different countries.

22.2 CUTTING FLUID SELECTION PROBLEM

Since the last two centuries, cutting fluids have been extensively utilized in metal cutting industries. At the beginning, simple oils were used with brushes to lubricate and cool down the machine tool, and seldom, lard, animal fat or whale oil was added to improve their lubricating ability. But, in response to dynamic and upsurging machining requirements, cutting fluids have evolved gradually from simple petroleum-based oils to complex blends of chemical additives, lubricants, and water-based oils. Due to varying demands of the manufacturing organizations, several types of cutting fluid with different compositions are prevalent in the market to meet the explicit machining needs. In order to exploit the fullest potential of cutting fluid while in application, it is always recommended to select the most apposite cutting fluid for the specific machining requirement. The performance of a cutting fluid does not only depend on its technological viability, but often also on several factors determined by the machine tool set-up, like chip removal, heat dissipation, lubrication of the machine parts and possibility of chip breakage. The mode of application of cutting fluid can also range from flood to minimum quantity lubrication. However, the application method employed is usually governed by the requirements of machining application and availability of the equipment. It is observed that a machine operator usually utilizes his knowledge and experience while selecting a particular mode of application of cutting fluid. Hence, a scrupulous/conscientious choice of cutting fluid for a specified machining requirement is critical while considering the following important aspects:

- Physical and chemical properties of the cutting fluid;
- Applicability of cutting fluids to various cutting tools and workpiece materials;
- Ability to perform different machining operations; and
- Economical considerations, such as purchase cost, storage cost, operating cost, disposal cost, etc.

Several other factors, like compatibility with machine tool, operator interaction, performance requirements and environment friendliness also govern the selection of a cutting fluid [9]. One specific cutting fluid is highly suitable for a given machining condition, but may not be equally appropriate for another application. Moreover, several of the above-listed criteria may be both qualitative and quantitative in nature, sometimes having complex interrelationship (minimization of cost and maximization of safety, minimization of corrosion and maximization of cooling effect) among them. Such an in-depth analysis, however, is time consuming and requires a clear understanding of the machining operation. The complexity and difficulty associated with determining the most suitable cutting fluid from a list of feasible alternatives based on some evaluation criteria necessitate adopting an MCDM tool, which can deal with such problems.

In this example, a decision-making problem regarding the selection of the most suitable cutting fluid for cutting spur gears fabricated of alloy steel is considered. The tool material used for cutting gears is carbide. The machining process used for cutting gears is hobbing, which is usually a slow speed machining operation where the cut is quite heavy and continuous. The increase in temperature at the workpiece and tool interface adversely affects the dimensional accuracy and desired surface finish of the spur gears. In such severe applications, wetting agents (generally fatty esters) increase the cutting fluid's ability to coat the cutting tool, workpiece, and metal fines, which thereby enhances its lubricating property and guards against microscopic welding. It has already been known that neat oils with extreme pressure additives (EPA), such as chlorine and sulfur, perform best in heavy duty machining operations on alloy steel. Cutting fluids with chlorine additive offers greater tool life at relatively low cutting speeds, while sulfur additives are used for machining of high alloy steels. Viscosity of a cutting fluid is also a vital property with respect to its performance and maintenance. Fluids with lower viscosity let micro-chip and dirt to settle out of the suspension. With increasing environmental regulation, trouble, and cost associated with disposal of the cutting fluids have increased. On these days, minimizing the environmental footprint has also become a significant objective for both the manufacturers and end users of cutting fluids. Therefore, in this example, seven neat oil alternatives are evaluated based on five properties of cutting fluid, i.e., viscosity at 40°C (in cSt), sulfur wt.%, chlorine wt.%, fatty ester wt.% and disposability which have significant effects on the given machining requirements. Disposability of a cutting fluid is measured in R scale, while comparing its physical and chemical properties that affect the environment. The values 5 and 1 used for R scale symbolize the highest and the lowest disposability of the cutting fluid alternatives, respectively. The higher value for disposability indicates that the effect of cutting fluid disposal on the environment is comparatively less harmful. The decision matrix comprising all the values related to the chosen criteria for different alternatives considered in this example is shown in Table 22.1. In addition, the last row of Table 22.1 provides the priority weights of all the evaluation criteria computed using entropy method to avoid any subjectivity in the decision-making process.

Next, following the steps of the modified similarity index-based approach, the selection of the most effective neat oil for cutting gears made of alloy steel from the identified candidate alternatives is carried out. At first, the normalized values for the decision matrix employing the vector normalization technique are attained within a scale of 0 to 1, as displayed in Table 22.2.

TABLE 22.1 Decision Matrix for the Neat Oil Selection Problem

Alternative	Evaluation Criteria				
	Viscosity at 40°C (C_1)	Sulfur wt.% (C_2)	Chlorine wt.% (C_3)	Fatty Ester wt.% (C_4)	Disposability (C_5)
A_1	12.5	2.8	1.8	0.5	4
A_2	31.6	3.5	1	3	5
A_3	13.5	2.3	1.1	5	5
A_4	24.1	2.5	2.7	5	4
A_5	45.1	2.4	2.6	5	3
A_6	38.5	2.2	2.4	4	3
A_7	21.6	1.1	3.5	7	4
Weight	0.0919	0.1621	0.2738	0.2748	0.1973

In the next step, the degree of conflict between each alternative and PIS (cos θ^+) and NIS (cos θ^-) is computed from the performance matrix. The columns of the normalized decision matrix are multiplied by the associated criteria weights to obtain the corresponding performance matrix. The degrees of similarity between the alternatives and PIS (S^+) and NIS (S^-) are then estimated. Finally, the overall performance score for each alternative is computed using Eqn. (22.10) and all the values obtained are arranged in Table 22.3.

TABLE 22.2 Normalized Decision Matrix for the Neat Oil Selection Problem

Alternative	Criteria				
	C_1	C_2	C_3	C_4	C_5
A_1	0.1627	0.4248	0.2939	0.0409	0.3714
A_2	0.4114	0.5310	0.1633	0.2456	0.4642
A_3	0.1757	0.3490	0.1796	0.4093	0.4642
A_4	0.3137	0.3793	0.4408	0.4093	0.3714
A_5	0.5871	0.3641	0.4245	0.4093	0.2785
A_6	0.5012	0.3338	0.3919	0.3274	0.2785
A_7	0.2812	0.1669	0.5715	0.5730	0.3714

TABLE 22.3 Weighted Normalized Matrix for Neat Oil Selection Problem

Alternative	C_1	C_2	C_3	C_4	C_5	Cos θ^+	Cos θ^-	S^+	S^-	P_i
A_1	0.0150	0.0689	0.0805	0.0112	0.0733	0.8187	0.8541	0.4169	0.8434	0.3308
A_2	0.0378	0.0861	0.0447	0.0675	0.0916	0.8624	0.8411	0.5205	0.7227	0.4186
A_3	0.0162	0.0566	0.0492	0.1125	0.0916	0.9279	0.7086	0.5958	0.8064	0.4249
A_4	0.0288	0.0615	0.1207	0.1125	0.0733	0.9948	0.7692	0.7509	0.6319	0.5430
A_5	0.0540	0.0590	0.1162	0.1125	0.0550	0.9714	0.7912	0.7170	0.6282	0.5330
A_6	0.0461	0.0541	0.1073	0.0900	0.0550	0.9723	0.8193	0.6330	0.6878	0.4793
A_7	0.0258	0.0271	0.1565	0.1575	0.0733	0.9704	0.6722	0.8990	0.5891	0.6041
A(+)	0.0150	0.0861	0.1565	0.1575	0.0916	–	–	–	–	–
A(–)	0.0540	0.0271	0.0447	0.0112	0.0550	–	–	–	–	–

When this selection problem is solved employing the modified similarity index-based method, A_7 emerges out as the most appropriate neat oil among all the considered alternatives for cutting gears based on the specified machining conditions. The alternatives A_4 and A_1 are identified as the second best and the least suitable choices, respectively. The appropriateness of neat oil alternative A_7 can be justified while having a closer look at its various chemical properties. Here, viscosity is the sole non-beneficial criterion where a lower value is desired and the other four are beneficial attributes preferred with their higher values. Although, A_7 contains comparatively less sulfur wt.%, it has higher values for the other three beneficial criteria, i.e., chlorine wt.%, fatty ester wt.% and disposability, and moderate value for viscosity. Hence, its selection with respect to the considered criteria can be well justified.

KEYWORDS

- **conflict index**
- **extreme pressure additives**
- **methodologies**
- **neat oil alternative**
- **negative-ideal solution**
- **positive-ideal solution**
- **similarity index**

REFERENCES

1. Deng, H., (2007). A similarity-based approach to ranking multi-criteria alternatives. In: *Proc. of the International Conference on Intelligent Computing, Lecture Notes in Artificial Intelligence* (Vol. 4682, pp. 253–262).
2. Safari, H., Khanmohammadi, E., Hafezamini, A., & Ahangari, S. S., (2013). A new technique for multi-criteria decision making based on modified similarity method. *Middle-East Journal of Scientific Research, 14*(5), 712–719.
3. Deng, H., (1999). Multicriteria analysis with fuzzy pairwise comparison. *International Journal of Approximate Reasoning, 21*(3), 215–231.
4. Safari, H., & Ebrahimi, E., (2014). Using modified similarity multiple criteria decision making technique to rank countries in terms of human development index. *Journal of Industrial Engineering and Management, 7*(1), 254–275.
5. Suresh, R. K., Krishnaiah, G., & Venkataramaiah, P., (2017). Selection of best novel MCDM method during turning of hardened AISI D3 tool steel under minimum quantity lubrication using bio-degradable oils as cutting fluids. *International Journal of Applied Engineering Research, 12*(19), 8082–8091.
6. Prasad, K., & Chakraborty, S., (2018). Application of the modified similarity-based method for cutting fluid selection. *Decision Science Letters, 7*, 273–286.

7. Mallick, B., Das, S., Sarkar, B., & Das, S., (2019). Application of the modified similarity-based method for multi-criteria inventory classification. *Decision Science Letters, 8*, 455–470.
8. Taşabat, S. E., (2019). A novel multicriteria decision-making method based on distance, similarity, and correlation: DSC TOPSIS. *Mathematical Problems in Engineering* (p. 20). Article ID: 9125754. https://doi.org/10.1155/2019/9125754.
9. De Chiffre, L., & Belluco, W., (2002). Investigations of cutting fluid performance using different machining operations. *Lubrication Engineering, 58*, 22–29.

CHAPTER 23

MULTI-ATTRIBUTIVE BORDER APPROXIMATION AREA COMPARISON (MABAC) METHOD

The multi-attributive border approximation area comparison (MABAC) method [1] is a recently developed MCDM tool, and has already been well accepted due to its computational simplicity and robustness. In this method, the performance of each criterion function is partitioned into two specific areas, i.e., upper approximation area consisting of the ideal (positive-ideal) alternatives and lower approximation area with the anti-ideal (negative-ideal) alternatives. It evaluates the corresponding strengths and weaknesses of each alternative over its competitors with reference to each criterion under consideration. This method is implemented based on the following procedural steps [2–5]:

- **Step 1:** Develop the initial decision matrix (X):

$$X = \begin{bmatrix} x_{11} & x_{12} & \cdots & x_{1n} \\ x_{21} & x_{22} & \cdots & x_{2n} \\ \cdots & \cdots & \cdots & \cdots \\ x_{m1} & x_{m2} & \cdots & x_{mn} \end{bmatrix} \quad (23.1)$$

  where; x_{ij} denotes the performance of ith alternative with respect to j^{th} criterion, m is the number of alternatives and n is the number of criteria.

- **Step 2:** Normalize the decision matrix:

$$N = \begin{bmatrix} n_{11} & n_{12} & \cdots & n_{1n} \\ n_{21} & n_{22} & \cdots & n_{2n} \\ \cdots & \cdots & \cdots & \cdots \\ n_{m1} & n_{m2} & \cdots & n_{mn} \end{bmatrix} \quad (23.2)$$

Multi-Criteria Decision-Making Methods in Manufacturing Environments: Models and Applications.
Shankar Chakraborty, Prasenjit Chatterjee, & Partha Protim Das (Authors)

In order to make all the elements of the initial decision matrix dimensionless and comparable, their values are normalized using Eqns. (23.3) and (23.4) depending on the type of the criterion considered. After normalization, the corresponding normalized decision matrix $N = [n_{ij}]_{m\times n}$ is obtained.

i. **For Beneficial Criterion:**

$$n_{ij} = [x_{ij} - \min(x_{ij})]/[\max(x_{ij}) - \min(x_{ij})] \quad (\mathrm{i} = 1, 2, \ldots, m; \mathrm{j} = 1, 2, \ldots, n) \tag{23.3}$$

ii. **For Cost Criterion:**

$$n_{ij} = [\max(x_{ij}) - x_{ij}]/[\max(x_{ij}) - \min(x_{ij})] \tag{23.4}$$

- **Step 3:** Compute the weighted normalized decision matrix (V):

$$V = \begin{bmatrix} v_{11} & v_{12} & \cdots & v_{1n} \\ v_{21} & v_{22} & \cdots & v_{2n} \\ \cdots & \cdots & \cdots & \cdots \\ v_{m1} & v_{m2} & \cdots & v_{mn} \end{bmatrix} = \begin{bmatrix} w_1 \cdot (n_{11}+1) & w_2 \cdot (n_{12}+1) & \cdots & w_n \cdot (n_{1n}+1) \\ w_1 \cdot (n_{21}+1) & w_2 \cdot (n_{22}+1) & \cdots & w_n \cdot (n_{2n}+1) \\ \cdots & \cdots & \cdots & \cdots \\ w_1 \cdot (n_{m1}+1) & w_2 \cdot (n_{m1}+1) & \cdots & w_n \cdot (n_{mn}+1) \end{bmatrix} \tag{23.5}$$

where; w_j ($j = 1, 2, \ldots, n$) is the weight of jth criterion.

- **Step 4:** Determine the border approximation area matrix (G):

$$G = [g_1, g_2, \ldots, g_n] \tag{23.6}$$

$$g_j = \left(\prod_{i=1}^{m} v_{ij} \right)^{1/m} \tag{23.7}$$

where; v_{ij} is the element of the weighted normalized matrix (V).

- **Step 5:** Compute the distance of the alternative from the border approximation area for the matrix elements (Q).

$$Q = \begin{bmatrix} q_{11} & q_{12} & \cdots & q_{1n} \\ q_{21} & q_{22} & \cdots & q_{2n} \\ \cdots & \cdots & \cdots & \cdots \\ q_{m1} & q_{m2} & \cdots & q_{mn} \end{bmatrix} \tag{23.8}$$

The distances of the alternatives from the border approximation area (q_{ij}) are estimated as the differences between the elements in the weighted matrix (V) and values of the border approximation area matrix (G).

$$Q = V - G = \begin{bmatrix} v_{11} & v_{12} & \cdots & v_{1n} \\ v_{21} & v_{22} & \cdots & v_{2n} \\ \cdots & \cdots & \cdots & \cdots \\ v_{m1} & v_{m2} & \cdots & v_{mn} \end{bmatrix} - \begin{bmatrix} g_1 & g_2 & \cdots & g_n \\ g_1 & g_2 & \cdots & g_n \\ \cdots & \cdots & \cdots & \cdots \\ g_1 & g_2 & \cdots & g_n \end{bmatrix}$$

$$= \begin{bmatrix} v_{11} - g_1 & v_{12} - g_2 & \cdots & v_{1n} - g_n \\ v_{21} - g_1 & v_{22} - g_2 & \cdots & v_{2n} - g_n \\ \cdots & \cdots & \cdots & \cdots \\ v_{m1} - g_1 & v_{m2} - g_2 & \cdots & v_{mn} - g_n \end{bmatrix} \quad (23.9)$$

where; g_j is the border approximation area for j^{th} criterion.

The position of an alternative A_i can be in the border approximation area (G), upper approximation area (G^+) or lower approximation area (G^-), i.e., $A_i \in \{G \vee G^+ \vee G^-\}$. The ideal alternative (A^+) is always positioned in the upper approximation area (G^+), while the location of the anti-ideal alternative (A^-) is in the lower approximation area (G^-). The position of alternative A_i in the approximation area (G, G^+, or G^-) is determined based on the following condition:

$$A_i \in \begin{cases} G^+ & \text{if } q_{ij} > 0 \\ G & \text{if } q_{ij} = 0 \\ G^- & \text{if } q_{ij} < 0 \end{cases} \quad (23.10)$$

It is observed that for an alternative A_i to be identified as the best, it has the maximum number of criteria belonging to the upper approximation area (G^+).

- **Step 6:** Calculate the criterion function value. Based on the summation of row elements in matrix (Q), the final scores of the criterion function for different alternatives are derived as follows:

$$S(A_i) = \sum_{j=1}^{n} q_{ij} \quad (i = 1, 2, \ldots, m) \quad (23.11)$$

The considered alternatives can now be ranked based on $S(A_i)$ values. The best alternative should have the highest $S(A_i)$ score.

23.1 REVIEW OF THE LITERATURE

Through six simple procedural steps, Božanić et al. [6] presented the application of MABAC as an effective MCDM tool to support decision-making on the use of force in a defensive operation. While identifying the most suitable NTM process for machining of cylindrical through holes on non-conductive ceramic materials, Chatterjee et al. [2] first applied FARE method to estimate the corresponding criteria weights, and then ranked the considered NTM processes using MABAC method based on their feasibility. Using the MABAC method, Biswas & Das [3] endeavored to single out the most apposite hybrid vehicle based on some important evaluation criteria, like cost, mileage, tail pipe emission, comfort, and high tank size volume. Roy et al. [4] combined AHP and MABAC methods in uncertain environment to assess and prioritize medical tourism destinations in India. The validity of the derived ranking solutions was later established while comparing with rough-TOPSIS and rough-VIKOR methods. Sharma et al. [7] proposed a hybrid MCDM approach while combining rough AHP and MABAC methods for prioritizing some shortlisted Indian railway stations. Rough AHP was implemented to calculate criteria weights based on the preferences of the participating decision makers, while the candidate railway stations were ranked using MABAC method. Biswas & Das [5] presented the application of MABAC method for selection and ranking of a group of seven battery electric vehicles based on various technical and operational criteria, like fuel economy, base model pricing, quick accelerating time, battery range and top speed. While solving a commercially available scooter selection problem, Biswas & Saha [8] estimated the weights of the considered criteria, i.e., kerb weight, mileage, top speed, fuel tank capacity and price using fuzzy AHP method. The candidate alternative scooters were finally ranked using MABAC method.

Ecer et al. [9] adopted MABAC method to assess sustainability performance of 12 selected OPEC countries according to 41 sustainability development indicators in 10 dimensions. The derived results identified United Arab Emirates as the most sustainable OPEC country, followed by Qatar, Kuwait, and Iran. The ranking solutions were also compared with those obtained using CoCoSo, WASPAS, CODAS, and VIKOR methods. Using social needs, resource supply and economic factors as the main criteria along with 11 sub-criteria, Wu et al. [10] developed a location decision framework for agroforestry biomass cogeneration project in China while employing the MABAC method. The weights of the considered subjective and objective evaluation criteria were evaluated using BWM and entropy method, respectively. In an FMS system, Ahmad et al. [11] solved a robot selection problem for pick and place operation while proposing the integrated application of CRITIC and MABAC methods. The weights of five criteria, i.e., load capacity (LC), memory capacity (MC), manipulator reach (MR), maximum tip speed (MTS) and repeatability were calculated using CRITIC method, and the candidate robots were ranked based on MABAC method. Alidrisi [12] proposed the application of MABAC method as a novel MCDM technique to evaluate eight total productive maintenance (TPM) pillars to identify the maintenance maturity level in a public service sector. While attempting to identify the most suitable NTM processes for the generation of standard through holes on glass and deep through cavities on titanium, Chakraborty et al. [13] first applied the rough-entropy method to determine the weights of the considered evaluation criteria. The candidate NTM processes were then ranked according to their suitability of operation using MABAC method.

With the help of the MABAC method, Isik et al. [14] measured and evaluated the logistics performance of 11 Central and Eastern European Countries in order to achieve sustainable competitive advantage. Application of statistical variance technique identified timeliness and infrastructure as the most and the least important significant performance criteria respectively during the evaluation process. Muravev & Mijic [15] presented the application of an integrated MCDM framework for selection of the third-party providers. The BWM was first implemented to estimate the criteria weights, while the competing providers were ranked using MABAC method. Its performance was finally validated with respect to other MCDM tools, like MAIRCA and VIKOR models. Chakraborty et al. [16] first adopted DEMATEL method to portray the cause-and-effect relationship among 11 evaluation criteria while attempting to identify the best Indian smart city. The initial dataset containing 98 Indian smart cities was subsequently reduced to only 23 efficient cities using *k*-means clustering algorithm. Those shortlisted smart cities were finally ranked from the best to the worst using MABAC. Aizawl became the best Indian smart city, followed by Mangalore; while Warangal and Bilaspur emerged out as the worst performing cities. Agarwal et al. [17] integrated rough numbers with MABAC approach while solving an arc welding robot selection problem in a group decision-making environment. The weights of the evaluation criteria were estimated employing rough-entropy method, which the competing robot alternatives were ranked based on the criterion function values derived from the application of MABAC method.

23.2 WELDING PROCESS CONSUMABLE SELECTION PROBLEM

In a welding process, different consumables, like filler alloy, electrode material and shielding gas play important roles while providing good weld quality. Baghel & Nagesh [18] considered a filler alloy material selection problem while welding dissimilar aluminum alloy combination of 5083-O and 6061-T651 for extrusion applications. The choice of the appropriate filler alloy would determine the transverse and longitudinal strength of the weld joint. In order to satisfy this requirement, the filler materials should have higher values of tensile strength and impact toughness at lower cost. Thus, there must be good correlations between the available filler materials and their properties to produce robust welding of aluminum alloys with the lowest overall cost. Baghel & Nagesh [18] considered five filler alloy materials, i.e., ER5356, ER5556, ER5183, ER5654, and ER5554 for performing welding operation of dissimilar aluminum materials. The performance of those filler alloys was evaluated based on eight criteria, i.e., yield strength (YS) (in MPa), tensile strength (TS) (in MPa), elongation (E) (in %), Brinell hardness (BH), cost (C) (in USD), longitudinal shear strength (LSS) (in MPa), transverse shear strength (TSS) (in MPa) and impact strength (IS) (in J). The decision matrix consisting of five alternative welding filler alloy materials and eight evaluation criteria is provided in Table 23.1. While employing AHP method, the corresponding criteria weights were determined as w_{YS} = 0.0326, w_{TS} = 0.0235, w_{E} = 0.0476, w_{BH} = 0.1059, w_{C} = 0.3313, w_{LSS} = 0.1572, w_{TSS} = 0.2306, and w_{IS} = 0.0709, which would be subsequently applied in the MABAC method-based analysis. In this criterion set, all the evaluation criteria except cost (C) are beneficial in nature requiring their higher values. Now, depending on the type of the considered criterion (beneficial or cost), the initial decision matrix of Table 23.1 is normalized using Eqns. (23.3) and (23.4). This normalized decision matrix is exhibited in Table 23.2. Based on Eqn. (23.5), the corresponding weighted normalized decision matrix is now developed after multiplying each element of the normalized decision matrix by the related criteria weights, as shown in Table 23.3. While following Eqns. (23.6) and (23.7), the border approximation area matrix (*G*) for this problem is formulated, as depicted in Table 23.4. Now, the distances of the alternative welding filler alloy materials from the border approximation area are computed as the differences between the elements in the weighted normalized matrix and values of the border approximation area matrix. These distances are exhibited in Table 23.5. Finally, based on Eqn. (23.11), the final scores of criterion function for all the five filler alloys are calculated, as provided in Table 23.6. Based on these scores, the candidate filler alloys are ranked as ER5356 → ER5183 → ER5556 → ER5554 → ER5654. It can be interestingly noted that Baghel & Nagesh [18] also derived the same ranking of the welding filler alloys using AHP-TOPSIS method. From Table 23.5, it can be unveiled that filler alloys ER5356, ER5183, and

ER5556 are positioned in the upper approximation area, while the locations of filler alloys ER5554 and ER5654 are in the lower approximation area. The performance of the most preferred filler alloy ER5356 is excellent with respect to all the considered evaluation criteria. On the contrary, the worst preferred filler alloy ER5654 has strengths with regard to YS, E, and BH, but it has major weaknesses with respect to TS, C, LSS, TSS, and IS. Thus, the application of the MABAC method not only helps in ranking of the alternative welding filler alloy materials, but also highlights the relative strengths and weaknesses of each of the considered materials.

TABLE 23.1 Filler Alloys and Their Properties [18]

Filler Alloy	YS	TS	E	BH	C	LSS	TSS	IS
ER5356	240	268.89	17	105	13.5	117.21	179.26	18.2
ER5556	158.579	317.15	14	65	16	137.89	206.84	15.7
ER5183	275	296.47	17	99	15.5	127.55	193.05	16
ER5654	220.63	227.52	17	83	17.2	82.73	124.1	13
ER5554	124.1	186	8	39	13	79.289	103.42	10

Source: Adapted from [18].

TABLE 23.2 Normalized Decision Matrix for the Filler Alloy Material Selection Problem

Filler Alloy	YS	TS	E	BH	C	LSS	TSS	IS
ER5356	0.7681	0.6320	1.0000	1.0000	0.8810	0.6471	0.7333	1.0000
ER5556	0.2285	1.0000	0.6667	0.3939	0.2857	1.0000	1.0000	0.6951
ER5183	1.0000	0.8423	1.0000	0.9091	0.4048	0.8236	0.8667	0.7317
ER5654	0.6397	0.3166	1.0000	0.6667	0.0000	0.0587	0.2000	0.3659
ER5554	0.0000	0.0000	0.0000	0.0000	1.0000	0.0000	0.0000	0.0000

TABLE 23.3 Weighted Normalized Decision Matrix for the Filler Alloy Selection Problem

Filler Alloy	YS	TS	E	BH	C	LSS	TSS	IS
ER5356	0.0576	0.0384	0.0952	0.2118	0.6232	0.2589	0.3997	0.1418
ER5556	0.0400	0.0470	0.0793	0.1476	0.4260	0.3144	0.4612	0.1202
ER5183	0.0652	0.0433	0.0952	0.2022	0.4654	0.2867	0.4305	0.1228
ER5654	0.0535	0.0309	0.0952	0.1765	0.3313	0.1664	0.2767	0.0968
ER5554	0.0326	0.0235	0.0476	0.1059	0.6626	0.1572	0.2306	0.0709

TABLE 23.4 Border Approximation Area Matrix

Criteria	YS	TS	E	BH	C	LSS	TSS	IS
G	0.0483	0.0355	0.0799	0.1639	0.4860	0.2276	0.3474	0.1075

TABLE 23.5 Distances of Filler Alloys from the Border Approximation Area

Filler Alloy	YS	TS	E	BH	C	LSS	TSS	IS
ER5356	0.0094	0.0028	0.0153	0.0479	0.1371	0.0313	0.0523	0.0343
ER5556	–0.0082	0.0115	–0.0006	–0.0162	–0.0601	0.0868	0.1138	0.0127
ER5183	0.0169	0.0077	0.0153	0.0383	–0.0206	0.0591	0.0830	0.0153
ER5654	0.0052	–0.0046	0.0153	0.0126	–0.1547	–0.0612	–0.0707	–0.0107
ER5554	–0.0157	–0.0120	–0.0323	–0.0580	0.1766	–0.0704	–0.1168	–0.0366

TABLE 23.6 Score Values and Corresponding Ranks for Each Alternative

Filler Alloy	Score	Rank by MABAC Method	Rank by AHP-TOPSIS [18]
ER5356	0.3304	1	1
ER5556	0.1396	3	3
ER5183	0.2150	2	2
ER5654	–0.2688	5	5
ER5554	–0.1653	4	4

KEYWORDS

- **Brinell hardness**
- **multi-attributive border approximation area comparison**
- **tensile strength**
- **total productive maintenance**
- **transverse shear strength**
- **yield strength**

REFERENCES

1. Pamučar, D., & Ćirovic, G., (2015). The selection of transport and handling resources in logistics centers using multi-attributive border approximation area comparison (MABAC). *Expert Systems with Applications, 42*(6), 3016–3028.

2. Chatterjee, P., Mondal, S., Boral, S., Banerjee, A., & Chakraborty, S., (2017). A novel hybrid method for non-traditional machining process selection using factor relationship and multi-attributive border approximation method. *FACTA Universitatis Series: Mechanical Engineering, 15*(3), 439–456.
3. Biswas, T. K., & Das, M. C., (2018). Selection of hybrid vehicle for green environment using multi-attributive border approximation area comparison method. *Management Science Letters, 8*(2), 121–130.
4. Roy, J., Chatterjee, K., Bandyopadhyay, A., & Kar, S., (2018). Evaluation and selection of medical tourism sites: A rough analytic hierarchy process based multi-attributive border approximation area comparison approach. *Expert Systems, 35*. https://doi.org/10.1111/exsy.12232.
5. Biswas, T. K., & Das, M. C., (2019). Selection of commercially available electric vehicle using fuzzy AHP-MABAC. *Journal of Institution of Engineers (India) Series C, 100*(3), 531–537.
6. Božanić, D. I., Pamučar, D. S., & Karović, S. M., (2016). Application the MABAC method in support of decision-making on the use of force in a defensive operation. *Technics-Management, 12*, 97–104.
7. Sharma, H. K., Roy, J., Kar, S., & Prentkovskis, O., (2018). Multi criteria evaluation framework for prioritizing Indian railway stations using modified rough AHP-MABAC method. *Transport and Telecommunication, 19*(2), 113–127.
8. Biswas, T. K., & Saha, P., (2019). Selection of commercially available scooters by new MCDM method. *International Journal of Data and Network Science, 3*, 137–144.
9. Ecer, F., Pamucar, D., Zolfani, S. H., & Eshkalag, M. K., (2019). Sustainability assessment of OPEC countries: Application of a multiple attribute decision making tool. *Journal of Cleaner Production, 241*, 118324.
10. Wu, Y., Yan, Y., Wang, S., F Liu, F., Xu, C., & Zhang, T., (2019). Study on location decision framework of agroforestry biomass cogeneration project: A case of China. *Biomass and Bioenergy, 127*, 105289.
11. Ahmad, S., Bingol, S., & Wakeel, S., (2020). A hybrid multi-criteria decision making method for robot selection in flexible manufacturing system. *Middle East Journal of Science, 6*(2), 68–77.
12. Alidrisi, H., (2020). Maintenance maturity level identification using MABAC method: An adaptation of TPM pillars in a public service sector. *Journal of Scientific & Industrial Research, 79*, 914–917.
13. Chakraborty, S., Dandge, S. S., & Agarwal, S., (2020). Non-traditional machining processes selection and evaluation: A rough multi-attributive border approximation area comparison approach. *Computers & Industrial Engineering, 139*, 106201.
14. Isik, O., Aydin, Y., & Kosaroglu, S. M., (2020). The assessment of the logistics performance index of CEE countries with the new combination of SV and MABAC methods. *Scientific Journal of Logistics, 16*(4), 549–559.
15. Muravev, D., & Mijic, N., (2020). A novel integrated provider selection multicriteria model: The BWM-MABAC model. *Decision-making: Applications in Management and Engineering, 3*(1), 60–78.
16. Chakraborty, S., Ghosh, S., Agarwal, S., & Chakraborty, S., (2021). An integrated performance evaluation approach for the Indian smart cities. *OPSEARCH, 58*(4), 906–941.

17. Agarwal, S., Chakraborty, S., Prasad, K., & Chakraborty, S., (2021). A rough multi-attributive border approximation area comparison approach for arc welding robot selection. *Jordan Journal of Mechanical and Industrial Engineering, 15*(2), 169–180.
18. Baghel, P. K., & Nagesh, D. S., (2016). Multiattribute assessment of consumables for TIG welding of aluminum alloys. *International Journal of Manufacturing Engineering* (p. 9). Article ID: 7291898. http://dx.doi.org/10.1155/2016/7291898.

CHAPTER 24

MULTI-ATTRIBUTIVE IDEAL-REAL COMPARATIVE ANALYSIS (MAIRCA) METHOD

As mentioned earlier, TOPSIS identifies the most preferred solution, which would be nearest to the positive-ideal solution (PIS) and farthest from the negative ideal solution. The major disadvantage of this method is that though it measures two target values, it does not project the significance of the distances from those two solutions. It advocates that the best alternative may not be always nearest to the PIS [1]. Keeping in mind this disadvantage of TOPSIS, the MAIRCA method was developed in 2014 at the University of Defense in Belgrade [2]. It is a simple mathematical tool with a high degree of stability with respect to changes in both the nature and characteristic of the criteria considered [3]. It depends on the evaluation of the gap between the ideal and the real assessments. Summing up of these gaps for each criterion for each alternative provides the total gap, and finally, the alternatives are ranked based on the increasing order of these gap values. The highest ranking is allotted to the alternative having the smallest total gap. The alternative with the minimum gap has most of its criteria closest to their ideal values. The steps related to the application of MAIRCA method are provided as below [4, 5]:

- **Step 1: Initiation of the Decision Matrix (X):**

$$X = [x_{ij}]_{m \times n} = \begin{bmatrix} x_{11} & x_{12} & \cdots & x_{1n} \\ x_{21} & x_{22} & \cdots & x_{2n} \\ \cdots & \cdots & \cdots & \cdots \\ x_{m1} & x_{m2} & \cdots & x_{mn} \end{bmatrix} \quad (24.1)$$

where; x_{ij} denotes the performance value of ith alternative on j^{th} criterion.

Multi-Criteria Decision-Making Methods in Manufacturing Environments: Models and Applications.
Shankar Chakraborty, Prasenjit Chatterjee, & Partha Protim Das (Authors)

➢ **Step 2: Choosing a Particular Alternative:** Preference to choose one of the m candidate alternatives is determined by evaluating the preference order P_{R_i}.

$$P_{R_i} = \frac{1}{m}; \sum_{i=1}^{m} P_{R_i} = 1; \quad (i = 1,2.,,,,m) \tag{24.2}$$

When the decision maker is neutral towards all the considered alternatives, the preferences are treated to be equal.

$$P_{R_1} = P_{R_2} = \ldots = P_{R_m} \tag{24.3}$$

➢ **Step 3: Calculation of the Theoretical Evaluation Matrix (T_E):** The theoretical evaluation matrix (T_E) is formulated having an $m \times n$ format (m is the number of alternatives and n is the number of criteria). The elements of the theoretical evaluation matrix ($t_{E_{ij}}$) are calculated after multiplying the preferences of the alternatives P_{R_i} by the corresponding criteria weights.

$$T_E = \begin{bmatrix} t_{E_{11}} & t_{E_{12}} & \cdots & t_{E_{1n}} \\ t_{E_{21}} & t_{E_{22}} & \cdots & t_{E_{2n}} \\ \cdots & \cdots & \cdots & \cdots \\ t_{E_{m1}} & t_{E_{m2}} & \cdots & t_{E_{mn}} \end{bmatrix} = \begin{bmatrix} P_{R_1} w_1 & P_{R_1} w_2 & \cdots & P_{R_1} w_n \\ P_{R_2} w_1 & P_{R_2} w_2 & \cdots & P_{R_2} w_n \\ \cdots & \cdots & \cdots & \cdots \\ P_{R_m} w_1 & P_{R_m} w_2 & \cdots & P_{R_m} w_n \end{bmatrix} \tag{24.4}$$

where; w_j is the relative weight of the jth criterion which can be determined using AHP or entropy method.

As the decision maker is neutral to the initial selection of the alternatives, the preferences (P_{R_i}) become equal for all the considered alternatives. Eqn. (25.4) can now be rewritten as follows:

$$T_E = \begin{bmatrix} t_{E_1} & t_{E_2} & \cdots & t_{E_n} \end{bmatrix} = \begin{bmatrix} P_{R_i} w_1 & P_{R_i} w_2 & \cdots & P_{R_i} w_n \end{bmatrix} \tag{24.5}$$

➢ **Step 4: Determination of the Real Evaluation Matrix (T_R):**

$$T_R = \begin{bmatrix} t_{R_{11}} & t_{R_{12}} & \cdots & t_{R_{1n}} \\ t_{R_{21}} & t_{R_{22}} & \cdots & t_{R_{2n}} \\ \cdots & \cdots & \cdots & \cdots \\ t_{R_{m1}} & t_{R_{m2}} & \cdots & t_{R_{mn}} \end{bmatrix} \tag{24.6}$$

The elements of the real evaluation matrix ($t_{R_{ij}}$) are computed while multiplying the theoretical evaluation matrix elements ($t_{E_{ij}}$) by the elements of the initial decision matrix (x_{ij}).

i. **For Beneficial Criterion:**

$$t_{R_{ij}} = t_{E_{ij}} \left(\frac{x_{ij} - x_i^-}{x_i^+ - x_i^-} \right) \tag{24.7}$$

ii. **For Cost or Non-Beneficial Criterion:**

$$t_{R_{ij}} = t_{E_{ij}} \left(\frac{x_{ij} - x_i^+}{x_i^- - x_i^+} \right) \tag{24.8}$$

where; x_i^+ and x_i^- can be defined as follows:

$x_i^+ = \max(x_1, x_2, ..., x_m)$ is the maximum value of the criterion in the decision matrix.

$x_i^- = \min(x_1, x_2, ..., x_m)$ is the minimum value of the criterion in the decision matrix.

- **Step 5: Development of the Total Gap Matrix (T_G):** The elements of the gap matrix can be estimated as the difference (gap) (g_{ij}) between the theoretical ($t_{E_{ij}}$) and real evaluations ($t_{R_{ij}}$), or by actually subtracting the elements of the real evaluation matrix (T_R) from the elements of the theoretical evaluation matrix (T_E).

$$T_G = T_E - T_R = \begin{bmatrix} g_{11} & g_{12} & \cdots & g_{1n} \\ g_{21} & g_{22} & \cdots & g_{2n} \\ \cdots & \cdots & \cdots & \cdots \\ g_{m1} & g_{m2} & \cdots & g_{mn} \end{bmatrix} = \begin{bmatrix} t_{E_{11}} - t_{R_{11}} & t_{E_{12}} - t_{R_{12}} & \cdots & t_{E_{1n}} - t_{R_{1n}} \\ t_{E_{21}} - t_{R_{21}} & t_{E_{22}} - t_{R_{22}} & \cdots & t_{E_{2n}} - t_{R_{2n}} \\ \cdots & \cdots & \cdots & \cdots \\ t_{E_{m1}} - t_{R_{m1}} & t_{E_{m2}} - t_{R_{m2}} & \cdots & t_{E_{mn}} - t_{R_{mn}} \end{bmatrix} \tag{24.9}$$

Based on Eqn. (24.9), it can be observed that the gap value (g_{ij}) belongs to the interval $g_i \in [0, (t_{E_{ij}} - t_{R_{ij}})]$. It is always preferable that g_{ij} should be zero because the alternative with the smallest difference between the theoretical ($t_{E_{ij}}$) and real evaluations ($t_{R_{ij}}$) needs to be selected. When an alternative A_i has a theoretical evaluation value equal to the real evaluation value, i.e., $t_{E_{ij}} = t_{R_{ij}}$, the corresponding gap for the alternative would become zero, identifying it as the best (ideal) choice. On the contrary, if for an alternative, the real evaluation value is zero, the corresponding gap would become $g_{ij} = t_{E_{ij}}$. In this case, the alternative with the maximum gap value would be the worst (anti-ideal) choice.

- **Step 6: Computation of the Criteria Functions (Q_i) for the Alternatives:** The values of criteria functions for the alternatives can now be estimated by simply taking the sum of the elements in the total gap matrix.

$$Q_i = \sum_{j=1}^{n} g_{ij}, \quad (i = 1, 2, ..., m) \tag{24.10}$$

24.1 LITERATURE SURVEY

Considering nine restrictions and six evaluation criteria, Gigović et al. [6] integrated GIS and MAIRCA methods to help the decision makers in selecting of the most apposite location for ammunition depots in Serbia. The corresponding criteria weights were calculated using DEMATEL-ANP technique. Pamučar et al. [7] presented the simultaneous application of FUCOM and MAIRCA methods for selecting a level crossing in Serbia with respect to investment in its security equipment. Sensitivity analysis was later carried out to prove the robustness of the proposed approach. Ulutaş [8] first employed the SWARA method to determine the weights of seven evaluation criteria, i.e., hygiene, taste, variety of meals, service time, references, service quality and price, and ranked four catering firms based on their performance using the MAIRCA method. Bakir et al. [9] endeavored to develop an MCDM model to assess the operational performance of different airlines operating in emerging markets. Considering a real-life example with 11 candidate airlines and eight evaluation criteria, PIPRECIA method was first adopted to estimate the corresponding criteria weights, while the considered alternative airlines were later ranked using MAIRCA method. Chatterjee & Chakraborty [10] solved a piston material selection problem consisting of eight candidate alternatives and eight evaluation criteria using MAIRCA method. The criteria weights were determined using entropy method, and AISI 4140 steel was identified as the most appropriate material, followed by AISI 8660 steel.

24.2 BATTERY SELECTION PROBLEM FOR ELECTRIC HYBRID VEHICLES

Most of the electric and plug-in hybrid electric vehicles employ a battery as the primary power source and an internal combustion engine as the secondary power source. In those hybrid vehicles, selection of the most suitable battery with high efficiency at low cost is an important task to achieve better fuel economy, reliable performance, and durability. In the present-day market, different types of batteries are available whose performance can be evaluated based on various parameters, like energy density, durability, W-hour/cycle, specific energy, nominal voltage, efficiency, self-discharge rate, cost, etc.

Panday & Bansal [11] considered a battery selection problem for hybrid electric vehicles, and solved the same using Ashby's principle, TOPSIS and VIKOR methods. In this problem, the performance of four alternative batteries, like Li-ion, Ni-MH, Ni-Cd, and Pb-acid is assessed based on seven evaluation criteria, like energy density (in Wh/l), specific energy (in Wh/kg), electrical efficiency (in %), durability (in cycles), self-discharge rate (in %/month), energy/cycle (in Wh) and cost (in USD). The related decision matrix for this battery selection problem is shown in Table 24.1. Among these criteria, self-discharge rate and cost are non-beneficial attributes, whereas, the remaining are beneficial attributes. Panday & Bansal [11] employed a ratio method to determine the corresponding criteria weights as 0.3669, 0.2508, 0.0060, 0.2478, 0.04867, 0.0464, and 0.0334, respectively. Now, based on the initial decision matrix and calculated criteria weights, this battery selection problem for hybrid electric vehicles is solved using MAIRCA method.

As this problem consists of four alternative batteries and the concerned decision maker is neutral towards all the alternatives, the preferences towards the considered alternative batteries are treated to be equal (i.e., 0.25). Following the procedural steps of MAIRCA method, the elements of the theoretical evaluation matrix are first computed while multiplying the preferences of the alternatives by the corresponding criteria weights. This theoretical evaluation matrix is exhibited in Table 24.2. Now, the elements of the real evaluation matrix, as shown in Table 24.3, are estimated after multiplying the theoretical evaluation matrix elements by the elements of the initial decision matrix. Now, employing Eqn. (24.9), elements of the gap matrix (T_G) are calculated as the difference (gap) between the theoretical and real evaluations, i.e., by subtracting elements of the real evaluation matrix (T_R) from the elements of the theoretical evaluation matrix (T_E). The corresponding gap matrix is provided in Table 24.4. In the last step, using Eqn. (24.10), values of the criteria functions for the alternative batteries are computed by simply taking the sum of the elements in the total gap matrix. In Table 24.5, all the considered batteries are ranked in ascending order of their criteria functions. The ranking of the batteries is achieved as Li-ion → Ni-Cd → Pb-acid → Ni-MH. On the other hand, using Ashby's principle, Panday & Bansal [11] identified Li-ion as the most preferred battery, followed by Ni-MH battery. While solving the same problem using TOPSIS and VIKOR methods, Panday & Bansal [11] determined the rankings of the alternative batteries as Li-ion → Ni-MH → Ni-Cd → Pb-acid and Li-ion → Ni-Cd → Ni-MH → Pb-acid, respectively. It can be observed from Table 24.1 that Li-ion battery has higher specific energy and energy density than Ni-based batteries. Thus, selection of Li-ion as the most appropriate battery for electric hybrid vehicles is quite justifiable.

TABLE 24.1 Decision Matrix for the Battery Selection Problem [11]

Battery	Energy Density	Specific Energy	Cell Voltage	Electrical Efficiency	Self-Discharge Rate	Cost	Energy/ Cycle
Li-ion	180	180	3.6	85	5	24	8.6
Ni-MH	140	70	1.2	66	30	18.5	7.5
Ni-Cd	100	50	1.2	90	10	11	4.5
Pb-acid	70	35	2	90	20	8.5	24

TABLE 24.2 Theoretical Evaluation Matrix (T_E) for the Battery Selection Problem

Criteria	Energy Density	Specific Energy	Cell Voltage	Electrical Efficiency	Self-Discharge Rate	Cost	Energy/ Cycle
t_{E_j}	0.0122	0.0116	0.0917	0.0627	0.0015	0.0620	0.00834825

TABLE 24.3 Real Evaluation Matrix (T_R) for the Battery Selection Problem

Battery	Energy Density	Specific Energy	Cell Voltage	Electrical Efficiency	Self-Discharge Rate	Cost	Energy/ Cycle
Li-ion	0.0917	0.0627	0.0015	0.0490	0.0122	0	0.0018
Ni-MH	0.0584	0.0151	0	0	0	0.0041	0.0013
Ni-Cd	0.0250	0.0065	0	0.0620	0.0097	0.0097	0
Pb-acid	0	0	0.0005	0.0620	0.0049	0.0116	0.0083

TABLE 24.4 Total Gap Matrix (T_G) for the Battery Selection Problem

Battery	Energy Density	Specific Energy	Cell Voltage	Electrical Efficiency	Self-Discharge Rate	Cost	Energy/ Cycle
Li-ion	0	0	0	0.0129	0	0.0116	0.0066
Ni-MH	0.0334	0.0476	0.0015	0.0620	0.0122	0.0075	0.0071
Ni-Cd	0.0667	0.0562	0.0015	0	0.0024	0.0019	0.0083
Pb-acid	0.0073	0.0000	0.0917	0.0627	0.0010	0	0

TABLE 24.5 Criteria Functions for the Alternative Batteries (Q_i)

Battery	Q_i	Rank
Li-ion	0.0311	1
Ni-MH	0.1711	4
Ni-Cd	0.1371	2
Pb-acid	0.1627	3

KEYWORDS

- **durability**
- **efficiency**
- **electric hybrid vehicles**
- **energy density**
- **ideal solution**
- **multi-attributive real-ideal comparative analysis**

REFERENCES

1. Chakraborty, S., & Chatterjee, P., (2013). Selection of materials using multi-criteria decision-making methods with minimum data. *Decision Science Letters, 2*(3), 135–148.
2. Pamučar, D., Vasin, L., & Lukovac, L., (2014). Selection of railway level crossings for investing in security equipment using hybrid DEMATEL-MARICA model. In: *Proc. of the XVI International Scientific-Expert Conference on Railways* (pp. 89–92). Serbia.
3. Gigović, L., Pamučar, D., Bajić, Z., & Milicević, M., (2016). The combination of expert judgment and GIS-MAIRCA analysis for the selection of sites for ammunition depot. *Sustainability, 8*(4), 1–30.
4. Chatterjee, K., Pamučar, D., & Zavadskas, E. K., (2018). Evaluating the performance of suppliers based on using the R'AMATEL-MAIRCA method for green supply chain implementation in electronics industry. *Journal of Cleaner Production, 184*, 101–129.
5. Pamučar, D., Mihajlović, M., Obradović, R., & Atanasković, P., (2017). Novel approach to group multi-criteria decision making based on interval rough numbers: Hybrid DEMATEL-ANP-MAIRCA model. *Expert Systems with Applications, 88*, 58–80.
6. Gigović, L., Pamučar, D., Bajić, Z., & Milićević, M., (2016). The combination of expert judgment and GIS-MAIRCA analysis for the selection of sites for ammunition depots. *Sustainability, 8*. doi: 10.3390/su8040372.
7. Pamučar, D., Lukovac, V., Božanić, D., & Komazec, N., (2018). Multi-criteria FUCOM-MAIRCA model for the evaluation of level crossings: Case study in the Republic of Serbia. *Operational Research in Engineering Sciences: Theory and Applications, 1*(1), 108–129.
8. Ulutaş, A., (2019). The selection of catering firm with SWARA and MAIRCA methods. *Business & Management Studies: An International Journal, 7*(4), 1467–1479.
9. Bakir, M., Akan, Ş., Kiraci, K., Karabasevic, D., Stanujkic, D., & Popovic, G., (2020). Multiple-criteria approach of the operational performance evaluation in the airline

industry: Evidence from the emerging markets. *Romanian Journal of Economic Forecasting, XXIII*(2), 149–172.
10. Chatterjee, S., & Chakraborty, S., (2021). A multi-attributive ideal-real comparative analysis-based approach for piston material selection. *OPSEARCH* https://doi.org/10.1007/s12597-021-00536-2.
11. Panday, A., & Bansal, H. O., (2016). Multi-objective optimization in battery selection for hybrid electric vehicle applications. *Journal of Electrical Systems, 12*(2), 325–343.

CHAPTER 25

MEASUREMENT ALTERNATIVES AND RANKING ACCORDING TO COMPROMISE SOLUTION (MARCOS) METHOD

The development of measurement alternatives and ranking according to compromise solution (MARCOS) method is based on specifying the relationship between the alternatives and reference values (ideal and anti-ideal alternatives). Based on these relationships, the corresponding utility functions of the considered alternatives are calculated, and a compromise ranking is formulated with respect to the ideal utility functions. Utility functions define the position of an alternative with regard to ideal and anti-ideal solutions (AAIs). The best alternative should be positioned nearest to the ideal solution and, at the same time, farthest from the anti-ideal reference point. The steps involved in the MARCOS method are enumerated below [1]:

- **Step 1: Develop the Initial Decision Matrix:** For a decision-making problem with m alternatives and n criteria, the following matrix (X) is formulated.

$$X = \begin{bmatrix} x_{11} & x_{12} & \dots & x_{1n} \\ x_{21} & x_{22} & \dots & x_{2n} \\ \dots & \dots & \dots & \dots \\ x_{m1} & x_{m2} & \dots & x_{mn} \end{bmatrix} \tag{25.1}$$

- **Step 2: Formulate the Extended Initial Matrix:** From the initial decision matrix, the corresponding extended decision matrix is formulated. In this step, extension of the initial matrix is performed by defining the ideal (AI) and anti-ideal (AAI) solutions.

Multi-Criteria Decision-Making Methods in Manufacturing Environments: Models and Applications.
Shankar Chakraborty, Prasenjit Chatterjee, & Partha Protim Das (Authors)

$$X' = \begin{bmatrix} x_{aa1} & x_{aa2} & \dots & x_{aan} \\ x_{11} & x_{12} & \dots & x_{1n} \\ x_{21} & x_{22} & \dots & x_{2n} \\ \dots & \dots & \dots & \dots \\ x_{m1} & x_{m2} & \dots & x_{mn} \\ x_{ai1} & x_{ai2} & \dots & x_{ain} \end{bmatrix} \tag{25.2}$$

The anti-ideal solution (*AAI*) is an alternative with worst characteristics, while the ideal solution (*AI*) is an alternative having the best characteristics. Depending on the type of the criterion, *AAI* and *AI* can be defined as follows:

$$AAI = \min_i x_{ij} \text{ if } j \in B \text{ and} \max_i x_{ij} \text{ if } j \in C \tag{25.3}$$

$$AI = \max_i x_{ij} \text{ if } j \in B \text{ and} \min_i x_{ij} \text{ if } j \in C \tag{25.4}$$

where; *B* denotes a group of beneficial criteria and *C* signifies a group of cost criteria.

- **Step 3: Normalize the Extended Initial Matrix (X′):** The elements of the normalized matrix $N = [n_{ij}]_{m \times n}$ can be derived based on the following equations:

$$n_{ij} = \frac{x_{ai}}{x_{ij}} \text{ if } j \in C \tag{25.5}$$

$$n_{ij} = \frac{x_{ij}}{x_{ai}} \text{ if } j \in B \tag{25.6}$$

- **Step 4: Calculate the Weighted Normalized Decision Matrix, $V = [v_{ij}]_{m \times n}$:** The weighted normalized decision matrix *V* can be derived after multiplying the elements of the normalized matrix *N* by the corresponding criteria weights.

$$v_{ij} = n_{ij} \times w_j \tag{25.7}$$

where; w_j is the weight of jth criterion under consideration.

- **Step 5: Compute the Utility Degrees of Alternatives (K_i):** Based on the following expressions, the utility degrees of an alternative with respect to the anti-ideal and ideal solutions are computed.

$$K_i^- = \frac{S_i}{S_{aai}} \tag{25.8}$$

$$K_i^+ = \frac{S_i}{S_{ai}} \tag{25.9}$$

where; S_i ($i = 1, 2, \ldots, m$) is the sum of the elements of the weighted normalized decision matrix (V), as given in Eqn. (25.10), S_{aai} is the sum of elements of weighted normalized matrix for the anti-ideal alternative and S_{ai} is the sum of elements of weighted normalized matrix for the ideal alternative.

$$S_i = \sum_{i=1}^{n} v_{ij} \tag{25.10}$$

- **Step 6: Determine the Utility Functions of the Alternatives $f(K_i)$:** The utility function is the compromise of an alternative with respect to the ideal and anti-ideal solutions.

$$f(K_i) = \frac{K_i^+ + K_i^-}{1 + \frac{1 - f(K_i^+)}{f(K_i^+)} + \frac{1 - f(K_i^-)}{f(K_i^-)}} \tag{25.11}$$

where; $f(K_i^-)$ is the utility function in relation to the anti-ideal solution; and $f(K_i^+)$ is the utility function related to the ideal solution. The utility functions in relation to the ideal and anti-ideal solutions can be estimated using the following two equations:

$$f(K_i^-) = \frac{K_i^+}{K_i^+ + K_i^-} \tag{25.12}$$

$$f(K_i^+) = \frac{K_i^-}{K_i^+ + K_i^-} \tag{25.13}$$

- **Step 7: Rank the Alternatives:** Based on the descending values of the utility function, the candidate alternatives are finally ranked. The best alternative should have the maximum utility function value.

25.1 LITERATURE REVIEW

Biswas [2] proposed the integrated application of PIPRECIA, MABAC, CoCoSo, and MARCOS methods for solving a supplier selection problem in an Indian healthcare unit. Ozdagoglu et al. [3] applied MARCOS method to rank 166 Turkish universities based on five evaluation criteria. The derived ranking results were also compared with those obtained from CoCoSo method. Puška et al. [4] attempted to prove the application potentiality of MARCOS method in selecting the best project management software from a pool of four alternatives, i.e., Smartsheet, Asana, Microsoft Project and

Basecamp based on six evaluation criteria. Stević & Brković [5] solved an evaluation problem of human resources in a transport system while considering 23 drivers and five criteria. The FUCOM method was first employed to estimate the significance of the criteria, while MARCOS method was applied to evaluate the potential solutions.

25.2 MANUFACTURING PROCESS SELECTION PROBLEM

Ghaleb et al. [6] applied various MCDM techniques, like AHP, TOPSIS, and VIKOR methods in order to identify the most suitable manufacturing process for a submersible water pump driven by a built-in submersible motor driving multistage impeller/diffuser set. The considered product consisted of two sub-assemblies enclosed with two end supports, within a stainless-steel tube. The function of the stainless-steel tube is to hold either motor-stator or multistage impeller/diffuser set. For manufacturing of this product, five processes, i.e., sand casting, gravity die casting, investment casting, pressure die casting and additive manufacturing were taken into account. The relative performance of those processes was evaluated with respect to productivity, accuracy, complexity, flexibility, material utilization, quality, and operation cost. In Table 25.1, the initial decision matrix containing five alternative manufacturing processes and seven criteria is presented. In this table, a relative scale of 1–9 was employed to assess the performance of the manufacturing processes with regard to each of the evaluation criteria. This table also provides the ideal (*AI*) and anti-ideal (*AAI*) solutions for this decision-making problem. While solving the same problem using AHP method, Ghaleb et al. [6] estimated the corresponding criteria weights, as shown in Table 25.2. The extended decision matrix of Table 25.1 and the criteria weight set of Table 25.2 are adopted here for MARCOS method-based evaluation and ranking of the considered manufacturing processes.

Following the procedural steps of MARCOS method, the elements of the extended decision matrix are normalized, as shown in Table 25.3. Now, employing the elements of this normalized decision matrix and criteria weights, the corresponding weighted normalized decision matrix is developed in Table 25.4. Based on Eqns. (25.8)–(25.9), the utility degrees of all the alternative manufacturing processes with respect to the anti-ideal and ideal solutions are calculated, as exhibited in Table 25.5. Using those utility degree values, the corresponding utility function for each of the alternatives is determined. These values are finally adopted to rank the alternative manufacturing processes depending on their capability to produce the considered

TABLE 25.1 Extended Initial Decision Matrix [6]

Alternative	Productivity	Accuracy	Complexity	Flexibility	Material Utilization	Quality	Operation Cost
AAI	2	3	3	6	3	2	6
Sand casting	5	3	9	9	3	2	6
Gravity die casting	9	8	3	6	6	9	8
Investment casting	4	9	9	8	9	9	9
Pressure die casting	9	9	4	8	8	9	9
Additive manufacturing	2	8	9	9	9	9	9
AI	9	9	9	9	9	9	9

Source: Adapted from [6].

TABLE 25.2 Criteria Weights for the Manufacturing Process Selection Problem

Criteria	Productivity	Accuracy	Complexity	Flexibility	Material Utilization	Quality	Operation Cost
Weight	0.264454	0.32392	0.104459	0.102623	0.086151	0.050267	0.068127

TABLE 25.3 Normalized Decision Matrix for the Manufacturing Process Selection Problem

Alternative	Productivity	Accuracy	Complexity	Flexibility	Material Utilization	Quality	Operation Cost
AAI	0.222	0.333	0.333	0.667	0.333	0.222	0.667
Sand casting	0.556	0.333	1.000	1.000	0.333	0.222	0.667
Gravity die casting	1.000	0.889	0.333	0.667	0.667	1.000	0.889
Investment casting	0.444	1.000	1.000	0.889	1.000	1.000	1.000
Pressure die casting	1.000	1.000	0.444	0.889	0.889	1.000	1.000
Additive manufacturing	0.222	0.889	1.000	1.000	1.000	1.000	1.000
AI	1.000	1.000	1.000	1.000	1.000	1.000	1.000

TABLE 25.4 Weighted Normalized Matrix for the Manufacturing Process Selection Problem

Alternative	**Productivity**	**Accuracy**	**Complexity**	**Flexibility**	**Material Utilization**	**Quality**	**Operation Cost**
AAI	0.059	0.108	0.035	0.068	0.029	0.011	0.045
Sand casting	0.147	0.108	0.104	0.103	0.029	0.011	0.045
Gravity die casting	0.264	0.288	0.035	0.068	0.057	0.050	0.061
Investment casting	0.118	0.324	0.104	0.091	0.086	0.050	0.068
Pressure die casting	0.264	0.324	0.046	0.091	0.077	0.050	0.068
Additive manufacturing	0.059	0.288	0.104	0.103	0.086	0.050	0.068
AI	0.264	0.324	0.104	0.103	0.086	0.050	0.068

TABLE 25.5 Results of MARCOS Method

Alternative	S_i	K_i^-	K_i^+	$f(K_i^-)$	$f(K_i^+)$	$f(K_i)$	Rank
AAI	0.355						
Sand casting	0.547	1.540	0.547	0.262	0.738	0.501	5
Gravity die casting	0.824	2.319	0.824	0.262	0.738	0.754	3
Investment casting	0.842	2.369	0.842	0.262	0.738	0.770	2
Pressure die casting	0.921	2.592	0.921	0.262	0.738	0.843	1
Additive manufacturing	0.758	2.134	0.758	0.262	0.738	0.694	4
AI	0.998	–	–				

product. It can be noticed that pressure die casting is the most appropriate process for this product, followed by investment casting. On the other hand, for manufacturing the same product, sand casting emerges out as the most unsuitable process. While solving the same problem using AHP, TOPSIS, and VIKOR methods, Ghaleb et al. [6] identified pressure die casting, investment casting, and sand casting as the first, second, and last ranked manufacturing processes respectively, for all the three adopted MCDM techniques, whereas, there were variations in the intermediate rankings for gravity die casting and additive manufacturing processes.

KEYWORDS

- **anti-ideal solution**
- **flexibility**
- **ideal solution**
- **ideal utility functions**
- **material utilization**
- **measurement alternatives**

REFERENCES

1. Stević, Ž., Pamučar, D., Puška, A., & Chatterjee, P., (2020). Sustainable supplier selection in healthcare industries using a new MCDM method: Measurement of alternatives and ranking according to compromise solution (MARCOS). *Computers & Industrial Engineering, 140*, 106231.
2. Biswas, S., (2020). Measuring performance of healthcare supply chains in India: A comparative analysis of multicriteria decision making methods. *Decision Making: Applications in Management and Engineering, 3*(2), 162–189.
3. Ozdagoglu, A., Ulutas, A., & Keles, M. K., (2020). The ranking of Turkish universities with COCOSO and MARCOS. *Journal of Economics, Business & Organization Research*, 374–392.
4. Puška, A., Stojanović, I., Maksimović, A., & Osmanović, N., (2020). Project management software evaluation by using the measurement of alternatives and ranking according to compromise solution (MARCOS) method. *Operational Research in Engineering Sciences: Theory and Applications, 3*(1), 89–102.
5. Stević, Ž., & Brković, N., (2020). A novel integrated FUCOM-MARCOS model for evaluation of human resources in a transport company. *Logistics, 4*. doi:10.3390/logistics4010004.

6. Ghaleb, A. M., Kaid, H., Alsamhan, A., Mian, S. H., & Hidri, L., (2020). Assessment and comparison of various MCDM approaches in the selection of manufacturing process. *Advances in Materials Science and Engineering* (p. 16). Article ID: 4039253. https://doi.org/10.1155/2020/4039253.

CHAPTER 26

COMBINED COMPROMISE SOLUTION (COCOSO) METHOD

The combined compromise solution (CoCoSo) method is developed while integrating WSM and WPM techniques [1], taking into consideration the compromise solutions. The application of CoCoSo method for ranking of the alternatives involves the following steps:

- **Step 1:** Develop the initial decision matrix:

$$X = \begin{bmatrix} x_{11} & x_{12} & \cdots & x_{1n} \\ x_{21} & x_{22} & \cdots & x_{2n} \\ \cdots & \cdots & \cdots & \cdots \\ x_{m1} & x_{m2} & \cdots & x_{mn} \end{bmatrix} \tag{26.1}$$

where; x_{ij} is the performance score of ith alternative with respect to j^{th} criterion, m is the number of alternatives and n is the number of criteria.

- **Step 2:** In order to make all the elements of the decision matrix dimensionless and comparable, they are normalized using the following linear normalization equations:

  i. **For Beneficial Criterion:**

$$r_{ij} = \frac{x_{ij} - \min x_{ij}}{\max x_{ij} - \min x_{ij}} \tag{26.2}$$

  ii. **For Cost Criterion:**

$$r_{ij} = \frac{\max x_{ij} - x_{ij}}{\max x_{ij} - \min x_{ij}} \tag{26.3}$$

where; r_{ij} is the normalized element of x_{ij}.

Multi-Criteria Decision-Making Methods in Manufacturing Environments: Models and Applications.
Shankar Chakraborty, Prasenjit Chatterjee, & Partha Protim Das (Authors)

- **Step 3:** Now, taking into account WSM and WPM techniques, the corresponding performance indexes S_i and P_i for each of the candidate alternatives are determined.

$$S_i = \sum_{j=1}^{n} w_j \times r_{ij} \tag{26.4}$$

$$P_i = \sum_{j=1}^{n} \left(r_{ij}\right)^{w_j} \tag{26.5}$$

where; w_j is the relative weight of jth criterion.

- **Step 4:** In this step, three different appraisal scores of the alternatives are computed using the following aggregation strategies.

$$k_{ia} = \frac{S_i + P_i}{\sum_{i=1}^{m}\left(S_i + P_i\right)} \tag{26.6}$$

$$k_{ib} = \frac{S_i}{\min_i S_i} + \frac{P_i}{\min_i P_i} \tag{26.7}$$

$$k_{ic} = \frac{\lambda \times S_i + (1-\lambda) \times P_i}{\lambda \times \max_i S_i + (1-\lambda) \times \max_i P_i}, \quad 0 \le \lambda \le 1 \tag{26.8}$$

It can be noted that Eqn. (26.6) is developed based on the arithmetic mean of sums of WSM and WPM scores, whereas, Eqn. (26.7) is formulated expressing the sum of relative scores of WSM and WPM compared to the best. On the other hand, Eqn. (26.8) represents the balanced compromise of WSM and WPM model-based scores. In Eqn. (26.8), the value of λ varies between 0 and 1, with 0.5 as the most preferred value set by the decision-makers.

- **Step 5:** The final ranking of the alternatives is determined using k_i value, with the best alternative having the maximum k_i value.

$$k_i = \left(k_{ia} \times k_{ib} \times k_{ic}\right)^{\frac{1}{3}} + \frac{1}{3}\left(k_{ia} + k_{ib} + k_{ic}\right) \tag{26.9}$$

26.1 SURVEY OF THE LITERATURE

Barua et al. [2] integrated Taguchi methodology with CoCoSo method for studying the mechanical behavior of hybrid natural fiber reinforced nano-SiC particle composites, considering ultimate tensile strength, density, and flexural strength as the responses. While solving a supplier selection problem for a construction industry, Yazdani et al. [3] first applied DEMATEL and BWM to derive the relative importance of each of the seven evaluation criteria. In the later stage of the evaluation process, gray numbers are integrated with CoCoSo method for ranking of the 10 competing suppliers. The derived ranking was finally compared with that obtained using COPRAS method to prove the efficacy of CoCoSo-G approach in solving supplier selection problems. Bagal et al. [4] adopted CoCoSo and CODAS methods as effective multi-objective optimization techniques to determine the most suitable parametric combination for mechanical characterization of M30 concrete grade while treating compressive strength, split tensile strength and flexural strength as the measurable responses. Kumar & Verma [5] applied CoCoSo method to determine the optimal settings of drill speed, feed, and percentage of graphene oxide for achieving the most desired values of average surface roughness, mean roughness depth and circularity error, while performing machining operation on graphene nanocomposites. Principal component analysis was also adopted to investigate the influences of drilling parameters on the considered responses.

26.2 CONCRETE PUMP SELECTION PROBLEM

In the manufacturing and industrial domain, equipment selection is an important decision-making process involving opting for an optimal choice amongst several feasible alternatives, considering both the technical and economic benefits. Involvement of qualitative criteria and multiple decision makers also make the selection problem more complicated. Tam et al. [6] considered a concrete pump selection problem for a construction site and solved the same using SIR method. Concrete pumps are usually employed to cast concrete slabs and other structural elements. This problem consists of 10 alternative concrete pumps and nine evaluation criteria, i.e., maximum pumping speed (MPS) (in m^3/h), maximum vertical pumping distance (MVPD) (in m), workability (W) (slump in cm), maximum horizontal pumping distance (MHPD) (in m), maximum suitable aggregate diameter (MSAD) (in mm), weight of the equipment (WOE) (in tons), pipe diameter (PD) (in inch), machine power (MP) (in kW) and price (P) (in USD). Among these considered criteria, WOE and P are the non-beneficial attributes, and the remaining are the beneficial attributes. In a similar direction, another decision matrix of Table 26.1 is also developed for solving a new concrete pump selection problem. Applying AHP method, Tam et al. [6] determined the corresponding criteria weights as w_{MPS} = 0.141, w_{MVPD} = 0.114, w_{W} =

0.055, $w_{MHPD} = 0.102$, $w_{MSAD} = 0.172$, $w_{WOE} = 0.015$, $w_{PD} = 0.082$, $w_{MP} = 0.127$ and $w_P = 0.191$. These criteria weights are considered here for the CoCoSo method-based analysis. While solving this concrete pump selection problem using CoCoSo method, the decision matrix is first linearly normalized within an interval [0,1] using Eqns. (26.2) and (26.3), depending on the type of the criterion under consideration. This normalized decision matrix is depicted in Table 26.2.

TABLE 26.1 Decision Matrix for the Concrete Pump Selection Problem

Pump	MPS	MVPD	W	MHPD	MSAD	WOE	PD	MP	P
P1	57.15	86.25	5	297	57.5	2.7	5	99	3,57,000
P2	87.97	86.25	6	495	46	5.76	4	132	4,33,695
P3	52.78	69	6	330	46	7.74	4	135	6,29,782
P4	52.78	69	6	247	57.5	5.22	5	132	5,08,759
P5	70.38	69	3	264	46	6.84	7	123	4,25,102
P6	52.78	86.25	3	495	57.5	5.04	5	160	5,08,759
P7	68.54	89.7	6	327	69	8.46	6	164	3,41,445
P8	87.97	69	3	247	46	7.74	4	135	3,74,908
P9	87.97	96.6	3	495	46	4.05	8	164	3,48,616
P10	52.78	28.75	3	495	57.5	12.42	5	164	4,07,323

TABLE 26.2 Normalized Decision Matrix for the Concrete Pump Selection Problem

Pump	MPS	MVPD	W	MHPD	MSAD	WOE	PD	MP	P
P1	0.1242	0.8475	0.6667	0.2016	0.5	1	0.25	0	0.9461
P2	1	0.8475	1	1	0	0.6852	0	0.5077	0.6801
P3	0	0.5932	1	0.3347	0	0.4815	0	0.5538	0
P4	0	0.5932	1	0	0.5	0.7407	0.25	0.5077	0.4197
P5	0.5001	0.5932	0	0.0685	0	0.5741	0.75	0.3692	0.7099
P6	0	0.8475	0	1	0.5	0.7593	0.25	0.9385	0.4197
P7	0.4479	0.8983	1	0.3226	1	0.4074	0.5	1	1
P8	1	0.5932	0	0	0	0.4815	0	0.5538	0.8839
P9	1	1	0	1	0	0.8611	1	1	0.9751
P10	0	0	0	1	0.5	0	0.25	1	0.7715

Now, taking into account SAW and WPM approaches of CoCoSo method, the corresponding weighted comparability (S_i) and power weighted comparability (P_i) indexes are respectively computed using Eqns. (26.4)

and (26.5), as shown in Table 26.3. Using three different aggregation strategies, as presented in Eqns. (26.6)–(26.8), values of k_{ia}, k_{ib}, and k_{ic} for all the alternatives are calculated in Table 26.4. Finally, considering λ as 0.5, the values of k_{ia}, k_{ib}, and k_{ic} are aggregated using Eqn. (26.9) to provide the performance score for each of the alternative concrete pumps. The candidate alternatives are then ranked based on the descending order of their respective performance scores, as portrayed in Figure 26.1. It can be unveiled from this figure that P7 is the most preferred concrete pump, followed by P9, whereas, P3 is the worst chosen pump.

TABLE 26.3 Weighted Comparability (S_i) and Power Weighted Comparability (P_i) Indexes

Pump	S_i	P_i
P1	0.4735	7.3234
P2	0.5993	6.8222
P3	0.2343	4.7534
P4	0.3849	6.4826
P5	0.3977	6.3961
P6	0.5159	6.5965
P7	0.7906	8.7031
P8	0.455	4.8357
P9	0.7652	6.993
P10	0.4829	4.7318

TABLE 26.4 Performance Scores of the Alternative Concrete Pumps

Pump	k_{ia}	k_{ib}	k_{ic}	k_i	Rank
P1	0.1134	3.5686	0.8213	2.1938	4
P2	0.108	3.9991	0.7817	2.3259	3
P3	0.0726	2.0046	0.5254	1.2919	10
P4	0.0999	3.0125	0.7234	1.8802	7
P5	0.0988	3.0491	0.7156	1.8875	6
P6	0.1035	3.5955	0.7492	2.1359	5
P7	0.1381	5.2131	1	3.0134	1
P8	0.077	2.9638	0.5573	1.7022	9
P9	0.1129	4.7433	0.8172	2.6502	2
P10	0.0759	3.0606	0.5493	1.732	8

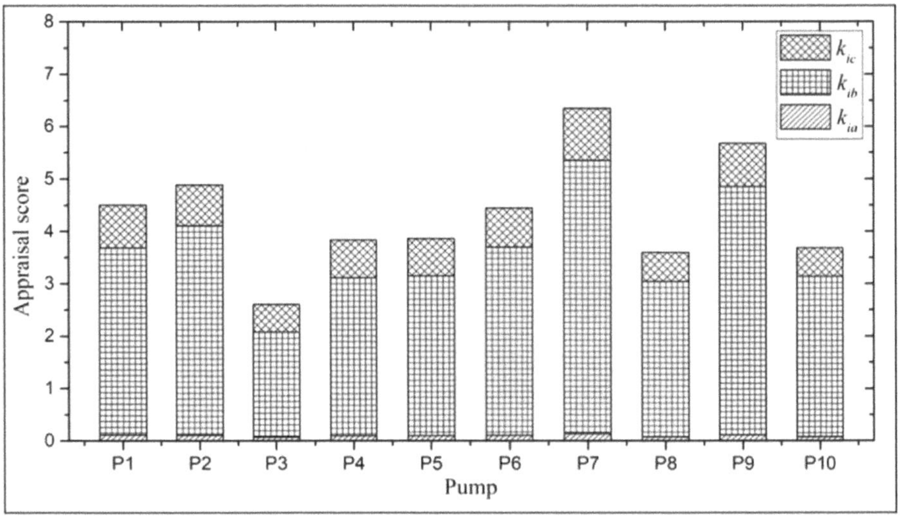

FIGURE 26.1 Ranking of the alternative concrete pumps.

KEYWORDS

- **combined compromise solution**
- **maximum horizontal pumping distance**
- **maximum suitable aggregate diameter**
- **maximum vertical pumping distance**
- **weighted product model**
- **weighted sum model**

REFERENCES

1. Yazdani, M., Zarate, P., Zavadskas, E. K., & Turskis, Z., (2019). A combined compromise solution (CoCoSo) method for multi-criteria decision-making problems. *Management Decision, 57*(9), 2501–2519.
2. Barua, A., Jeet, S., Bagal, D. K., Satapathy, P., & Agrawal, P. K., (2019). Evaluation of mechanical behavior of hybrid natural fiber reinforced nano SiC particles composite using hybrid Taguchi-Cocoso method. *International Journal of Innovative Technology and Exploring Engineering, 8*(10), 3341–3345.

3. Yazdani, M., Wen, Z., Liao, H., Banaitis, A., & Turskis, Z., (2019). A grey combined compromise solution (COCOSO-G) method for supplier selection in construction management. *Journal of Civil Engineering and Management, 25*(8), 858–874.
4. Bagal, D. K., Naik, B., Parida, B., Barua, A., Jeet, S., Singh, S. K., & Pattanaik, A. K., (2020). Comparative mechanical characterization of M30 concrete grade by fractional replacement of Portland pozzolana cement with industrial waste using CoCoSo and CODAS methods. *IOP Conf. Series: Materials Science and Engineering, 970*, 012015.
5. Kumar, J., & Verma, R. K., (2021). A novel methodology of combined compromise solution and principal component analysis (CoCoSo-PCA) for machinability investigation of graphene nanocomposites. *CIRP Journal of Manufacturing Science and Technology, 33*, 143–157.
6. Tam, C. M., Tong, T. K. L., & Wong, Y. W., (2004). Selection of concrete pump using the superiority and inferiority ranking method. *Journal of Construction Engineering and Management, 130*, 827–834.

CHAPTER 27

RANKING OF ALTERNATIVES THROUGH FUNCTIONAL MAPPING OF CRITERION SUB-INTERVALS INTO A SINGLE INTERVAL (RAFSI) METHOD

It has been observed that most of the MCDM techniques suffer from the rank reversal problem, i.e., when an existing alternative is removed from the set or a new alternative is included into the set, the final rankings of all the alternatives usually alter. To avoid this rank reversal problem, Žižović et al. [1] proposed a novel MCDM method in the form of RAFSI. Its application steps are detailed out as below [2]:

- **Step 1:** Develop the following decision matrix consisting of m alternatives and n criteria.

$$D = \begin{bmatrix} a_{11} & a_{12} & \dots & a_{1n} \\ a_{21} & a_{22} & \dots & a_{2n} \\ \dots & \dots & \dots & \dots \\ a_{m1} & a_{m2} & \dots & a_{mn} \end{bmatrix}$$

  where; a_{ij} is the performance score of ith alternative with respect to j^{th} criterion.
- **Step 2:** Define the ideal and the anti-ideal values. In the decision matrix, two values, i.e., a_{I_j} and a_{N_j} are defined for each criterion C_j ($j = 1, 2, \dots, n$), where a_{I_j} is the ideal value of C_j criterion and a_{N_j} is the anti-ideal value of C_j criterion. It is quite obvious that for beneficial criterion $a_{I_j} > a_{N_j}$ and for non-beneficial criterion $a_{I_j} < a_{N_j}$.
- **Step 3:** Map each of the elements of the decision matrix into corresponding criteria intervals. These criteria intervals can be defined as follows:

Multi-Criteria Decision-Making Methods in Manufacturing Environments: Models and Applications.
Shankar Chakraborty, Prasenjit Chatterjee, & Partha Protim Das (Authors)

a. $C_j \in \{a_{N_j}, a_{I_j}\}$ for beneficial criterion;
b. $C_j \in \{a_{I_j}, a_{N_j}\}$ for non-beneficial criterion.

To make all the criteria of the initial decision matrix equal or convert them into a criteria interval $[n_1, n_{2k}]$, a sequence of numbers is formed where $(k - 1)$ number of points are inserted between the highest and the lowest values of the criteria interval.

$$n_1 < n_2 \leq n_3 < n_4 \leq n_5 < n_6 \leq \ldots \leq n_{2k-1} < n_{2k} \quad (27.1)$$

It can be noted that this criteria interval remains constant for all criteria having n_1 and n_{2k} fixed points. The sub-intervals of the criteria are then mapped into the corresponding criteria intervals. Thus, the map of minimum value a_{N_j} (for beneficial criterion) and a_{I_j} (for non-beneficial criterion) is n_1. In the similar direction, the map of maximum value a_{I_j} (for beneficial criterion) and a_{N_j} (for non-beneficial criterion) is n_{2k}. It is always recommended that the ideal value is at least six times better than the anti-ideal value, i.e., $n_1 = 1$ and $n_{2k} = 6$. Other values of n_1 and n_{2k}, like $n_1 = 1$ and $n_{2k} = 9$ may also be considered.

A function $f_s(x)$ is now defined in Eqn. (27.2) to map sub-intervals into criteria interval $[n_1, n_{2k}]$. The endpoints of this interval $[n_1, n_{2k}]$ would determine the ratio (to be determined by the concerned decision maker) of a barely acceptable alternative to the ideal alternative.

$$f_s(x) = \frac{n_{2k} - n_1}{a_{I_j} - a_{N_j}} x + \frac{a_{I_j}.n_1 - a_{N_j}.n_{2k}}{a_{I_j} - a_{N_j}} \quad (27.2)$$

where; n_{2k} and n_1 denote the relation to depict the extent to which the ideal value is preferred over the anti-ideal value.

Eqn. (27.2) can be used to map a part of the interval $[a_{N_j}, a_{I_j}]$ into an interval $[n_1, n_{2k}]$. Here, all these parts, i.e., all the functions $f_1(x)$, $f_2(x)$, …, $f_n(x)$ denote a function $f_s(x)$ to map the entire criterion interval into a defined numerical interval. Besides representing a function to map a part of an interval, Eqn. (27.2) can also map a complete criterion interval into the corresponding numerical interval. Thus, the numbers a_{I_j} and a_{N_j} can either represent values from inside the criterion interval or endpoints of the criterion interval.

Thus, a standardized decision matrix $S = [s_{ij}]_{m \times n}$ ($i = 1, 2, \ldots, m$; $j = 1, 2, \ldots, n$) is formulated where all of its elements are mapped into the interval $[n_1, n_{2k}]$. After functional mapping of the elements of the decision matrix into criteria interval $[n_1, n_{2k}]$, the condition $n_1 \leq s_{ij} \leq n_{2k}$ is achieved for every i, j.

The elements of $S = [s_{ij}]_{m\times n}$ are obtained employing Eqn. (27.2), i.e., $s_{ij} = f_{A_i}(C_j)$. It can be noted that:

a. In case of beneficial criterion, for a_{x_j} (where $a_{x_j} < a_{I_j}$), the equality $f(a_{x_j}) = f(a_{I_j})$ holds.
b. in case of non-beneficial criterion, for a_{x_j} (where $a_{x_j} > a_{I_j}$), the equality $f(a_{x_j}) = f(a_{I_j})$ holds.

➢ **Step 4:** Calculate the arithmetic and harmonic means. Based on Eqns. (27.3) and (27.4), the corresponding arithmetic and harmonic means are respectively computed for minimum and maximum sequence of the elements n_1 and n_{2k}.

$$A = \frac{n_1 + n_{2k}}{2} \tag{27.3}$$

$$H = \frac{2}{\frac{1}{n_1} + \frac{1}{n_{2k}}} \tag{27.4}$$

➢ **Step 5:** Formulate the normalized decision matrix, $(i = 1, 2, \ldots, m; j = 1, 2, \ldots, n)$. Now, employing Eqns. (27.5) and (27.6), all the elements of the standardized decision matrix S are normalized, while transferring them into the interval [0,1].

i. For Benefit Criterion:

$$\hat{s}_{ij} = \frac{s_{ij}}{2A} \tag{27.5}$$

ii. For cost criterion:

$$\hat{s}_{ij} = \frac{H}{2s_{ij}} \tag{27.6}$$

Thus, from the standardized decision matrix, a new normalized decision matrix is developed, as shown below:

$$\hat{S} = \begin{bmatrix} \hat{s}_{11} & \hat{s}_{12} & \cdots & \hat{s}_{1n} \\ \hat{s}_{21} & \hat{s}_{22} & \cdots & \hat{s}_{2n} \\ \cdots & \cdots & \cdots & \cdots \\ \hat{s}_{m1} & \hat{s}_{m2} & \cdots & \hat{s}_{mn} \end{bmatrix} \tag{27.7}$$

where;

For the elements of the normalized decision matrix, the following relations must apply.

i. **For Beneficial Criterion:**

$$0 < \frac{n_1}{2A} \le \hat{s}_{ij} \le \frac{n_{2k}}{2A} < 1 \tag{27.8}$$

ii. **For beneficial criterion:**

$$0 < \frac{H}{2n_{2k}} \le \hat{s}_{ij} \le \frac{H}{2n_1} < 1 \tag{27.9}$$

- **Step 6:** Compute the criteria function values of all the alternatives $V(A_i)$ using the following expression:

$$V(A_i) = w_1 \times \hat{s}_{i1} + w_2 \times \hat{s}_{i2} + \ldots + w_n \times \hat{s}_{in} \tag{27.10}$$

The candidate alternatives are then ranked from the best to the worst based on the descending order of the calculated $V(A_i)$ values.

As RAFSI is an entirely new method, developed only in 2020, its application in the real-time manufacturing environment is limited in the literature. Alossta et al. [3] applied AHP and RAFSI methods for solving a location selection problem for emergency medical services considering four criteria and five alternatives. The corresponding criteria weights were determined using AHP method, whereas, the alternative locations were ranked using RAFSI method.

27.1 WELDING PROCESS SELECTION PROBLEM

Welding is one of the most important primary production processes intended to join two or more pieces of similar or dissimilar materials so as to achieve complete coalescence. Using the welding process, almost all materials can be joined, but not always by the same welding process. The application a particular welding process for the joining operation mainly depends on the mechanical properties of the constituent materials and requirements to fulfill the desired quality of weld joint at minimum cost. In some cases, depending on the requirements of the end products, there may not be any other choice, and only a specific welding process needs to be considered without any other alternative. But there are other instances where a suitable welding process needs to be identified from a set of equally feasible alternatives in producing the end product based on a number of evaluation criteria.

Let us consider an MCDM problem to select the most appropriate arc welding process to join mild steel (0.2% C) of 6 mm thickness. The performance of three alternative arc welding processes, e.g., W_1 (gas tungsten arc welding), W_2 (gas metal arc welding) and W_3 (shielded metal arc welding) is evaluated based on six criteria, i.e., C_1 (weld quality), C_2 (availability of consumables), C_3 (operator fatigue), C_4 (skill required), C_5 (cleaning required after welding) and C_6 (initial preparation required). All these evaluation criteria are subjective and are converted into their corresponding crisp (objective) values using a fuzzy conversion scale. The relative importance of those six criteria is determined as 0.303, 0.091, 0.182, 0.121, 0.242, and 0.061, respectively. Thus, C_1 is the most important criteria and C_6 is the least important one. The decision matrix of this arc welding process selection problem is provided in Table 27.1. Among these criteria, C_1 and C_2 are benefit attributes, and C_3, C_4, C_5, and C_6 are cost attributes. Now, this arc welding process selection problem is solved using RAFSI method.

TABLE 27.1 Decision Matrix for the Arc Welding Process Selection Problem

Welding Process	C_1	C_2	C_3	C_4	C_5	C_6
W_1	0.865	0.745	0.865	0.590	0.590	0.745
W_2	0.665	0.865	0.745	0.665	0.745	0.865
W_3	0.590	0.590	0.590	0.745	0.865	0.590

Now, based on the procedural steps of this method, two values, i.e., a_{I_j} and a_{N_j} are defined for each criterion C_j ($j = 1, 2, \ldots, n$), as shown below:

$$a_{I_j} = [0.9\ 0.9\ 0.5\ 0.5\ 0.5\ 0.5]$$
$$a_{N_j} = [0.5\ 0.5\ 0.9\ 0.8\ 0.9\ 0.9]$$

Thus, it leads to the ranges of values for all the considered criteria as $C_1 \in [0.5, 0.9]$; $C_2 \in [0.5, 0.9]$; $C_3 \in [0.5, 0.9]$; $C_4 \in [0.5, 0.8]$; $C_5 \in [0.5, 0.9]$; and $C_6 \in [0.5, 0.9]$. Now, Eqn. (27.2) is adopted to map a part of the interval $[a_{I_j}, a_{N_j}]$ into an interval $[n_1, n_{2k}]$. Here, all these parts, i.e., all the functions $f_1(x), f_2(x), \ldots, f_n(x)$ denote a function $f_s(x)$ to map the entire criterion interval into a defined numerical interval. Table 27.2 shows this mapping of elements of the initial decision matrix into the corresponding criteria intervals. In this calculation process, values of the arithmetic mean (A) and harmonic mean (H) are estimated as 3.50 and 1.71, respectively.

$$f_i(C_1) = 12.5 \times C_1 - 5.25$$
$$f_i(C_2) = 12.5 \times C_2 - 5.25$$
$$f_i(C_3) = 12.5 \times C_3 - 5.25$$
$$f_i(C_4) = 16.667 \times C_4 - 7.333$$
$$f_i(C_5) = 12.5 \times C_5 - 5.25$$
$$f_i(C_6) = 12.5 \times C_6 - 5.25$$

TABLE 27.2 Mapping of Elements of the Initial Decision Matrix into Criteria Intervals

Welding Process	C_1	C_2	C_3	C_4	C_5	C_6
W_1	5.5625	2.1250	5.5625	2.5000	4.0625	4.0625
W_2	3.0625	4.0625	4.0625	3.7500	5.5625	5.5625
W_3	2.1250	5.5625	2.1250	5.0833	2.1250	2.1250

Using Eqns. (27.5) and (27.6), the normalized decision matrix for this arc welding process selection problem is subsequently developed, as exhibited in Table 27.3. Finally, the criteria function values of all the alternative processes are computed while applying Eqn. (27.10). These criteria function values are provided in Table 27.4. Based on these values, it can be revealed that W_3 (shielded metal arc welding) emerges out as the most appropriate welding process to perform the considered joining operation. On the other hand, W_2 (gas metal arc welding) is the least preferred welding process.

TABLE 27.3 Normalized Decision Matrix for the Arc Welding Process Selection Problem

Welding Process	C_1	C_2	C_3	C_4	C_5	C_6
W_1	0.7946	0.3036	0.1541	0.3429	0.2110	0.2110
W_2	0.4375	0.5804	0.2110	0.2286	0.1541	0.1541
W_3	0.3036	0.7946	0.4034	0.1686	0.4034	0.4034

TABLE 27.4 Calculated Criteria Functions of the Arc Welding Processes

Welding Process	V_i	Rank
W_1	0.4158	2
W_2	0.3625	3
W_3	0.4394	1

KEYWORDS

- **criterion interval**
- **functional mapping**
- **mechanical properties**
- **multi-criteria decision making**
- **single interval**
- **welding process**

REFERENCES

1. Žižović, M., Pamučar, D., Albijanić, M., Chatterjee, P., & Pribićević, I., (2020). Eliminating rank reversal problem using a new multi-attribute model - the RAFSI method. *Mathematics, 8*. doi: 10.3390/math8061015.
2. Pamučar, D., Žižović, M., Marinković, D., Doljanica, D., Jovanović, S. V., & Brzaković, P., (2020). Development of a multi-criteria model for sustainable reorganization of a healthcare system in an emergency situation caused by the COVID-19 pandemic. *Sustainability, 12*, doi: 10.3390/su12187504.
3. Alossta, A., Elmansouri, O., & Badi, I., (2021). Resolving a location selection problem by means of an integrated AHP-RAFSI approach. *Reports in Mechanical Engineering, 2*(1), 135–142.

CHAPTER 28

SIMULTANEOUS EVALUATION OF CRITERIA AND ALTERNATIVES (SECA) METHOD

The SECA method deals with evaluating a multi-objective non-linear mathematical model to determine the overall performance scores of the alternatives and the weights of the criteria simultaneously [1]. It is based on two reference points defined for criteria weights. The first reference point is defined using the within-criterion variation information obtained as the standard deviation of each criterion, and the second reference point is defined based on between-criteria variation information derived as the correlation measure. Thus, the corresponding mathematical model consists of three objective functions, i.e., (a) maximizing the overall performance of the alternatives; (b) minimizing the deviation of criteria weights from the reference point based on within-criterion variation information; and (c) minimizing the deviation of criteria weights from the reference point based on between-criterion variation information. The procedural steps of SECA method are explained as below:

- **Step 1: Development of the Initial Decision Matrix:** In this step, a decision matrix is formed based on the performance information of *m* alternatives against *n* criteria.

$$X = \begin{bmatrix} x_{11} & x_{12} & \cdots & x_{1n} \\ x_{21} & x_{22} & \cdots & x_{2n} \\ \cdots & \cdots & \cdots & \cdots \\ x_{m1} & x_{m2} & \cdots & x_{mn} \end{bmatrix} \quad (28.1)$$

where; x_{ij} denotes the performance value of ith alternative on j^{th} criterion and $x_{ij} > 0$.

Multi-Criteria Decision-Making Methods in Manufacturing Environments: Models and Applications.
Shankar Chakraborty, Prasenjit Chatterjee, & Partha Protim Das (Authors)

- **Step 2: Normalization of the Decision Matrix:** The elements of the initial decision matrix are normalized to develop the normalized decision matrix. Depending on the type of criterion considered, any of the following two equations can be employed.

$$n_{ij} = \frac{x_{ij}}{\max_i x_{ij}} \quad \text{if } j \in \text{beneficial criteria} \tag{28.2}$$

$$n_{ij} = \frac{\min_i x_{ij}}{x_{ij}} \quad \text{if } j \in \text{non-beneficial criteria} \tag{28.3}$$

where; n_{ij} represents the normalized value of each element of the decision matrix.

- **Step 3: Calculation of the Standard Deviation and Degree of Conflict:** The vector of each criterion as $V_j = [n_{ij}]_{m\times 1}$ is now determined. Based on the elements of the normalized decision matrix, within-criterion, and between-criteria variation information are estimated as standard deviation (σ_j) and correlation (r_{jl}) between each pair of vectors of criteria, respectively. Thus, the degree of conflict (π_j) between criterion j and other criteria can be calculated as:

$$\pi_j = \sum_{l=1}^{n} (1 - r_{jl}) \tag{28.4}$$

In SECA method, standard deviation (σ_j) indicates the dispersion of data and is used to calculate the contrast intensity of interaction among the alternatives for each criterion. On the other hand, the correlation (r_{jl}) index denotes the strength of the linear relationship between two criteria. It is utilized to calculate the degree of conflict between a criterion with respect to other criteria.

- **Step 4: Determination of Reference Points for Criteria Weights:** The reference points of within-criterion and between-criteria for weights are obtained as the normalized values of σ_j and π_j.

$$\sigma_j^N = \frac{\sigma_j}{\sum_{l=1}^{n} \sigma_l} \tag{28.5}$$

$$\pi_j^N = \frac{\pi_j}{\sum_{l=1}^{n} \pi_l} \tag{28.6}$$

where; σ_j^N and π_j^N are the normalized values of σ_j and π_j, respectively.

- **Step 5: Development of the Multi-Objective Linear Programming (LP) Model:** A multi-objective linear programming (LP) model comprising the following objective functions, i.e., maximization of overall performance (S_i) of alternatives and minimization of the deviation of criteria weights from reference points for each criterion along with the required constraints is now formulated.

$$\max\ S_i = \sum_{j=1}^{n} w_j n_{ij}, \quad \forall i \in 1, 2, \ldots, m \tag{28.7}$$

$$\min\ \varphi_b = \sum_{j=1}^{n} (w_j - \sigma_j^N)^2, \tag{28.8}$$

$$\min\ \varphi_c = \sum_{j=1}^{n} (w_j - \pi_j^N)^2, \tag{28.9}$$

$$\text{s.t.}\ \sum_{j=1}^{n} w_j = 1,$$

$$w_j \leq 1, \quad \forall j \in 1, 2, \ldots, n$$

$$w_j \geq \varepsilon, \quad \forall j \in 1, 2, \ldots, n$$

where; φ_b and φ_c are the deviations of criteria weights from the reference points of within-criterion and between-criteria, respectively; w_j is the weight of jth criterion; $\varepsilon \leq w_j \leq 1$ sets the range of w_j between ε and 1; and ε is the minimum positive value of the criterion weight (which is generally considered as 0.001).

- **Step 6: Optimization of the Multi-Objective Non-Linear Programming Model:** The optimization model, stated above, can be transformed into a single-objective optimization problem as shown below:

$$\max\ Z = \varphi_a - \beta(\varphi_b + \varphi_c), \tag{28.10}$$

$$\text{s.t. Si}\ \varphi_a \leq S_i, \quad \forall i \in 1, 2, \ldots, m$$

$$S_i = \sum_{j=1}^{n} w_j n_{ij}, \quad \forall i \in 1, 2, \ldots, m$$

$$\varphi_b = \sum_{j=1}^{n} (w_j - \sigma_j^N)^2,$$

$$\varphi_c = \sum_{j=1}^{n} (w_j - \pi_j^N)^2,$$

$$\sum_{j=1}^{n} w_j = 1,$$

$$w_j \leq 1, \quad \forall j \in 1, 2, \ldots, n$$

$$w_j \geq \varepsilon, \quad \forall j \in 1, 2, \ldots, n$$

The objective function Z should be maximized, whereas, the deviations of criteria weights from the reference points need to be minimized. Hence, the deviation of criteria weights, i.e., φ_b and φ_c with coefficient β is subtracted from the performance of alternatives φ_a. The coefficient β plays an important role in determining the criteria weights and overall performance of the alternatives, and is selected based on the expert's opinion. The overall performance score of each alternative (S_i) and objective weight of each criterion (w_j) can be determined while solving the above-developed mathematical model in LINGO 18 software. A higher S_i value indicates superiority of ith alternative over the others.

Based on 15 economic, socio-cultural, technological, and environmental criteria, and six alternatives, Azbari et al. [2] derived the optimal ranking of wastewater allocation alternatives. A detailed sensitivity analysis was also carried out to investigate the effects of different β values on the criteria weights and derived ranking results. Baradari et al. [3] applied SECA method to prioritize information technology infrastructure libraries based on some identified key performance indicators. Bazrafshan et al. [4] solved an asymmetric traveling salesman problem using the SECA method while considering the number of constraints, number of variables, type of variables, time of solving and, differences between the optimum and the relaxed value for comparing these constraints as the evaluation criteria. In an attempt to appraise the relative performance of alternative battery electric vehicles, Ecer [5] first identified price, permitted load and energy consumption as the most significant criteria affecting the selection decision. Several MCDM techniques, like SECA, MARCOS, MAIRCA, COCOSO, ARAS, and COPRAS were later employed to rank the considered alternatives from the best to the worst. Tesla Model S emerged out as the best choice. Bahrami & Rastegar [6] first applied fuzzy BWM to estimate the weights of three criteria, i.e., voltage drop, losses, and feeder loading margin while solving a critical power distribution feeder selection problem. Those criteria weights were subsequently utilized to prioritize the considered feeders using the VIKOR method. Finally, SECA method was employed to weight the evaluation criteria and rank the feeders without considering the opinions of the participating experts.

28.1 3D PRINTER NOZZLE MATERIAL SELECTION PROBLEM

The 3D printing technology has propelled the growth of additive manufacturing with an ever-increasing range of products based on various available processes [7]. Designing and fabrication of 3D printer's components significantly influence the development of the end product. In fused deposition modeling (FDM), a common rapid prototyping process, inside the extruder, there is a motor which feeds the filament to the hot end where the filament melts and the nozzle helps to deposit the molten filament to form layers of the desired product in a controlled way. Continuous lifting of the nozzle or lowering of the platform after each layer results in the building of the product layer upon layer [8].

The nozzle is the last component that the filament material contacts before it gets extruded. It needs to be exchangeable according to the requirements. Size, shape, and material of the nozzle are observed to be the influencing factors to be considered while selecting a nozzle. Dimension of the nozzle diameter directly affects horizontal resolution and thickness of the 3D-printed product. Another major determinant of the performance of the nozzle is its constituent material. Brass is a soft material, relatively cheap, good conductor of heat and easy to machine, but it wears out easily if abrasive filaments are used. Hardened steel and stainless steel are better for abrasive filaments, but they are not good conductors of heat. Ruby, plated copper and aluminum alloys are also employed to make harder nozzles that can withstand constant abrasion. Brass and aluminum start to become extremely weak at higher temperature. Nickel layer plated over copper is preferred for higher corrosion resistance and thermal conductivity. It has already been acknowledged that while designing a mechanical component, the properties of the constituent materials play significant roles. Due to the huge availability of the candidate materials, and considering the constraints and criteria requirements imposed by the designers, it has now become extremely difficult to find out the best material from a set of alternatives for a particular application. Conflicting criteria requirements along with the presence of several candidate alternatives lead to the deployment of an MCDM method.

It has already been mentioned that nozzle is the end-most part of the extruder of a 3D printer through which the molten filament material flows. Hence, to have an improved additive manufacturing experience with better finished products, it has become crucial to choose the most apposite nozzle material under the given conditions, resulting in minimum wear along with easy flow of the molten filament at minimum cost. Table 28.1 shows the corresponding decision matrix, consisting of nine important evaluation criteria and eight nozzle material alternatives. In this decision matrix, all the criteria values, except corrosion resistance, wear resistance and machinability, are expressed in absolute units. A 9-point relative scale with 1 – very low, 3 – low, 5 – medium, 7 – high, and 9 – very high is incorporated here to express values of the three qualitative material properties. The details of those nine criteria are provided in Table 28.2 along with their weights (measured using entropy method). Table 28.3 shows the list of eight nozzle materials alternatives. It is worthwhile to mention here that among all the criteria under consideration, C_1 and C_6 are non-beneficial in nature requiring their lower values.

TABLE 28.1 Decision Matrix for the 3D Printer Nozzle Material Selection Problem

Alternative	Criteria								
	C_1	C_2	C_3	C_4	C_5	C_6	C_7	C_8	C_9
A_1	8.49	124	1	65.1	1	5.44	255	430	7
A_2	7.5	46.9	3	262	3	4.35	810	1,010	7
A_3	7.81	16.7	9	252	3	3.12	666	939	3
A_4	15.7	110	7	1,006	9	46.17	415	344	1
A_5	4.43	6.7	9	334	5	30.47	880	950	5
A_6	2.71	179	7	88.3	1	1.52	257	301	7
A_7	2.81	153	7	135	1	5.80	370	444	9
A_8	8.64	190	9	145	1	6.53	383	511	5

TABLE 28.2 Nozzle Material Selection Criteria

Symbol	Properties of Nozzle Material	Weight
C_1	Density (g/cc)	0.0218
C_2	Thermal conductivity (W/m-K)	0.1735
C_3	Corrosion resistance	0.0777
C_4	Hardness (BHN)	0.2479
C_5	Wear resistance	0.2203
C_6	Cost (USD/kg)	0.0514
C_7	Yield strength (MPa)	0.0667
C_8	Ultimate tensile strength (MPa)	0.0659
C_9	Machinability	0.0750

While solving this 3D printer nozzle material selection problem using SECA method, the original decision matrix of Table 28.1 is first normalized depending on the type of the considered criterion. The corresponding normalized matrix is provided in Table 28.4. Using the procedural steps of SECA method, the standard deviation between each pair of vectors of criteria is computed from the normalized decision matrix, as shown in Table 28.5.

The correlation between each pair of vectors of criteria and degree of conflict between criteria are calculated using Eqn. (28.4), and are also exhibited in Table 28.5. Thus, the reference points of within-criterion and between-criteria for criteria weights, i.e., normalized standard deviation and degree of conflict are obtained using Eqns. (28.5) and (28.6), respectively, as depicted in Table 28.5.

TABLE 28.3 Nozzle Material Alternatives

Symbol	Nozzle Material
A_1	Brass
A_2	Hardened steel
A_3	Stainless steel
A_4	Tungsten carbide
A_5	Titanium alloy (TiAl6V4)
A_6	Aluminum 6065
A_7	Aluminum 7075-T6
A_8	Nickel plated copper (cold drawn)

TABLE 28.4 Normalized Decision Matrix for the 3D Printer Nozzle Material Selection Problem

Alternative	C_1	C_2	C_3	C_4	C_5	C_6	C_7	C_8	C_9
A_1	0.3192	0.6526	0.1111	0.0647	0.1111	0.2794	0.2898	0.4257	0.7778
A_2	0.3613	0.2468	0.3333	0.2604	0.3333	0.3494	0.9205	1	0.7778
A_3	0.347	0.0879	1	0.2505	0.3333	0.4872	0.7568	0.9297	0.3333
A_4	0.1726	0.5789	0.7778	1	1	0.0329	0.4716	0.3406	0.1111
A_5	0.6117	0.0353	1	0.332	0.5556	0.0499	1	0.9406	0.5556
A_6	1	0.9421	0.7778	0.0878	0.1111	1	0.292	0.298	0.7778
A_7	0.9644	0.8053	0.7778	0.1342	0.1111	0.2621	0.4205	0.4396	1
A_8	0.3137	1	1	0.1441	0.1111	0.2328	0.4352	0.5059	0.5556

The normalized decision matrix and calculated reference points are now fed into LINGO 18 software. By solving the developed multi-objective non-linear programming model and substituting different β values (i.e., 0.1, 0.2, 0.3, 0.4, 0.5, 1, 2, 3, 4, and 5), the corresponding criteria weights and overall performance of the candidate material alternatives are obtained. For varying values of β (β = 0.1, 0.2, 0.3, 0.4, 0.5, 1, 2, 3, 4, and 5), 10 sets of criteria weights are derived, as shown in Table 28.6. From this table, it can be observed that the criteria weights vary for different β values, and these variations of criteria weights with changing β values are portrayed in Figure 28.1. This figure reveals that the criteria weights become more stable for $\beta \geq 4$. Thus, the criteria weights at $\beta = 4$ are considered here as the final weights. Accordingly, criterion C_2 receives the maximum weight as 0.1498, and criteria C_3 and C_4 have the minimum weight as 0.0949. On the other hand, on solving the same model using LINGO 18 while considering different values of β (β = 0.1, 0.2, 0.3, 0.4, 0.5, 1, 2, 3, 4, and 5), the overall performance scores of the alternative 3D printer nozzle materials are also derived in Table 28.7. The value of β for which the performance scores of the alternative materials are not the same, is considered for the final ranking of the alternatives. For instance, for $\beta = 1$, the performance scores of A_1 and A_4 alternatives are the same (0.4372) and hence, it is not possible to rank the candidate alternatives practically at that β value. Thus, from $\beta \geq 2$, the performance scores of the alternatives are taken into account for their subsequent rankings. From Table 28.7, it can be noticed that for $\beta \geq 2$, alternative A_6 (Aluminum 6065) receives the highest performance score indicating it to be the best choice among all the 3D printer nozzle material alternatives. Rankings of the alternative materials based on their performance scores considering different β values (β = 0.1, 0.2, 0.3, 0.4, 0.5, 1, 2, 3, 4, and 5) are provided in Table 28.8. From this table, it can be unveiled that the rankings of the alternative materials for $\beta \geq 2$ are completely the same and stable. Hence, based on the derived results, alternative A_6 (Aluminum 6065) with the maximum overall performance score can be chosen as the most appropriate material for 3D printer nozzle.

TABLE 28.5 Standard Deviation, Degree of Conflict, and Reference Points

Criteria	σ_j	π_j	σ_j^N	π_j^N
C_1	0.315	7.6956	0.1122	0.1014
C_2	0.3785	9.516	0.1348	0.1254
C_3	0.3307	7.6086	0.1178	0.1003
C_4	0.3036	8.7809	0.1081	0.1157
C_5	0.3143	8.6072	0.1119	0.1134
C_6	0.3064	8.4234	0.1091	0.111
C_7	0.2799	7.9663	0.0997	0.105
C_8	0.2946	8.3519	0.1049	0.1101
C_9	0.2848	8.9344	0.1014	0.1177

TABLE 28.6 Criteria Weights for Varying Values of β

Criteria	β									
	0.1	**0.2**	**0.3**	**0.4**	**0.5**	**1**	**2**	**3**	4	5
C_1	0.001	0.001	0.012	0.043	0.012	0.0937	0.1046	0.1053	0.1057	0.1059
C_2	0.3327	0.2797	0.2581	0.2408	0.2581	0.2024	0.1696	0.1564	0.1498	0.1459
C_3	0.1678	0.1226	0.1054	0.0912	0.1054	0.0668	0.0808	0.0902	0.0949	0.0977
C_4	0.001	0.0263	0.0408	0.0499	0.0408	0.0657	0.0779	0.0892	0.0949	0.0984
C_5	0.0632	0.0758	0.0752	0.076	0.0752	0.0768	0.0845	0.0939	0.0986	0.1014
C_6	0.001	0.001	0.001	0.0211	0.001	0.0842	0.1029	0.1053	0.1065	0.1072
C_7	0.0408	0.0874	0.1006	0.0981	0.1006	0.0912	0.0964	0.0984	0.0994	0.1
C_8	0.1895	0.1758	0.1699	0.1542	0.1699	0.1227	0.1186	0.1149	0.1130	0.1119
C_9	0.203	0.2304	0.237	0.2257	0.237	0.1965	0.1647	0.1464	0.1372	0.1316

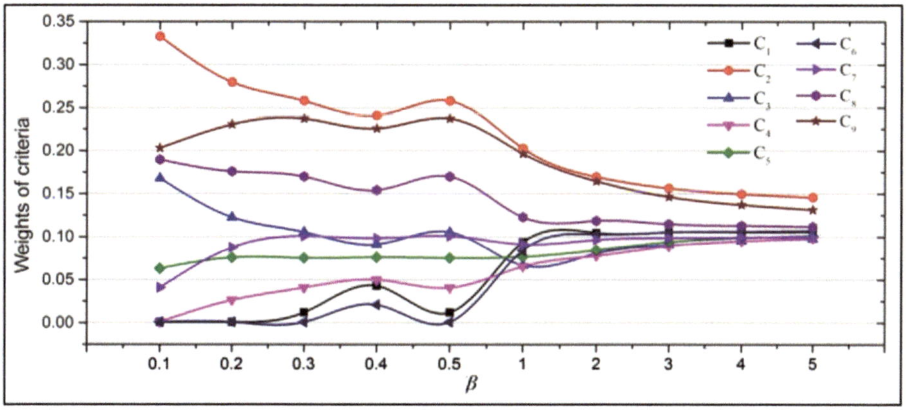

FIGURE 28.1 Variations of criteria weights for different β values.

TABLE 28.7 Overall Performance Scores of the Alternatives for Varying β Values

Alternative	β									
	0.1	**0.2**	**0.3**	**0.4**	**0.5**	**1**	**2**	**3**	**4**	**5**
A_1	0.4939	0.4863	0.4811	0.4682	0.4811	0.4372	0.4028	0.3826	0.3725	0.3665
A_2	0.545	0.5782	0.5861	0.5711	0.5861	0.5377	0.5265	0.5174	0.5128	0.5101
A_3	0.4939	0.4863	0.4811	0.4682	0.4811	0.4488	0.468	0.4755	0.4793	0.4816
A_4	0.4939	0.4863	0.4811	0.4682	0.4811	0.4372	0.449	0.4673	0.4764	0.4819
A_5	0.5474	0.5647	0.5693	0.5544	0.5693	0.5157	0.5282	0.535	0.5384	0.5404
A_6	0.6793	0.6288	0.6145	0.6249	0.6145	0.6509	0.6379	0.6232	0.6158	0.6114
A_7	0.7103	0.6783	0.6695	0.6618	0.6695	0.6335	0.6045	0.5859	0.5766	0.5711
A_8	0.7346	0.6701	0.6432	0.6122	0.6432	0.5472	0.5212	0.5097	0.504	0.5005

TABLE 28.8 Rankings of the 3D Printer Nozzle Material Alternatives for Different β Values

Alternative	β									
	0.1	0.2	0.3	0.4	0.5	1	2	3	4	5
A_1	7	7	7	7	7	7.5	8	8	8	8
A_2	5	4	4	4	4	4	4	4	4	4
A_3	7	7	7	7	7	6	6	6	6	7
A_4	7	7	7	7	7	7.5	7	7	7	6
A_5	4	5	5	5	5	5	3	3	3	3
A_6	3	3	3	2	3	1	1	1	1	1
A_7	2	1	1	1	1	2	2	2	2	2
A_8	1	2	2	3	2	3	5	5	5	5

KEYWORDS

- **3D printer nozzle material**
- **entropy method**
- **environmental criteria**
- **multi-objective linear**
- **sensitivity analysis**
- **simultaneous evaluation**

REFERENCES

1. Keshavarz-Ghorabaee, M., Amiri, M., Zavadskas, E. K., Turskis, Z., & Antucheviciene, J., (2018). Simultaneous evaluation of criteria and alternatives (SECA) for multi-criteria decision-making. *Informatica, 29*(2), 265–280.
2. Azbari, K. E., Ashofteh, P. S., Golfam, P., Vijay, P., & Singh, V. P., (2021). Optimal wastewater allocation with the development of an SECA multi-criteria decision-making method. *Journal of Cleaner Production, 321*, 129041.
3. Baradari, I., Shoar, M., Nezafati, N., & Mohammadreza, M. M., (2021). A new approach for KPI ranking and selection in ITIL processes: Using simultaneous evaluation of criteria and alternatives (SECA). *Journal of Industrial Engineering and Management Studies, 8*(1), 152–179.
4. Bazrafshan, R., Zolfani, S. H., & Al-e-hashem, S. M. J. M., (2021). Comparison of the sub-tour elimination methods for the asymmetric traveling salesman problem applying the SECA method. *Axioms, 10*. https://doi.org/10.3390/axioms10010019.

5. Ecer, F., (2021). A consolidated MCDM framework for performance assessment of battery electric vehicles based on ranking strategies. *Renewable and Sustainable Energy Reviews, 143*, 110916.
6. Bahrami, S., & Rastegar, M., (2022). Security-based critical power distribution feeder identification: Application of fuzzy BWM-VIKOR and SECA. *Electrical Power and Energy Systems, 134*, 107395.
7. Lu, B., Li, D., & Tian, X., (2015). Development trends in additive manufacturing and 3D printing. *Engineering, 1*, 85–89.
8. Xia, H., Lu, J., & Tryggvason, G., (2018). Fully resolved numerical simulations of fused deposition modeling. Part II - solidification, residual stresses and modeling of the nozzle. *Rapid Prototyping Journal, 24*, 973–987.

CHAPTER 29

R METHOD

The R method is a recently developed MCDM technique [1], which ranks the alternatives based on their performance scores with respect to each of the evaluation criteria. Furthermore, it also ranks the considered criteria based on the opinion of the concerned decision-maker. These assigned ranks are subsequently converted into corresponding weights, and composite scores are evaluated using these weights, leading to a final ranking of the alternatives based on their composite scores. The procedural steps of R method are presented as below [1, 2]:

- **Step 1:** Construct the decision matrix based on the performance scores of the alternatives against each criterion.
- **Step 2:** Assign ranks (1, 2, 3, …, etc.), to the criteria based on their significance and perception of the decision maker. Assign average rank to equally significant criteria.
- **Step 3:** Assign ranks (1, 2, 3, …, etc.), to the candidate alternatives based on their performance scores against each criterion. Allocate average rank to those alternatives having equal performance score against a specific criterion.
- **Step 4:** Transform the ranks assigned to both the alternatives and criteria into corresponding weights based on the information, as provided in Table 29.1. However, to compute weights from the assigned ranks, the following equation can be employed.

$$w_j = \frac{[1/\sum_{k=1}^{j}(1/r_k)]}{\sum_{j=1}^{n}[1/\sum_{k=1}^{j}(1/r_k)]} \quad (29.1)$$

where; w_j is the weight of jth alternative or criterion (j = 1, 2, …, n), r_k is the rank assigned to k^{th} alternative or criterion (k = 1, 2, …, j) and n is the number of alternatives or criteria.

Multi-Criteria Decision-Making Methods in Manufacturing Environments: Models and Applications.
Shankar Chakraborty, Prasenjit Chatterjee, & Partha Protim Das (Authors)

- **Step 5:** Calculate the composite scores of the candidate alternatives by adding up the products of the criteria weights and the corresponding weights of the alternatives.
- **Step 6:** Award ranks to the alternatives based on their composite scores. The alternative with the maximum composite score is the best option.

Table 29.1 provides values of the weights calculated from different ranks assigned to the alternatives or criteria based on Eqn. (29.1). To illustrate the calculation steps involved in Table 29.1, the computation procedure of weights for four criteria or alternatives (column values under rank 4) is shown as below [2]:

1/Reciprocal of rank 1: 1/(1/1) = 1.0000
1/Reciprocal of rank up to 2: 1/(1/1 + 1/2) = 0.6666
1/Reciprocal of rank up to 3: 1/(1/1 + 1/2 + 1/3) = 0.5454
1/Reciprocal of rank up to 4: 1/(1/1 + 1/2 + 1/3 + 1/4) = 0.4800
Total = 1.000 + 0.6666 + 0.5454+ 0.4800 = 2.6921

Therefore, weight is assigned to rank 1 = 1/2.6921 = 0.3714, similarly for rank 2 = 0.6666/2.6921 = 0.2476, for rank 3 = 0.5454/2.6921 = 0.2026 and for rank 4 = 0.48/2.6921 = 0.1783.

TABLE 29.1 Weight Calculations for Different Assigned Ranks

Rank	Number of Criteria or Alternatives								
	2	**3**	**4**	**5**	**6**	**7**	**8**	**9**	**10**
	Calculated Weights								
1	0.6	0.452	0.371	0.319	0.283	0.255	0.233	0.215	0.201
2	0.4	0.301	0.248	0.213	0.188	0.17	0.155	0.144	0.134
3	–	0.247	0.203	0.174	0.154	0.139	0.127	0.117	0.109
4	–	–	0.178	0.153	0.136	0.122	0.112	0.103	0.096
5	–	–	–	0.140	0.124	0.112	0.102	0.094	0.088
6	–	–	–	–	0.115	0.104	0.095	0.088	0.082
7	–	–	–	–	–	0.098	0.09	0.083	0.077
8	–	–	–	–	–	–	0.086	0.079	0.074
9	–	–	–	–	–	–	–	0.076	0.071
10	–	–	–	–	–	–	–	–	0.068

Rao & Lakshmi [1] first developed the R method, and proved its potentiality, efficacy, and robustness while solving vendor selection, industrial robot selection, material selection, and FMS selection problems. This newly developed MCDM technique provided excellent ranking solutions to all the considered real-time problems. Rao & Lakshmi [2] also adopted R method to rank the optimal Pareto solutions for an EDM process, design variables of a compression ignition biodiesel engine with an exhaust gas recirculation system and design variables of a solar-assisted Brayton engine system. The best solutions for the considered multi-objective and many-objective optimization problems were later derived using the same method.

29.1 CONVEYOR SELECTION PROBLEM

To demonstrate the procedural steps involved in R method, a conveyor selection problem [3] is considered here. This problem consists of four alternative conveyors to be evaluated based on six criteria. The corresponding decision matrix is provided in Table 29.2. Among these criteria, fixed cost per hour (FIC) (in €/h) and variable cost per hour (VAC) (in €/h) are non-beneficial attributes requiring with their lower values. On the other hand, higher values for speed of conveyor (SPC) (in m/min), item width (ITWI) (jn cm), item weight (ITWEI) (in kg) and flexibility (FLEX) are usually preferred. Among the four beneficial criteria, FLEX was expressed subjectively, and Rao [4] adopted a linguistic scale to convert the subjective values of FLEX criterion into quantitative measures. Now, following the procedural steps of R method, it is first required to assign the corresponding ranks to the considered alternatives and evaluation criteria. Karande & Chakraborty [5] applied AHP to determine the weights of those six criteria as $w_{FIC} = 0.1049$, $w_{VAC} = 0.1260$, $w_{SPC} = 0.1260$, $w_{ITWI} = 0.2402$, $w_{ITWEI} = 0.2245$, and $w_{FLEX} = 0.1782$. Based on these criteria weights, rank 1 is assigned to ITWI criterion, followed by ITWEI, FLEX, SPC, VAC, and FIC, as exhibited in Table 29.3. As VAC and SPC have the same priority weight, they are assigned with their average rank of 4.5. When there is no prior information regarding the criteria weights, the opinion of a decision maker may be sought to assign the corresponding ranks to the set of evaluation criteria. In Table 29.3, based on the type of the criterion (beneficial or non-beneficial), ranks are also assigned to the four alternative conveyors based on their performance against each criterion. For example, in Table 29.3, VAC being a non-beneficial criterion, its lowest value is always desirable. Thus, for this criterion, conveyor C_B is the best choice and is assigned with a rank of 1. As the conveyors C_A and C_C have the same value for VAC, they are assigned with an average rank of 2.5 (i.e., average of 2 and 3). Conveyor C_D, having the highest VAC value, is allotted with a rank of 4. Similarly, for beneficial criteria, the alternative conveyor with the highest value for the corresponding criterion is assigned with a rank of 1, and so on. Now, using Eqn. (29.1) and Table 29.1, the weights for different assigned ranks are calculated for all the considered alternative conveyors and evaluation criteria, as exhibited in Table 29.4. Since, in this decision-making problem, there are four alternatives, the set of weights to be assigned to each alternative is {Rank 1 (0.371), Rank 2 (0.248), Rank 3 (0.203), Rank 4 (0.178)}. In the similar direction, for the six evaluation criteria, the set of weights to be assigned is {(Rank 1 (0.283), Rank 2 (0.188), Rank 3 (0.154), Rank 4.5 (0.13), Rank 6 (0.115)}. To estimate weight for an average rank, average of the weights for the corresponding ranks is considered. For example, conveyors C_A and C_C have the same average rank for VAC criterion. Hence, both of them have assigned the average weight of 0.2251 (average of 0.248 and 0.203). The composite score for each of the alternative conveyors is finally calculated by adding the products of criteria weights and corresponding alternative weights, and the candidate conveyors are ranked based on the descending values of this composite score. Thus, conveyor C_C with the maximum composite score emerges out as the best choice for the considered handling task, followed by conveyors C_B, C_A, and C_D, respectively. Thus, the complete ranking of the conveyors is derived as $C_C \rightarrow C_B \rightarrow C_A \rightarrow C_D$. Conveyor C_C also appeared to be the first choice when the same problem was solved using other MCDM techniques, like GTMA, weighted utility additive theory, VIKOR, PROMETHEE, ELECTRE, EDAS, CODAS, WASPAS, and MOORA [4–6]. Due to differences in the mathematical treatments in all these MCDM techniques, there are slight variations in the intermediate rankings of the conveyors, but, the top ranked conveyor (C_C) exactly matches. It proves the potentiality of R method in identifying the best alternative from a set of feasible options for a given decision-making problem.

TABLE 29.2 Decision Matrix for the Conveyor Selection Problem [3]

Alternative	Criteria					
	FIC	VAC	SPC	ITWI	ITWEI	FLEX
C_A	2	0.45	12	15	10	Very good (0.745)
C_B	2.3	0.44	13	20	10	Excellent (0.955)
C_C	2.25	0.45	11	30	20	Excellent (0.955)
C_D	2.4	0.46	10	25	15	Very good (0.745)

Source: Reprinted with permission from Ref. [3]. © 2021 Elsevier

TABLE 29.3 Ranks Assigned to the Alternatives and Criteria for the Conveyor Selection Problem

Alternative	Criteria					
	FIC	VAC	SPC	ITWI	ITWEI	FLEX
C_A	1	2.5	2	4	3.5	2.5
C_B	3	1	1	3	3.5	1.5
C_C	2	2.5	3	1	1	1.5
C_D	4	4	4	2	2	2.5
Criteria rank	6	4.5	4.5	1	2	3

TABLE 29.4 Assigned Weights to the Alternatives and Criteria for the Conveyor Selection Problem

Alternative	Criteria						Composite Score	Rank
	FIC	VAC	SPC	ITWI	ITWEI	FLEX		
C_A	0.3714	0.2251	0.2476	0.1783	0.1904	0.22512	0.2251	3
C_B	0.2026	0.3714	0.3714	0.2026	0.1904	0.30954	0.2607	2
C_C	0.2476	0.2251	0.2026	0.3714	0.3714	0.30954	0.3067	1
C_D	0.1783	0.1783	0.1783	0.2476	0.2476	0.22512	0.2182	4
Criteria weight	0.1150	0.1300	0.1300	0.2830	0.1880	0.1540	–	

KEYWORDS

- **criteria weight**
- **flexibility**
- **item weight**
- **item width**
- **multi-criteria decision making**
- **speed of conveyor**

REFERENCES

1. Rao, R., & Lakshmi, J., (2021). R-method: A simple ranking method for multi-attribute decision-making in the industrial environment. *Journal of Project Management, 6*, 223–230.

2. Rao, R. V., & Lakshmi, R. J., (2021). Ranking of pareto-optimal solutions and selecting the best solution in multi-and many-objective optimization problems using R-method. *Soft Computing Letters, 3*, 100015.
3. Kulak, O., (2005). A decision support system for fuzzy multi-attribute selection of material handling equipments. *Expert Systems with Applications, 29*, 310–319.
4. Rao, R. V., (2007). *Decision Making in the Manufacturing Environment Using Graph Theory and Fuzzy Multiple Attribute Decision Making Methods*. Springer, UK.
5. Karande, P., & Chakraborty, S., (2013). Material handling equipment selection using weighted utility additive theory. *Journal of Industrial Engineering.* Article ID: 268708. http://dx.doi.org/10.1155/2013/268708.
6. Mathew, M., & Sahu, S., (2018). Comparison of new multi-criteria decision making methods for material handling equipment selection. *Management Science Letters, 8*, 139–150.

CHAPTER 30

COMPROMISE RANKING OF ALTERNATIVES FROM DISTANCE TO IDEAL SOLUTION (CRADIS) METHOD

The compromise ranking of alternatives from distance to ideal solution (CRADIS) method [1] is a recently developed MCDM technique based on the determination of the deviations of the alternatives from the ideal and anti-ideal solutions. It basically combines the application steps of ARAS, MARCOS, and TOPSIS methods. It employs the ideal solutions representing the maximum values of the alternatives. The procedural steps of CRADIS method are presented as below:

- **Step 1:** Develop the initial decision matrix $D = [x_{ij}]_{m \times n}$ with m alternatives and n criteria, where x_{ij} denotes the performance of ith alternative with respect to j^{th} criterion.
- **Step 2:** Normalize the decision matrix. For normalization, the following two equations are adopted.

$$n_{ij} = \frac{x_{ij}}{x_{max}} \text{ (for beneficial criterion)} \quad (30.1)$$

$$n_{ij} = \frac{x_{min}}{x_{ij}} \text{ (for non-beneficial criterion)} \quad (30.2)$$

- **Step 3:** Develop the weighted normalized decision matrix (v_{ij}) after multiplying the elements of the normalized decision matrix by the corresponding criteria weights.

$$v_{ij} = n_{ij} \times w_j \quad (30.3)$$

- **Step 4:** Identify the ideal and anti-ideal solutions. The ideal solution has the maximum v_{ij} value in the weighted normalized decision

Multi-Criteria Decision-Making Methods in Manufacturing Environments: Models and Applications.
Shankar Chakraborty, Prasenjit Chatterjee, & Partha Protim Das (Authors)

matrix, while the anti-ideal solution possesses the minimum v_{ij} value in weighted normalized decision matrix.

$$t_i = \max v_{ij} \tag{30.4}$$

$$t_{ai} = \min v_{ij} \tag{30.5}$$

- **Step 5:** Determine the deviations from the ideal and anti-ideal solutions.

$$d^+ = t_i - v_{ij} \tag{30.6}$$

$$d^- = v_{ij} - t_{ai} \tag{30.7}$$

- **Step 6:** Calculate the grades of the deviation of individual alternatives from the ideal and anti-ideal solutions.

$$s_i^+ = \sum_{j=1}^{n} d^+ \tag{30.8}$$

$$s_i^- = \sum_{j=1}^{n} d^- \tag{30.9}$$

- **Step 7:** Estimate the utility function for each alternative in relation to the deviations from the optimal alternatives.

$$K_i^+ = \frac{s_0^+}{s_i^+} \tag{30.10}$$

$$K_i^- = \frac{s_i^-}{s_0^-} \tag{30.11}$$

where; s_0^+ is the optimal alternative having the minimum distance from the ideal solution, while s_0^- is the optimal alternative having the maximum distance from the anti-ideal solution.

- **Step 8:** Rank the candidate alternatives from the best to the worst according to the descending order of the calculated utility degree. The utility degree (Q_i) is calculated while taking the average deviation of the alternatives based on the developed utility function. The best alternative should have the maximum Q_i value.

$$Q_i = \frac{K_i^+ + K_i^-}{2} \tag{30.12}$$

30.1 CONVEYOR BELT MATERIAL SELECTION PROBLEM

To illustrate the numerical computation of CRADIS method in solving MCDM problems, the example of the conveyor belt material (aramid) selection problem of Chapter 6 is considered here. The corresponding decision matrix having 10 alternative materials and five criteria is again exhibited in Table 30.1. The same entropy method-based criteria weight set as $w_{MWT} = 0.1450$, $w_{CT} = 0.1233$, $w_{ME} = 0.2737$, $w_{BW} = 0.2197$, and $w_{C} = 0.2383$ is also employed here to determine the most appropriate conveyor belt material. Now, based on the procedural steps of CRADIS method, and applying Eqns. (30.1) and (30.2), all the elements of the initial decision matrix are normalized. The normalized decision matrix is shown in Table 30.2. To develop the weighted normalized decision matrix (as provided in Table 30.3), the elements of the normalized decision matrix are multiplied by the respective criteria weights. This step is also known as aggravation of the decision matrix. From this aggravated decision matrix, the ideal solution (maximum value in the matrix) and anti-ideal solution (minimum in the matrix) are subsequently identified in Table 30.4. Using these ideal and anti-ideal solutions, their deviations from the elements of the weighted normalized decision matrix are then evaluated, using Eqns. (30.6) and (30.7). These values are subsequently utilized in calculating grades of the deviation of individual alternatives from the ideal and anti-ideal solutions. The utility functions for each alternative in relation to the deviations from the optimal alternatives are then computed, which are finally employed to determine the corresponding utility degrees of all the candidate conveyor belt materials. When the candidate alternatives are ranked based on the utility degree, alternative M10 emerges out as the best aramid material for design of conveyor belt. All these calculations are exhibited in Table 30.5.

TABLE 30.1 Quantitative Data for the Aramid Belt Materials

Designation of Belt	Belt Material	MWT (N/mm)	CT) (mm)	ME (KN/mm)	BW (kg/m²)	C (INR)
DP 500	M1	63	2.5	23	11.7	2,255
DP 630	M2	80	2.6	27	11.8	2,345
DP 800	M3	100	2.7	31	12.0	2,436
DP 1000	M4	125	3.5	37	12.8	3,157
DP 1250	M5	160	3.6	44	16.2	3,247
DP 1400	M6	175	3.7	49	16.4	3,338
DP 1600	M7	200	3.8	54	16.5	3,428
DP 1800	M8	225	4.2	60	16.9	3,789
DP 2000	M9	250	4.5	66	17.3	4,059
DP 2500	M10	315	5.0	80	17.9	4,510

TABLE 30.2 Normalized Decision Matrix for the Conveyor Belt Material Selection Problem

Belt Material	MWT	CT	ME	BW	C
M1	0.2000	1.0000	0.2875	1.0000	1.0000
M2	0.2540	0.9615	0.3375	0.9915	0.9616
M3	0.3175	0.9259	0.3875	0.9750	0.9257
M4	0.3968	0.7143	0.4625	0.9141	0.7143
M5	0.5079	0.6944	0.5500	0.7222	0.6945
M6	0.5556	0.6757	0.6125	0.7134	0.6756
M7	0.6349	0.6579	0.6750	0.7091	0.6578
M8	0.7143	0.5952	0.7500	0.6923	0.5951
M9	0.7937	0.5556	0.8250	0.6763	0.5556
M10	1.0000	0.5000	1.0000	0.6536	0.5000

TABLE 30.3 Aggravation of the Decision Matrix

Belt Material	MWT	CT	ME	BW	C
M1	0.0290	0.1233	0.0787	0.2197	0.2383
M2	0.0368	0.1186	0.0924	0.2178	0.2292
M3	0.0460	0.1142	0.1061	0.2142	0.2206
M4	0.0575	0.0881	0.1266	0.2008	0.1702
M5	0.0737	0.0856	0.1505	0.1587	0.1655
M6	0.0806	0.0833	0.1676	0.1567	0.1610
M7	0.0921	0.0811	0.1847	0.1558	0.1568
M8	0.1036	0.0734	0.2053	0.1521	0.1418
M9	0.1151	0.0685	0.2258	0.1486	0.1324
M10	0.1450	0.0617	0.2737	0.1436	0.1192

TABLE 30.4 Identification of the Ideal and Anti-Ideal Solutions

Value	MWT	CT	ME	BW	C
t_i	0.1450	0.1233	0.2737	0.2197	0.2383
t_{ai}	0.0290	0.0617	0.0787	0.1436	0.1192

TABLE 30.5 Ranking of the Alternative Conveyor Belt Materials

Alternative	S_i^+	S_i^-	K_i^+	K_i^-	Q_i	Rank
M1	0.3110	0.2569	0.8260	0.8260	0.8260	5
M2	0.3053	0.2627	0.8416	0.8445	0.8431	3
M3	0.2989	0.2690	0.8594	0.8648	0.8621	2
M4	0.3568	0.2111	0.7201	0.6789	0.6995	9
M5	0.3660	0.2019	0.7019	0.6491	0.6755	10
M6	0.3508	0.2171	0.7324	0.6982	0.7153	8
M7	0.3295	0.2384	0.7796	0.7665	0.7730	7
M8	0.3238	0.2441	0.7933	0.7848	0.7890	6
M9	0.3096	0.2583	0.8296	0.8304	0.8300	4
M10	0.2569	0.3110	1.0000	1.0000	1.0000	1

KEYWORDS

- **anti-numerical computation ideal solution**
- **compromise ranking**
- **entropy method**
- **ideal solution**
- **multi-criteria decision making**
- **numerical computation**

REFERENCE

1. Puška, A., Stević, Ž., & Pamučar, D., (2021). Evaluation and selection of healthcare waste incinerators using extended sustainability criteria and multi-criteria analysis methods. *Environment, Development, and Sustainability*. https://doi.org/10.1007/s10668-021-01902-2.

PART III

Normalization in MCDM

CHAPTER 31

NORMALIZATION TECHNIQUES

All the MCDM techniques aim to evaluate and rank a group of alternatives that satisfies a set of given criteria [1]. In MCDM problems, criteria represent a set of independent evaluation attributes that must be satisfied by the considered alternatives. In these problems, each criterion may have different units or dimensions, e.g., kilograms or meters; but all of them have to be normalized/pre-processed to define a common dimensionless range/scale in order to allow their combination in the form of a single score for each alternative. Furthermore, if the normalization technique is not suitable for the given decision problem or for the selected MCDM technique, the best decision solution may often be overlooked [1]. Hence, normalization/pre-processing of data is considered to be an integral part of any decision-making process mainly for the reason that it converts the input information into comparable arithmetical data, thus permitting MCDM techniques to assess and rank the alternatives [2–5]. It has already been pointed out that although the normalization process scales the criteria values to be roughly around equivalent magnitude, different normalization methods can produce different solutions, which may lead in deviation from the initially recommended solutions. Different normalization procedures as recommended for dealing with MCDM problems are enlisted as below:

a. **Linear: Max:**

$$n_{ij} = \frac{x_{ij}}{x_{ij}^{\max}} \text{ (for beneficial criterion)} \quad (31.1)$$

$$n_{ij} = 1 - \frac{x_{ij}}{x_{ij}^{\max}} \text{ (for cost criterion)} \quad (31.2)$$

where; x_{ij} is the performance score of ith alternatives with respect to j^{th} criterion; $x_{ij}^{\max}$ is the maximum value of the considered criterion; and n_{ij} is the normalized value of x_{ij}.

Multi-Criteria Decision-Making Methods in Manufacturing Environments: Models and Applications.
Shankar Chakraborty, Prasenjit Chatterjee, & Partha Protim Das (Authors)

b. Linear: Max-Min:

$$n_{ij} = \frac{x_{ij} - x_{ij}^{\min}}{x_{ij}^{\max} - x_{ij}^{\min}} \text{ (for beneficial criterion)} \quad (31.3)$$

$$n_{ij} = \frac{x_{ij}^{\max} - x_{ij}}{x_{ij}^{\max} - x_{ij}^{\min}} \text{ (for cost criterion)} \quad (31.4)$$

where; $x_{ij}^{\min}$ is the minimum value of the criterion under consideration.

c. Linear: Sum:

$$n_{ij} = \frac{x_{ij}}{\sum_{j=1}^{n} x_{ij}} \text{ (for beneficial criterion)} \quad (31.5)$$

$$n_{ij} = \frac{1/x_{ij}}{\sum_{j=1}^{n} \left(1/x_{ij}\right)} \text{ (for cost criterion)} \quad (31.6)$$

where; n is the number of criteria.

d. Vector Normalization:

$$n_{ij} = \frac{x_{ij}}{\sqrt{\sum_{j=1}^{n} x_{ij}^2}} \text{ (for beneficial criterion)} \quad (31.7)$$

$$n_{ij} = 1 - \frac{x_{ij}}{\sqrt{\sum_{j=1}^{n} x_{ij}^2}} \text{ (for cost criterion)} \quad (31.8)$$

e. Logarithmic Normalization:

$$n_{ij} = \frac{\ln(x_{ij})}{\ln\left(\prod_{j=1}^{n} x_{ij}\right)} \text{ (for beneficial criterion)} \quad (31.9)$$

$$n_{ij} = \frac{1 - \dfrac{\ln(x_{ij})}{\ln\left(\prod_{j=1}^{n} x_{ij}\right)}}{n-1} \text{ (for cost criterion)} \quad (31.10)$$

f. Non-Linear Normalization:

$$n_{ij} = \left(\frac{x_{ij}}{x_{ij}^{\max}}\right)^2 \text{ (for beneficial criterion)} \quad (31.11)$$

$$n_{ij} = \left(\frac{x_{ij}^{\min}}{x_{ij}}\right)^2 \text{ (for cost criterion)} \tag{31.12}$$

g. Jüttler's-Körth's Normalization (JKN):

$$n_{ij} = 1 - \left|\frac{x_{ij}^{\max} - x_{ij}}{x_{ij}^{\max}}\right| \text{ (for beneficial criterion)} \tag{31.13}$$

$$n_{ij} = 1 - \left|\frac{x_{ij}^{\min} - x_{ij}}{x_{ij}^{\min}}\right| \text{ (for cost criterion)} \tag{31.14}$$

KEYWORDS

- **arithmetical data**
- **beneficial criterion**
- **criteria**
- **Jüttler's-Körth's normalization**
- **multi-criteria decision making**
- **normalization process**

REFERENCES

1. Chatterjee, P., & Chakraborty, S., (2014). Investigating the effect of normalization norms in flexible manufacturing system selection using multi-criteria decision-making methods. *Journal of Engineering Science and Technology Review, 7*(3), 141–150.
2. Lakshmi, T. M., Joshuva, A., Vetriselvi, A. K., & Venkatesan, V. P., (2016). A study on different types of normalization methods in adaptive technique for order preference by similarity to ideal solution (TOPSIS). *International Journal of Engineering Research & Technology, 4*(5), 1–7.
3. Vafaei, N., Ribeiro, R. A., & Camarinha-Matos, L. M., (2018). Data normalization techniques in decision making: Case study with TOPSIS method data normalization techniques in decision making: Case study with TOPSIS method. *International Journal of Information and Decision Sciences, 10*(1), 19–38.
4. Acuña-Soto, C., Liern, V., & Pérez-Gladish, B., (2021). Normalization in TOPSIS-based approaches with data of different nature: Application to the ranking of mathematical videos. *Annals of Operations Research, 296*, 541–569.
5. Ersoy, N., (2021). Selecting the best normalization technique for ROV method: Towards a real-life application. *Gazi University Journal of Science, 34*(2), 592–609.

PART IV

Criteria Weight Measurement Techniques

CHAPTER 32

CRITERIA WEIGHT MEASUREMENT TECHNIQUES

32.1 ENTROPY METHOD

Entropy weights are the measures of uncertainties in the information formulated using probability theory [1]. It indicates that a wide distribution represents more uncertainty than does a sharply peaked one. Entropy weights measure the information content in the attribute values of the alternatives, thereby evaluating each attribute's usefulness in detecting differences in the data. For example, if an attribute has the same value for each of the alternatives, then it would provide no information to distinguish the alternatives. On the other hand, an attribute having different values for each alternative has high information content and is useful when comparing and contrasting the alternatives. The steps of calculating criteria weights using entropy method are enlisted as below [2]:

- **Step 1:** Normalize each criterion of the decision matrix.

$$r_{ij} = \frac{x_{ij}}{\left[\sum_{i=1}^{m} x_{ij}^2\right]^{1/2}} \quad (i = 1, 2, \ldots, m; j = 1, 2, \ldots, n) \tag{32.1}$$

  Given the initial decision matrix $X = [x_{ij}]_{m \times n}$ consisting of m alternatives and n criteria, all of its elements are normalized using vector normalization procedure.

- **Step 2:** Calculate the entropy value (e_j) of each criterion.

$$e_j = -k \sum_{i=1}^{m} \left(r_{ij} \times \ln r_{ij} \right) \tag{32.2}$$

  where; $k = 1/(\ln m)$ is a constant to assure the condition $0 \le e_j \le 1$.

Multi-Criteria Decision-Making Methods in Manufacturing Environments: Models and Applications.
Shankar Chakraborty, Prasenjit Chatterjee, & Partha Protim Das (Authors)

➢ **Step 3:** Determine the objective weight (w_j) for each criterion.

$$w_j = \frac{1-e_j}{\sum_{j=1}^{n}(1-e_j)} \tag{32.3}$$

These objective weights can be considered as the criteria weights for a given decision-making problem. Manshadi et al. [3] considered a cryogenic storage tank material selection problem and solved the same using a modified digital logic technique. This problem consists of seven material alternatives whose performance is evaluated based on seven criteria, i.e., toughness index (calculated based on ultimate tensile strength, yield strength and ductility), yield strength (in MPa), Young's modulus (in GPa), density (in g/cm^2), thermal expansion (in 10^{-6}/°C), thermal conductivity (in cal/cm^2/cm/°C/s) and specific heat (in cal/g/°C). The corresponding decision matrix is exhibited in Table 32.1. This cryogenic storage tank material selection problem is considered here to demonstrate the applicability of the entropy method in determining the related criteria weights. At first, using Eqn. (32.1), all the elements of the original decision matrix are normalized in the interval [0,1], as depicted in Table 32.2. Now, based on Eqn. (32.2), values of $r_{ij} \times \ln(r_{ij})$ are computed, as shown in Table 32.3. These values are finally employed to determine the corresponding criteria weights for this cryogenic storage tank material selection problem, as shown in Table 32.4. It can be noticed from this table that in order to fulfill the requirements of the most appropriate material for the cryogenic storage tank, thermal conductivity and toughness index emerge out as the two most important evaluation criteria. On the other hand, thermal expansion and specific heat are the two least important criteria.

TABLE 32.1 Decision Matrix for the Cryogenic Storage Tank Material Selection Problem [3]

Material	Toughness Index	Yield Strength	Young's Modulus	Density	Thermal Expansion	Thermal Conductivity	Specific Heat
Al 2024-T6	75.5	420	74.2	2.8	21.4	0.37	0.16
Al 5052-O	95	91	70	2.68	22.1	0.33	0.16
SS 301-FH	770	1,365	189	7.9	16.9	0.04	0.08
SS 310-3AH	187	1,120	210	7.9	14.4	0.03	0.08
Ti-6Al-4V	179	875	112	4.43	9.4	0.016	0.09
Inconel 718	239	1,190	217	8.51	11.5	0.31	0.07
70Cu-30Zn	273	200	112	8.53	19.9	0.29	0.06

Source: Reprinted with permission from Ref. [3]. © 2021 Elsevier.

TABLE 32.2 Normalized Decision Matrix for Entropy Weight Calculation

Material	Toughness Index	Yield Strength	Young's Modulus	Density	Thermal Expansion	Thermal Conductivity	Specific Heat
Al 2024-T6	0.0841	0.1787	0.1841	0.1604	0.4720	0.5651	0.5636
Al 5052-O	0.1058	0.0387	0.1737	0.1535	0.4874	0.5040	0.5636
SS 301-FH	0.8575	0.5808	0.4690	0.4526	0.3727	0.0611	0.2818
SS 310-3AH	0.2083	0.4765	0.5211	0.4526	0.3176	0.0458	0.2818
Ti-6Al-4V	0.1993	0.3723	0.2779	0.2538	0.2073	0.0244	0.3170
Inconel 718	0.2662	0.5063	0.5385	0.4876	0.2536	0.4734	0.2466
70Cu-30Zn	0.3040	0.0851	0.2779	0.4887	0.4389	0.4429	0.2113

TABLE 32.3 Calculated $r_{ij} \times \ln(r_{ij})$ Values

Material	Toughness Index	Yield Strength	Young's Modulus	Density	Thermal Expansion	Thermal Conductivity	Specific Heat
Al 2024-T6	–0.2082	–0.3077	–0.3116	–0.2936	–0.3544	–0.3225	–0.3232
Al 5052-O	–0.2376	–0.1259	–0.3040	–0.2877	–0.3503	–0.3453	–0.3232
SS 301-FH	–0.1318	–0.3156	–0.3551	–0.3588	–0.3678	–0.1708	–0.3569
SS 310-3AH	–0.3268	–0.3532	–0.3397	–0.3588	–0.3643	–0.1413	–0.3569
Ti-6Al-4V	–0.3215	–0.3679	–0.3559	–0.3480	–0.3262	–0.0907	–0.3642
Inconel 718	–0.3523	–0.3446	–0.3333	–0.3502	–0.3479	–0.3540	–0.3452
70Cu-30Zn	–0.3620	–0.2097	–0.3559	–0.3499	–0.3614	–0.3607	–0.3285

TABLE 32.4 e_j Values Along with Criteria Weights

Criteria	Toughness Index	Yield Strength	Young's Modulus	Density	Thermal Expansion	Thermal Conductivity	Specific Heat
e_j	0.6848	0.7146	0.8314	0.8284	0.8726	0.6301	0.8464
Weight	0.1980	0.1793	0.1060	0.1078	0.0800	0.2324	0.0965

32.2 DECISION-MAKING TRIAL AND EVALUATION LABORATORY (DEMATEL) METHOD

The decision-making trial and evaluation laboratory (DEMATEL) method is used to perceive intricate relationships and build a network relation map between criteria. It was mainly developed by the Battelle memorial association of the Geneva research center [4, 5] to study complicated problems concerning about race, hunger, environmental protection, energy, etc. It is based on a concept of pair-wise comparison of decision-making attributes (alternatives, criteria). The attributes are compared with respect to the relative influence of each over the other. The main objective of DEMATEL method is to directly compare the interaction relationship between different variables of a complicated system to determine direct and indirect causal relationships and influence levels between the variables. A visual structural matrix and a causal diagram are developed to express the causal relationships and influence levels between the considered variables and assist in making appropriate decisions. The evaluation criteria are usually consisted of many complicated aspects, including financial and non-financial, as well as qualitative aspects, and are either directly or indirectly mutually related. DEMATEL method is based on the notion of digraphs, which can separate the considered criteria into cause-and-effect groups to visually observe the inside of a complex system/problem. It assumes a system containing a set of components $C = \{C_1, C_2, \ldots, C_n\}$, with pair-wise relations that can be evaluated. The procedural steps of DEMATEL method are represented as below [6–8]:

- **Step 1: Develop the Direct-Relation Matrix (A):** At first, the decision makers or experts evaluate the relationship between the sets of paired criteria to indicate the direct effect that they believe each ith criterion exerts on each jth criterion, as indicated by an integer scale (score) ranging from 0 to 4, representing no influence (0), low influence (1), medium influence (2), high influence (3) and very high influence (4). Then, as a result of these evaluations, the initial data is obtained as a direct-relation matrix (A) which is in the form of an $n \times n$ matrix, in which the individual element (a_{ij}) denotes the degree to which i^{th} criterion affects j^{th} criterion and n denotes the total number of criteria.

$$A = \begin{bmatrix} 0 & a_{12} & \cdots & a_{1j} & \cdots & a_{1n} \\ a_{21} & 0 & \cdots & a_{2j} & \cdots & a_{2n} \\ \cdots & \cdots & \cdots & \cdots & \cdots & \cdots \\ \cdots & \cdots & \cdots & \cdots & \cdots & \cdots \\ a_{n1} & a_{n2} & \cdots & a_{nj} & \cdots & 0 \end{bmatrix}$$

  The digraph portrays contextual relationships between the elements of a system, as shown in Figure 32.1. For example, an arrow from element b to element a signifies that b affects a, and the influence level is 4. In the similar direction, an arrow from a to d denotes that a affects d with an influence level of 3.

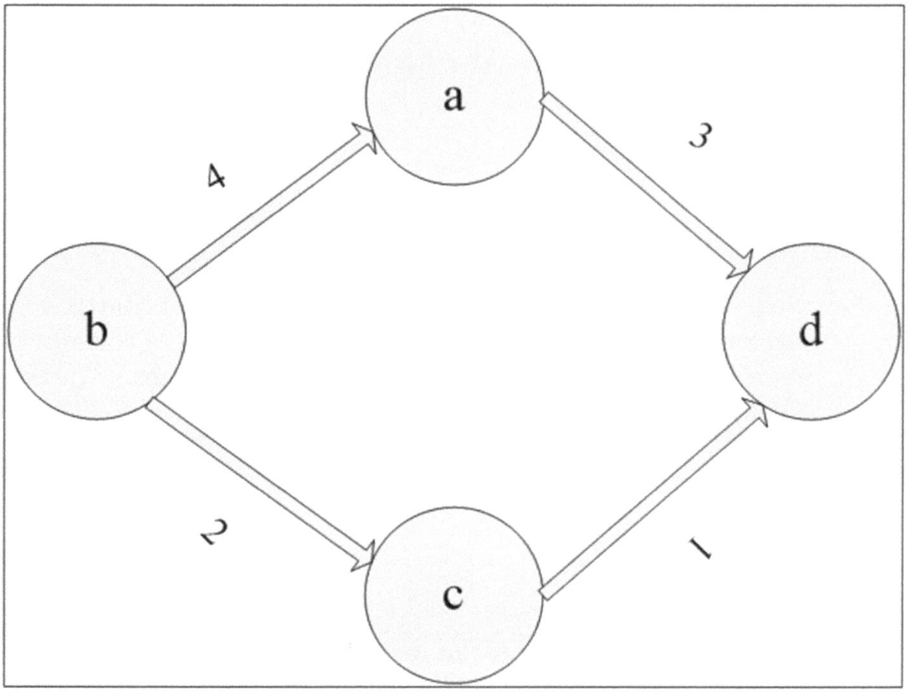

FIGURE 32.1 A typical digraph.

- **Step 2: Develop the Normalized Direct-Relation Matrix (*X*):** Once the direct-relation matrix is formulated, the corresponding normalized matrix (*X*) is derived employing Eqns. (32.4) and (32.5). Each element in matrix *X* should range between 0 and 1.

$$X = k.\,A \tag{32.4}$$

where;

$$k = \frac{1}{\max\limits_{1 \le i \le n}\left(\sum_{j=1}^{n} a_{ij}\right)} \quad (i, j = 1, 2, \ldots, n) \tag{32.5}$$

- **Step 3: Compute the Total-Relation Matrix (*T*):** The total-relation matrix (*T*) is determined using Eqn. (32.6), in which *I* represent the identity matrix. The element t_{ij} denotes the indirect effects that i^{th} criterion has on j^{th} criterion, and the matrix *T* reflects the total relationship between each pair of system criteria.

$$T = [t_{ij}]_{n\times n},\ i, j = 1, 2, \ldots, n$$

$$T = X + X^2 + X^3 + \ldots + X^k = X(I + X + X^2 + \ldots + X^{k-1})$$

$$[(I - X)(I - X)^{-1}] = X(I - X^k)(I - X)^{-1}$$

Then,

$$T = X(I - X)^{-1},\ \text{when } k \to \infty,\ X^k = [0]_{n\times n}$$

$$T = X(I - X)^{-1} \tag{32.6}$$

- **Step 4: Determine Sums of Rows and Columns of Matrix *T*:** In the total-relation matrix *T*, the sum of rows and sum of columns are represented by vectors *D* and *R*, as derived using Eqns. (32.7) and (32.8), respectively.

$$D_i = \left[\sum_{j=1}^{n} t_{ij}\right]_{n\times 1} = [t_i]_{n\times 1},\ i = 1, 2, \ldots, n \tag{32.7}$$

$$R_j = \left[\sum_{i=1}^{n} t_{ij}\right]_{1\times n} = [t_j]_{n\times 1},\ j = 1, 2, \ldots, n \tag{32.8}$$

- **Step 5: Set a Threshold Value (α):** This threshold value (α) is obtained from the average of the elements in matrix *T*, as computed using Eqn. (32.9), where *N* is the total number of elements in matrix *T*. This calculation is aimed to eliminate some minor effect elements in matrix *T*.

$$\alpha = \frac{\sum_{i=1}^{n}\sum_{j=1}^{n}[t_{ij}]}{N} \tag{32.9}$$

Since matrix *T* provides information on how one factor affects another, it is necessary for the decision maker to set up a threshold value to filter out some negligible effects. While doing so, only the effects greater than the set threshold value are chosen and shown in a digraph.

- **Step 6: Develop the Causal Diagram:** The horizontal axis vector ($D_k + R_k$) named 'prominence' is computed by adding *R* to *D* ($k = i = j = 1$) which reveals how much importance the criterion has. Similarly, the vertical axis ($D_k - R_k$) named 'relation' is calculated by subtracting *R* from *D*, which divides the criteria into a cause group and an effect group. If D_i is the sum of the ith row in matrix *T*, it summarizes both direct and indirect effects given by the ith criterion over other criteria. Similarly, R_j indicates the sum of the jth column in matrix *T*, and it shows both direct and indirect effects given by j^{th}

criterion over other criteria. If $j = 1$, it indicates the total effects given and received by i^{th} criterion. Thus, $(D_k + R_k)$ shows the degree of importance that i^{th} criterion plays in the entire system. In the contrary, $(D_k - R_k)$ determines the net effect that i^{th} criterion contributes to the entire system.

Generally, when $(D_k - R_k)$ is positive, the criterion belongs to the cause group. Otherwise, if $(D_k - R_k)$ is negative, the criterion belongs to the effect group. Therefore, a causal diagram can be developed while mapping the dataset of $(D_k + R_k, D_k - R_k)$, providing valuable insight for making decisions. Therefore, the decision maker can use the causal relationship of the variables and their interaction influence levels to identify the driving variables of the core problem in a complicated system, and plan for suitable decision to solve the problem in accordance with attribute type and influence level. Figure 32.2 portrays the procedural steps of DEMATEL method.

- **Step 7: Calculation of the Criteria Weights:** Now, in order to determine the relative importance (weight) of each of the considered criteria, the following formulations are adopted [9].

$$t_i^+ = \sum_{j=1}^{n} t_{ij} + \sum_{j=1}^{n} t_{ji} \tag{32.10}$$

$$t_i^- = \sum_{j=1}^{n} t_{ij} - \sum_{j=1}^{n} t_{ji} \tag{32.11}$$

$$t_i^{average} = \frac{(t_i^+ + t_i^-)}{2} \tag{32.12}$$

$$w_i = \frac{t_i^{average}}{\sum_{i=1}^{n} t_i^{average}} \tag{32.13}$$

where; w_i is the weight of ith criterion.

An efficient enterprise resource planning (ERP) system would provide a seamless information flow throughout the organization and help to integrate and automate the business information and processes within and across diversified functional areas in an organization. Gupta & Naqvi [10] attempted to identify the major critical success factors for successful implementation of an ERP system in a given organization. Those critical success factors were listed as cost, consultant, quality, vendor, ERP fit, process aspect, client, and technical aspect. Based on the opinions of 10 experts, the performance of three alternative ERP systems was evaluated, and the best system was singled out using fuzzy-TOPSIS method. This ERP system selection problem is considered here to illustrate the application of DEMATEL method in portraying the causal relationships among the considered criteria while estimating their relative weights.

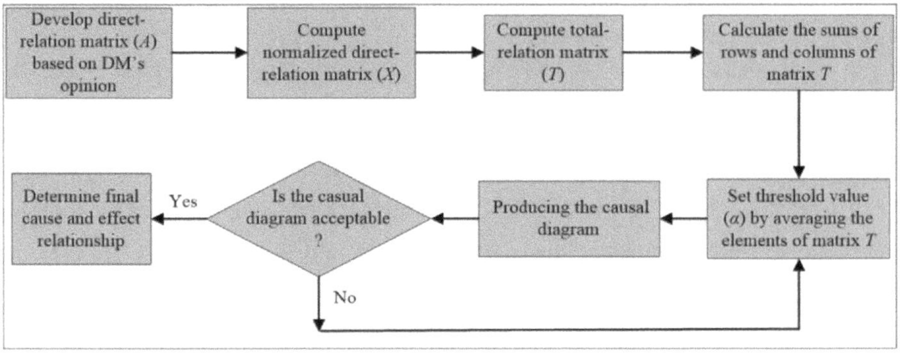

FIGURE 32.2 Process algorithm of DEMATEL method.

Using an integer scale ranging from 0 to 4, with no influence (0), low influence (1), medium influence (2), high influence, (3) and very high influence (4), and based on the opinions of a decision maker, the corresponding direct-relation matrix for this ERP selection problem is formulated, as shown in Table 32.5. In this matrix, the direct effect of the ith criterion on j^{th} criterion is pair-wise highlighted. For example, cost has high influence on consultant, it has very high influence on quality and so on. From the direct-relation matrix, the corresponding normalized matrix is constructed, while converting all of its elements in an interval [0,1]. This normalized direct-relation matrix is depicted in Table 32.6. In the next step, based on Eqn. (32.6), the total-relation matrix is formulated in Table 32.7. From the total-relation matrix T, the sum of rows and the sum of columns, represented as vectors D and R, are calculated using Eqns. (32.7) and (32.8), respectively. In the similar direction, application of Eqn. (32.9) determines the corresponding threshold value (α) as 1.3912 which would be helpful in developing the causal relationships amongst the criteria in the form of a digraph. The values of t_{ij} higher than 1.3912 are incorporated to draw the unidirectional or bidirectional arrows in the digraph. For example, in Table 32.7, the value of t_{12} = 1.6470 signifies that there would be an arrow joining criteria nodes 'cost' and 'consultant,' highlighting the significant influence of 'cost' criterion on 'consultant' criterion. The developed digraph for this ERP selection problem is exhibited in Figure 32.3.

TABLE 32.5 Initial Direct-Relation Matrix for the Criteria for the ERP Selection Problem

Criteria	Cost	Consultant	Quality	Vendor	ERP Fit	Process Aspect	Client	Technical Aspect
Cost	0	3	4	3	2	2	3	2
Consultant	3	0	3	3	2	3	3	2
Quality	4	3	0	2	3	2	1	3
Vendor	3	3	2	0	2	2	3	2
ERP fit	2	2	3	2	0	3	2	3
Process aspect	2	3	2	2	3	0	2	3
Client	3	3	1	3	2	1	0	2
Technical aspect	2	2	3	2	3	3	2	0

TABLE 32.6 Normalized Direct-Relation Matrix for the Criteria

Criteria	Cost	Consultant	Quality	Vendor	ERP Fit	Process Aspect	Client	Technical Aspect
Cost	0	0.1579	0.2105	0.1579	0.1053	0.1053	0.1579	0.1053
Consultant	0.1579	0	0.1579	0.1579	0.1053	0.1579	0.1579	0.1053
Quality	0.2105	0.1579	0	0.1053	0.1579	0.1053	0.0526	0.1579
Vendor	0.1579	0.1579	0.1053	0	0.1053	0.1053	0.1579	0.1053
ERP fit	0.1053	0.1053	0.1579	0.1053	0	0.1579	0.1053	0.1579
Process aspect	0.1053	0.1579	0.1053	0.1053	0.1579	0	0.1053	0.1579
Client	0.1579	0.1579	0.0526	0.1579	0.1053	0.0526	0	0.1053
Technical aspect	0.1053	0.1053	0.1579	0.1053	0.1579	0.1579	0.1053	0

Table 32.8 shows the computed values of D and R vectors for this problem. In Table 32.9, the corresponding values of $(D + R)$ and $(D - R)$ are provided. A positive value of $(D - R)$ for a specific criterion identifies it to belong to the cause group, whereas, a criterion belonging to the effect group has a negative value of $(D - R)$. Thus, it can be noticed from Table 32.9 that for this ERP selection problem, 'quality,' 'ERP fit,' 'process aspect' and 'technical aspect' are in the cause group of criteria, and 'cost,' 'consultant,' 'vendor' and 'client' belong to the effect group of criteria. Figure 32.4 provides the corresponding causal diagram showing the belongingness of different criteria in the cause-and-effect groups. It can be unveiled from this figure that 'process aspect' having the maximum $(D - R)$ value is the most significant and contributing criterion having the highest intensity of relation

with the other criteria. Thus, it exerts the greatest effect on the remaining ERP selection criteria. On the other hand, 'client' with the lowest value of $(D - R)$ is remarkably influenced by the other ERP selection criteria. Therefore, the decision maker can employ the causal relationship of the considered criteria and their interaction influence levels to identify the driving criterion for this ERP selection problem, and plan for suitable decision to solve the problem in accordance with attribute type and influence level. Finally, based on Eqns. (32.10)–(32.13), the relative weights of the ERP selection criteria are determined, as shown in Table 32.10. It can be interestingly noted that for this problem, almost all the criteria have equal importance in arriving at the most suitable ERP selection decision.

TABLE 32.7 Total-Relation Matrix for the Criteria

Criteria	Cost	Consultant	Quality	Vendor	ERP Fit	Process Aspect	Client	Technical Aspect
Cost	1.5216*	1.6470*	1.6317*	1.5096*	1.4561*	1.3915*	1.4355*	1.4561*
Consultant	1.6500*	1.5065*	1.5862*	1.5058*	1.4523*	1.4290*	1.4337*	1.4523*
Quality	1.6381*	1.5899*	1.4149*	1.4186*	1.4511*	1.3541	1.3088	1.4511*
Vendor	1.5090*	1.5025*	1.4121*	1.2443	1.3241	1.2684	1.3165	1.3241
ERP fit	1.4626*	1.4562*	1.4514*	1.3306	1.233	1.3121	1.2647	1.3693
Process aspect	1.4631*	1.4979*	1.4137*	1.3346	1.3694	1.1791	1.2703	1.3694
Client	1.3681	1.3622	1.2381	1.2555	1.1959	1.1078	1.0629	1.1959
Technical aspect	1.4626*	1.4562*	1.4514*	1.3306	1.3693	1.3121	1.2647	1.2330

$^*t_{ij} > 1.3912.$

TABLE 32.8 Computation of Vectors D and R

Criteria	D	R
Cost	12.049	12.0751
Consultant	12.0159	12.0185
Quality	11.6266	11.5994
Vendor	10.9011	10.9297
ERP fit	10.8798	10.8513
Process aspect	10.8975	10.3541
Client	9.7864	10.357
Technical aspect	10.8798	10.8513

TABLE 32.9 Total and Net Effects of the Responses

Criteria	D + R	D – R	Criteria Group
Cost	24.1241	–0.0261	Effect
Consultant	24.0343	–0.0026	Effect
Quality	23.226	0.0273	Cause
Vendor	21.8308	–0.0285	Effect
ERP fit	21.7311	0.0285	Cause
Process aspect	21.2516	0.5434	Cause
Client	20.1434	–0.5706	Effect
Technical aspect	21.7311	0.0285	Cause

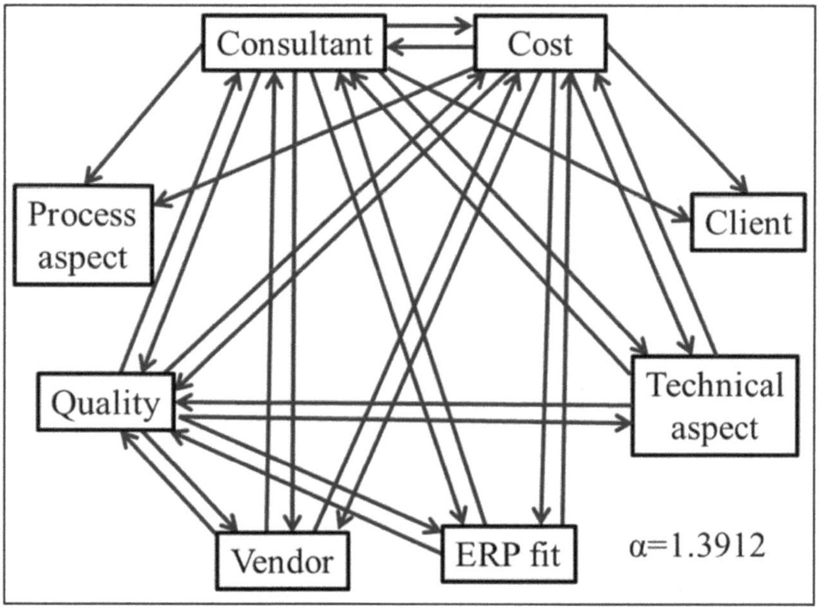

FIGURE 32.3 DEMATEL diagraph for the ERP selection problem.

32.3 STEP-WISE WEIGHT ASSESSMENT RATIO ANALYSIS (SWARA) METHOD

The SWARA method has now become a popular approach for estimating criteria weights in MCDM problems. The steps involved in this method are exhibited as below [11]:

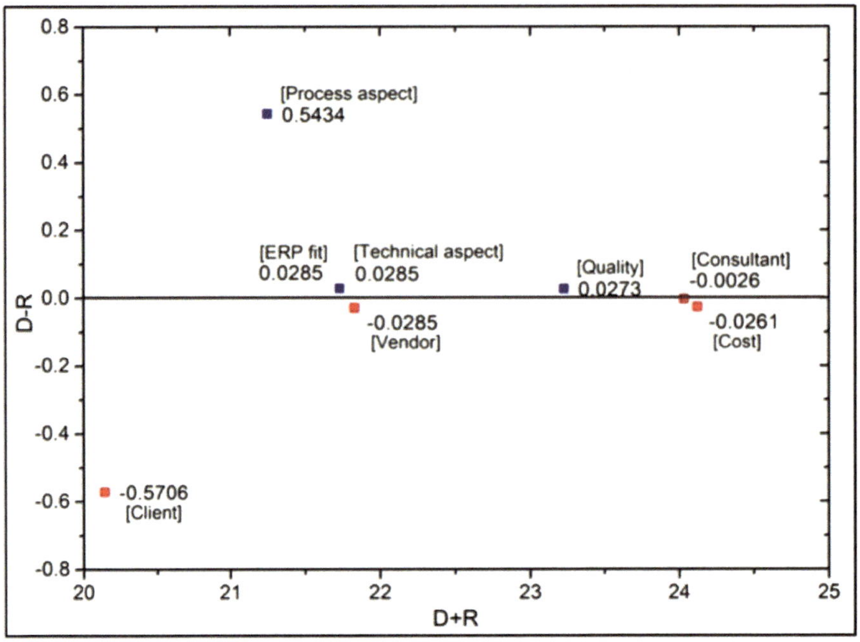

FIGURE 32.4 DEMATEL causal diagram for the ERP selection problem.

TABLE 32.10 Calculation of Criteria Weights Using DEMATEL Method

Criteria	t^+	t^-	taverage	w
Cost	24.1241	–0.0261	12.049	0.1353
Consultant	24.0343	–0.0026	12.0159	0.1350
Quality	23.226	0.0273	11.6266	0.1306
Vendor	21.8308	–0.0285	10.9011	0.1224
ERP fit	21.7311	0.0285	10.8798	0.1222
Process aspect	21.2516	0.5434	10.8975	0.1224
Client	20.1434	–0.5706	9.7864	0.1099
Technical aspect	21.7311	0.0285	10.8798	0.1222

- **Step 1:** The considered criteria are first sorted in descending order according to their expected significance.
- **Step 2:** Starting from the second criterion, the concerned decision maker expresses the relative importance of jth in relation to $(j-1)^{th}$ criterion, for each particular criterion. This ratio is known as the comparative importance of average value, s_j [12].

➢ **Step 3:** Determine the coefficient k_j using the following expression:

$$k_j = \begin{cases} 1, & j=1 \\ s_j + 1, & j>1 \end{cases} \tag{32.14}$$

➢ **Step 4:** Estimate the recalculated weight q_j as follows:

$$q_j = \begin{cases} 1, & j=1 \\ \dfrac{q_{j-1}}{k_j}, & j>1 \end{cases} \tag{32.15}$$

➢ **Step 5:** Calculate the relative criteria weights.

$$w_j = \frac{q_j}{\sum_{k=1}^{n} q_k} \tag{32.16}$$

where; w_j is the weight (relative importance) of the jth criterion.

It is worthwhile to mention here that both AHP and SWARA methods are quite capable of accurately estimating the criteria weights for any decision-making problem. Both of them employ pair-wise comparisons to denote the relative importance of the decision-making elements in a hierarchy, which is known as the comparative importance of average value in SWARA method. For a decision-making problem consisting of n number of evaluation criteria, AHP requires a significantly greater number of pair-wise comparisons as compared to SWARA method, which employs significantly less, i.e., $(n - 1)$ comparisons. Thus, while dealing with a high-dimensional problem using AHP method, the number of pair-wise comparisons significantly increases, negatively affecting the consistency of the derived solutions. But, in AHP method, there is an effective procedure to check the consistency of the performed comparisons, which helps in identifying biasness in the decision-making process. The SWARA method does not have such a procedure. For performing pair-wise comparisons in AHP method, a 9-point scale is usually adopted. On the contrary, in the SWARA method, the decision makers have greater freedom to express their judgments using a wider scale (any value between 0 and 1). Less number of pair-wise comparisons and easiness in expressing the judgments make SWARA method more attractive than AHP technique.

In an attempt to demonstrate the mathematical calculations involved in SWARA method, an alternative fuel selection problem [13] for electrical power generator is considered here. While identifying the most apposite fuel for electrical power generator, the corresponding decision matrix was developed with seven alternatives and seven evaluation criteria. In this method, using a 1–9 point scale, all the considered criteria need to be first sorted based on their expected relative significance. From Table 32.11, it can be noticed that among these criteria, maximum significance is provided to 'calorific value' criterion, followed by 'pour point.' On the other hand, 'viscosity' has the least significance. This table also segregates all the criteria into beneficial and cost categories. Table 32.12 shows the detailed calculations for criteria weight measurement using SWARA method. In this table, the concerned decision maker expresses the relative importance of the jth criterion in relation to $(j-1)^{th}$ criterion, leading to the estimation of comparative importance of average value (s_j). Now, employing Eqns. (32.14) and (32.15), values of coefficient (k_j) and recalculated weight (q_j) are respectively computed. Finally, the relative criteria weights are determined using Eqn. (32.16). It can be revealed that 'calorific value' has the maximum importance, followed by 'pour point.' Viscosity has the minimum relative weight. While solving the same problem applying AHP and fuzzy AHP techniques, Durairaj et al. [13] also determined the same importance of the criteria under consideration.

TABLE 32.11 List of the Considered Criteria Along with Their Relative Importance [13]

Criteria	Relative Importance	Type
Calorific value (C_1) (kJ/kg)	9	Beneficial
Density (C_2) (kg/m^3)	4	Beneficial
Cetane number (C_3)	3	Beneficial
Flashpoint (C_4) (°C)	5	Beneficial
Viscosity (C_5) (mm^2/sec)	1	Cost
Cloud point (C_6) (°C)	2	Cost
Pour point (C_7) (°C)	7	Cost

TABLE 32.12 Calculation of Criteria Weights Using SWARA Method

Criteria	Comparative Importance of Average Value s_j	Coefficient $k_j = 1 + s_j$	Recalculated Weight q_j	Weight w_j
C_1	–	1	1	0.1990
C_7	0.15	1.15	0.8696	0.1731
C_4	0.04	1.04	0.8361	0.1664
C_2	0.29	1.29	0.6481	0.1290
C_3	0.02	1.02	0.6354	0.1265
C_6	0.20	1.2	0.5295	0.1054
C_5	0.05	1.05	0.5043	0.1003

32.4 PIVOT PAIRWISE RELATIVE CRITERIA IMPORTANCE ASSESSMENT (PIPRECIA) METHOD

The PIPRECIA method, evolved from SWARA, is also a criteria weight measurement technique based on subjective judgments of the decision-makers in a group decision-making environment reflecting their cognitive attitudes [11]. In SWARA method, the considered evaluation criteria are required to be ranked according to their relative significance. However, it has been observed that PIPRECIA method is more effective than the SWARA method in complex decision-making situations although it does not require the prior ranking of the criteria based on their relative importance. Application of PIPRECIA method for criteria weight determination adopts the following steps [14, 15]:

- **Step 1:** Consider a set of relevant evaluation criteria. These criteria may be ranked in descending order of their significance (not mandatory).
- **Step 2:** Determine the relative importance (s_j) of the criteria. Beginning with the second criterion, the coefficient signifying the relative importance of each criterion (s_j) is evaluated.

$$s_j = \begin{cases} >1, & \text{if } C_j \succ C_{j-1} \\ =1, & \text{if } C_j = C_{j-1} \\ <1, & \text{if } C_j \prec C_{j-1} \end{cases} \tag{32.17}$$

- **Step 3:** Compute the coefficient k_j for each criterion using the following expression:

$$k_j = \begin{cases} 1, & \text{if } j = 1 \\ 2 - s_j, & \text{if } j > 1 \end{cases} \tag{32.18}$$

- **Step 4:** Estimate the recalculated weight q_j for the jth criterion.

$$q_j = \begin{cases} 1, & \text{if } j = 1 \\ \dfrac{q_{j-1}}{k_j}, & \text{if } j > 1 \end{cases} \tag{32.19}$$

- **Step 5:** Determine the relative weights of the evaluation criteria based on the judgment of an individual decision maker.

$$w_j = \frac{q_j}{\sum_{k=1}^{n} q_k} \tag{32.20}$$

where; w_j is the relative weight assigned to the jth criterion.

➤ **Step 6:** Calculate the group relative weights of the considered criteria.

In a group decision-making situation, if there are M number of decision makers, the following equations can be employed to determine the aggregated criteria weights.

$$w_j^* = \left(\prod_{m=1}^{M} w_j^m \right)^{\frac{1}{M}} \tag{32.21}$$

$$w_j = \frac{w_j^*}{\sum_{j=1}^{n} w_j^*} \tag{32.22}$$

Kazancoglu & Burmaoglu [16] considered an ERP software selection problem taking into account six alternatives and 10 evaluation criteria, and later solved the same using TODIM method. In order to exhibit the mathematical calculations involved in PIPRECIA method, this ERP software selection problem is considered here. The list of evaluation criteria is provided in Table 32.13. Its application starts with sorting the criteria in descending order, based on their expected significance. Now, using the preference of the participating decision maker, the comparative importance of average value for each criterion (s_j) is determined, as depicted in Table 32.14. Following the procedural steps of PIPRECIA method, values of the coefficient k_j and recalculated weight q_j are respectively estimated, while applying Eqns. (32.18) and (32.19). The final criteria weights are computed using Eqn. (32.20). From Table 32.14, it can be unveiled that 'strategy-fit' is assigned with the highest importance, followed by 'technology' criterion. For this problem, 'cost' and 'benefits' criteria have the minimum importance. Under a group decision-making environment, Kazancoglu & Burmaoglu [16] also determined the same order of preference for the 10 considered criteria.

32.5 CRITERIA IMPORTANCE THROUGH INTERCRITERIA CORRELATION (CRITIC) METHOD

In any decision-making problem, the basic objective is to identify the best alternative or course of action based on a set of conflicting subjective and objective criteria. Amongst these evaluation criteria, the determination of the weights for subjective criteria mainly depends on the knowledge, expertise, and perception of the participating decision makers. But involvement of

TABLE 32.13 Evaluation Criteria for the ERP Software Selection Problem

Symbol	Criteria
C_1	Strategy-fit
C_2	Technology
C_3	Change management
C_4	Risk
C_5	Implement ability
C_6	Business functionality
C_7	Vendor credentials
C_8	Flexibility
C_9	Cost
C_{10}	Benefits

TABLE 32.14 Calculation of Criteria Weights Using PIPRECIA Method

Criteria	Comparative Importance of Average Value s_j	Coefficient $k_j = 2 - s_j$	Recalculated Weight q_j	Weight w_j
C_1	–	1.00	1.00	0.29
C_2	0.70	1.30	0.77	0.23
C_3	0.65	1.35	0.57	0.17
C_4	0.60	1.40	0.41	0.12
C_5	0.50	1.50	0.27	0.08
C_6	0.45	1.55	0.18	0.05
C_7	0.35	1.65	0.11	0.03
C_8	0.30	1.70	0.06	0.02
C_9	0.25	1.75	0.04	0.01
C_{10}	0.20	1.80	0.02	0.01

multiple decision makers while solving a specific decision-making problem makes the solution approach complex and time-consuming. To avoid this complexity, objective weighting approaches are usually preferred. The CRITIC method, proposed by Diakoulaki et al. [17], is such an objective method for criteria weight measurement, involving intensity of the contrast and conflict present in the original decision matrix itself. It belongs to the class of correlation methods relying on the performance scores of the considered alternatives with respect to a set of evaluation criteria. In this method, the determination of the criteria weights depends on the standard deviations of the normalized criteria values and correlation coefficients between all pairs of the considered criteria. Its procedural steps are detailed out as below [18–21]:

- **Step 1:** Develop the initial decision matrix, $X = [x_{ij}]_{m\times n}$ where; x_{ij} is the performance measure of ith alternative with respect to j^{th} criterion, m is the number of alternatives and n is the number of criteria.
- **Step 2:** Normalize the initial decision matrix using the following expression:

$$r_{ij} = \frac{x_{ij} - x_j^{min}}{x_j^{max} - x_j^{min}} \tag{32.23}$$

where; $x_j^{max} = \max(x_{ij}, i = 1, 2, \ldots, m)$ and $x_j^{min} = \min(x_{ij}, i = 1, 2, \ldots, m)$.

- **Step 3:** In this step, the weight of a specific criterion is determined based on its standard deviation value and its correlation with other evaluation criteria. Thus, the weight of jth criterion w_j can be derived as:

$$w_j = \frac{C_j}{\sum_{j=1}^{n} C_i} \tag{32.24}$$

where; C_j is the quantity of information contained in j^{th} criterion which can be estimated based on the following expression:

$$C_j = \sigma_j \sum_{i=1}^{m} \left(1 - r_{ij}\right) \tag{32.25}$$

where; σ_j is the standard deviation of the jth criterion; and r_{ij} is the correlation coefficient between the ith and j^{th} criteria. Thus, it can be revealed that higher C_j value represents a greater amount of information extracted from a specific criterion, implying its higher relative significance for the given decision-making problem.

To illustrate the mathematical steps of CRITIC method, a spar material selection problem for wing structure of a human-powered aircraft [22] is considered here. The related decision matrix consists of 12 alternative materials and six evaluation criteria. This initial decision matrix is provided in Table 32.15. In this matrix, 'creep resistance' is expressed qualitatively. These qualitative ratings are subsequently converted into quantitative values using a relative scale as 'Excellent' = 0.895, 'Good' = 0.695, 'Average' = 0.495 and 'Weak' = 0.295. Table 32.16 exhibits the transformed decision matrix with all the evaluation criteria expressed quantitatively. Now, depending on the type of the evaluation criterion considered, the decision matrix of Table 32.16 is normalized. The corresponding normalized decision matrix is provided in Table 32.17. While determining criteria weights using CRITIC method, standard deviation values of the considered criteria and correlation coefficients between them play important roles. The standard deviation values of the six criteria are shown in Table 32.18, while Table 32.19 provides the correlation coefficients between these criteria. Finally, using Eqn. (32.24), the criteria weights for this spar material selection problem are calculated, as shown in Table 32.20. For this material selection problem, 'price' and 'density' of the alternative materials are identified as the two most important evaluation criteria.

TABLE 32.15 Data for the Spar Material Selection Problem for Wing Structure of a Human-Powered Aircraft [22]

Material	Price	Tensile Strength (Mpa)	Young's Modulus (Gpa)	Density (g/cm3)	Compressive Strength (Mpa)	Creep Resistance (25°C)
Al 7075-T6	3.5	581	70	2.6	581	Good
Al 2024-T4	3.5	425	72.5	2.6	425	Good
Ti-6Al-4 V	21	1,008	112	4.4	1,008	Excellent
Ti-2Fe-3Al-10 V	22	1,295	120	4.5	1,295	Excellent
E-glass 73%-epoxy	2.6	1,642	55.9	2.17	410	Average
E-glass 56%-epoxy	2.5	1,028	42.8	1.97	290	Weak
E-glass 65%-polyester	2.5	340	19.6	1.8	90	Weak
S-glass 70%-epoxy continuous fibers	9	2,100	62.3	2.11	550	Average
S-glass 70%-epoxy fabric	8	680	22	2.11	180	Average
Carbon 63%-epoxy	45	1,725	158.7	1.61	900	Average
Aramid 62%-epoxy	20	1,311	82.7	1.38	300	Average
Balsa	6	28.5	7	0.22	17.5	Average

Source: Reprinted with permission from Ref. [22]. © 2021 Elsevier

32.6 FACTOR RELATIONSHIP (FARE) METHOD

The FActor RElationship (FARE) method for determination of criteria weights was first proposed by Ginevicius [23]. In this method, the following steps are adopted for estimating criteria weights [24, 25]:

TABLE 31.16 Quantitative Data for Application of CRITIC Method

Material	Price	Tensile Strength	Young's Modulus	Density	Compressive Strength	Creep Resistance
Al 7075-T6	3.5	581	70	2.6	581	0.695
Al 2024-T4	3.5	425	72.5	2.6	425	0.695
Ti-6Al-4 V	21	1,008	112	4.4	1,008	0.895
Ti-2Fe-3Al-10 V	22	1,295	120	4.5	1,295	0.895
E-glass 73%-epoxy	2.6	1,642	55.9	2.17	410	0.495
E-glass 56%-epoxy	2.5	1,028	42.8	1.97	290	0.295
E-glass 65%-polyester	2.5	340	19.6	1.8	90	0.295
S-glass 70%-epoxy continuous fibers	9	2,100	62.3	2.11	550	0.495
S-glass 70%-epoxy fabric	8	680	22	2.11	180	0.495
Carbon 63%-epoxy	45	1,725	158.7	1.61	900	0.495
Aramid 62%-epoxy	20	1,311	82.7	1.38	300	0.495
Balsa	6	28.5	7	0.22	17.5	0.495

TABLE 32.17 Normalized Matrix for Application of CRITIC Method

Material	Price	Tensile Strength	Young's Modulus	Density	Compressive Strength	Creep Resistance
Al 7075-T6	0.9765	0.2667	0.4153	0.4439	0.441	0.667
Al 2024-T4	0.9765	0.1914	0.4318	0.4439	0.319	0.667
Ti-6Al-4 V	0.5647	0.4728	0.6922	0.0234	0.775	1.000
Ti-2Fe-3Al-10 V	0.5412	0.6114	0.7449	0	1.000	1.000
E-glass 73%-epoxy	0.9976	0.7789	0.3223	0.5444	0.307	0.333
E-glass 56%-epoxy	1.0000	0.4825	0.2360	0.5911	0.213	0.000
E-glass 65%-polyester	1.0000	0.1504	0.0831	0.6308	0.057	0
S-glass 70%-epoxy continuous fibers	0.8471	1.0000	0.3645	0.5584	0.417	0.333
S-glass 70%-epoxy fabric	0.8706	0.3145	0.0989	0.5584	0.127	0.333
Carbon 63%-epoxy	0	0.8190	1.0000	0.6752	0.691	0.333
Aramid 62%-epoxy	0.5882	0.6191	0.4990	0.7290	0.221	0.333
Balsa	0.9176	0.	0	1.0000	0	0.333

TABLE 32.18 Standard Deviation Values for the Criteria

Criteria	Price	Tensile Strength	Young's Modulus	Density	Compressive Strength	Creep Resistance
Standard deviation	0.302	0.302	0.295	0.278	0.304	0.328

TABLE 32.19 Correlation Coefficient Values of the Paired Criteria

Criteria	Price	Tensile Strength	Young's Modulus	Density	Compressive Strength	Creep Resistance
Price	1.0000	–0.4712	–0.8465	0.1707	–0.6358	–0.2838
Tensile strength	–0.4712	1.0000	0.5621	–0.2067	0.4886	0.0275
Young's modulus	–0.8465	0.5621	1.0000	–0.5183	0.8716	0.5539
Density	0.1707	–0.2067	–0.5183	1.0000	–0.8009	–0.7829
Compressive strength	–0.6358	0.4886	0.8716	–0.8009	1.000	0.7642
Creep resistance	–0.2838	0.0275	0.5539	–0.7829	0.7642	1.000

TABLE 32.20 Determination of Criteria Weights Using CRITIC Method

Criteria	Price	Tensile Strength	Young's Modulus	Density	Compressive Strength	Creep Resistance
C_j	2.132	1.391	1.293	1.981	1.321	1.550
w_j	0.221	0.144	0.134	0.205	0.136	0.160

- **Step 1:** Determine the potential impact of the considered criteria/attributes. The potential impact of the criteria is estimated using the following expression:

$$P = S(n - 1) \tag{32.26}$$

where; P is the potential of the system's criteria impact; S is the maximum value of the evaluation scale employed, as given in Table 32.21; and n is the number of criteria under consideration.

- **Step 2:** Ranking of the criteria and assessment of their interrelationship. In this step, all the considered criteria are now ranked depending on their importance based on the relationship among them. This pair-wise comparison is performed using the scale provided in Table 32.22. Here, any criterion with a lower rank has less significant impact on other criteria with higher ranks, and it would consequently transmit a larger part of its potential impact to others.

TABLE 32.21 Scale of Quantitative Evaluation of Interrelationship between the System's Criteria

Type of the Effect	Rating of the Effect Produced by Interrelationship
Almost none	1
Very weak	2
Weak	3
Lower than average	4
Average	5
Higher than average	6
Strong	7
Very strong	8
Almost absolute	9
Absolute	10

TABLE 32.22 Measurement Scale for Pair-Wise Comparison

Verbal Judgment/Preference	Numerical Rating
Extremely preferred	9
Very strongly to extremely preferred	8
Very strongly preferred	7
Strongly to very strongly preferred	6
Strongly preferred	5
Moderately to strongly preferred	4
Moderately preferred	3
Equally to moderately preferred	2
Equally to moderately preferred	1

- **Step 3:** Determine the impact of the criteria on the main criterion. The impact of criterion c_j on the main criterion is estimated and this impact is converted using the following expression:

$$c_{1j} = S - \tilde{c}_{1j} \tag{32.27}$$

where; c_{1j} is the impact of the jth criterion on the first criterion; and $\tilde{c}_{1j}$ is the part of jth criterion's potential impact transmitted to the main criterion.

- **Step 4:** Estimation of the total impact. Using Eqn. (32.28), the total impact and consistency of any criterion are estimated. The estimation is considered to be reliable, consistent, and accurate if the total impact of its criteria with a positive sign becomes equal to the total impact with a negative sign, i.e., their summation would always be zero.

$$P_j = \sum_{j=1}^{n} c_{ij}, \quad j \neq i \tag{32.28}$$

In the same way, the total impact can be calculated using Eqn. (32.29). The total impact of a criterion represents its dominance over the other criteria. Thus, the most significant criterion should be the first one with the maximum total dominance.

$$P_j = P_1 - n.c_{1j} \tag{32.29}$$

where; P_j is the total impact of jth criterion.

- **Step 5:** Calculation of the criteria weights. In this step, weights of the considered criteria are computed using Eqn. (32.30).

$$w_j = \frac{P_j^f}{P_S} = \frac{P_1 - nc_{1j} + S(n-1)}{nS(n-1)} \tag{32.30}$$

where; P_S is the total potential of a set of criteria, determined employing Eqn. (32.32); and P_j^f is the actual total impact of jth criterion on the system, determined using Eqn. (32.32).

$$P_S = n.P = n.S(n-1) \tag{32.31}$$

$$P_j^f = P_1 - nc_{1j} + S(n-1) = P_j + P \tag{32.32}$$

where; P_j is the total impact generated by the jth criterion of the system representing its total dependence on the other criteria and P is the potential impact of the criteria.

In order to demonstrate the application of FARE method in determining criteria weights of a given MCDM problem, an NTM process selection problem [26] is considered here. In this problem, the most suitable NTM process needs to be identified for machining of cylindrical through holes on non-conductive ceramic materials. The performance of the alternative NTM processes is evaluated based on seven machining characteristics, i.e., tolerance (TL), surface finish (SF), surface damage (SD), taper (T), material removal rate (MRR), work material (M) and cost (C). Yurdakul & Cogun [26] applied AHP method to determine the corresponding criteria weights as $w_{TL} = 0.32$, $w_{SF} = 0.19$, $w_{SD} = 0.04$, $w_T = 0.04$, $w_{MRR} = 0.19$, $w_M = 0.11$ and $w_C = 0.11$. As this problem consists of seven evaluation criteria and a 10-point rating scale is adopted to depict the interrelationships, the potential impact of the criteria is calculated as 10×(7 – 1) = 60, using Eqn. (32.26). Based on the estimated criteria weights [26], the initial priority of the NTM process selection criteria can be set as TL > SF > MRR > M > C > SD > T. It indicates that TL and T are respectively the most and the least important criteria for this decision-making problem. As in this problem, TL is identified as the most important criterion, the impact of other criteria would be transferred through it and their direct impacts on this NTM process selection problem would decrease, as shown in Table 32.23. It is worthwhile to mention here that this matrix of potential equilibrium would follow the condition, i.e., $a_{ij} = -a_{ji}$. For example, if the priority of TL to T is 4, the priority of T to TL would be automatically set as – 4. In this matrix, a row or column total determines the total effect or dependence of a particular criterion on other criteria as compared to it.

TABLE 32.23 Summary Matrix of Potential Equilibrium for the NTM Process Selection Criteria

Criteria	TL	SF	SD	T	MRR	M	C	Total Effect
TL	–	3	3	4	4	3	5	22
SF	–3	–	2	3	3	4	4	13
SD	–3	–2	–	2	–4	–3	–3	–13
T	–4	–3	–2	–	–2	–2	–2	–15
MRR	–4	–3	4	2	–	4	4	7
M	–3	–4	3	2	–4	–	4	–2
C	–5	–4	3	2	–4	–4	–	–12
Total effect	–22	–13	13	15	–7	2	12	–

Now, in Table 32.24, the total impact (P_j) for a specific criterion is computed while taking the summation of each row. It can be revealed from this table that the total impact of the criteria set having a positive sign becomes equal to the total impact of the criteria with a negative sign. Thus, the criteria set is entirely consistent. In this example, the total potential (P_S) value is calculated as 7×60 = 420, which would be subsequently employed for estimating the corresponding criteria weights. Based on these P_S values, the individual potentials of all the NTM process selection criteria are evaluated, along with the determination of the final criteria weights, as provided in Table 32.25. It can be interestingly noticed that FARE method also identifies TL and T as the most and the least significant criteria for this NTM process selection problem, which closely match with the observations of Yurdakul & Cogun [26].

TABLE 32.24 Total Impact of NTM Process Selection Criteria

Criteria	TL	SF	SD	T	MRR	M	C	Sum
TL	–	7	7	6	6	7	5	38
SF	–7	–	8	7	7	6	6	27
SD	–7	–8	–	8	–6	–7	–7	–27
T	–6	–7	–8	–	–8	–8	–8	–29
MRR	–6	–7	6	8	–	6	6	13
M	–7	–6	7	8	–6	–	6	2
C	–5	–6	7	8	–6	–6	–	–8

TABLE 32.25 Individual Potential and Criteria Weights

Criteria	P_j^f	$P_S = n.P = n.S$ $(n-1)$	$w_j = \frac{P_j^f}{P_S}$
TL	98	7 × 60 = 420	98/420 = 0.23
SF	87		87/420 = 0.21
SD	33		33/420 = 0.08
T	31		31/420 = 0.07
MRR	73		73/420 = 0.17
M	62		62/420 = 0.15
C	52		52/420 = 0.12

32.7 BEST WORST METHOD (BWM)

It has already been mentioned that while solving a given MCDM problem, determination of the corresponding criteria weights plays a pivotal role in arriving at the most accurate ranking of the candidate alternatives. The BWM, developed by Rezaei [27], is a new mathematical approach for criteria weight estimation, which is based on the comparison of the best criterion to all other criteria and all other criteria to the worst criterion. Its application thus depends on a comparison system consisting of two comparison vectors. While identifying the best and the worst criteria before the pair-wise comparisons between the criteria, this method provides a clear understanding to the concerned decision-makers with respect to the range of evaluation, thus helping in deriving more accurate and consistent solutions. In a similar direction, consideration of only two pair-wise comparison vectors with two opposed references (best and worst) in a single decision-making framework considerably reduces the chances of biasness in the decision-making process. Unlike AHP method, it does not take into account pair-wise comparisons between the secondary criteria. Thus, it has been proved itself as a more time-efficient approach with the ability to validate consistency of the pair-wise comparisons. In this method, there is a possibility of deriving multiple optimal solutions which would provide more flexibility in the decision-making process involving multiple decision makers. The procedural steps of BWM are elucidated as below [28–30]:

- **Step 1:** Consider a set of evaluation criteria in a decision-making problem. At first, a set of criteria (c_1, c_2, ..., c_n) is taken into consideration based on which the performance of the candidate alternatives is assessed.
- **Step 2:** Identify the best (most important) and the worst (least important) criteria from the given set of decision criteria.

  In this step, the concerned decision makers are asked to identify the best and the worst criteria depending on their level of expertise and knowledge.
- **Step 3:** Appraise the preference of the best criterion over all other criteria. Based on a scale ranging between 1 and 9, the preference of the best criterion over all the other criteria is evaluated, resulting in the development of the following best-to-others (A_B) vector:

  $$A_B = (a_{B1}, a_{B2}, \ldots, a_{Bn})$$

  where; a_{Bj} represents the preference of the best criterion B over j^{th} criterion. It is always true that $a_{BB} = 1$.
- **Step 4:** Assess the preference of all the other criteria over the worst criterion. Using the same scale, the preference of all the other criteria over the worst criterion is appraised, formulating the following others-to-worst (A_W) vector.

  $$A_W = (a_{1W}, a_{2W}, \ldots, a_{nW})^{\text{T}}$$

  where; a_{jW} denotes the preference of jth criterion over the worst criterion W. It becomes quite obvious that $a_{WW} = 1$.
- **Step 5:** Computation of the optimal criteria weights. To compute the optimal criteria weights for a decision-making problem, the maximum absolute differences $\{|w_B - a_{Bj}w_j|, |w_j - a_{jW}w_W|\}$ for all j require to be minimized. The corresponding mathematical expression can be stated as below:

$$\min \max_{j} \left\{ \left| w_B - a_{Bj} w_j \right|, \left| w_j - a_{jW} w_W \right| \right\} \tag{32.33}$$

Subject to:

$$\sum_{j} w_j = 1, w_j \geq 0 \text{ for all } j.$$

The above problem can now be solved by converting it into the following linear programming (LP) problem:

$$\min \xi$$

Subject to:

$$\left| \frac{w_B}{w_j} - a_{Bj} \right| \leq \xi, \quad \text{for all } j$$

$$\left| \frac{w_j}{w_W} - a_{jW} \right| \leq \xi, \quad \text{for all } j \tag{32.34}$$

$$\sum_{j} w_j = 1, w_j \geq 0 \text{ for all } j.$$

The solution of the above LP problem would provide the optimal criteria weights (w_1^*, w_2^*,..., w_n^*,) and the objective function value (ξ^*). The ξ^* is designated as the consistency ratio (CR) of the comparison method. A value of ξ^* approaching zero represents more consistency in the decision-making process.

In order to illustrate the formulation and application of BWM in estimating criteria weights, a group decision-making problem having eight criteria (C_1, C_2, ..., C_8) and involvement of four decision makers is considered. Now, based on the opinions of the participating decision makers, the following best-to-others and others-to-worst matrixes are subsequently developed, as provided in Table 32.26.

Now, using Eqn. (32.34), the corresponding LP model for the decision maker 1 is formulated, as shown in Table 32.27. While solving this developed LP model, the weights of the considered criteria are determined, as exhibited in Table 32.28. In a similar direction, the criteria weights assigned by the remaining decision makers are also estimated. These criteria weights are now aggregated together using the geometric mean method to compute the final weights for the considered decision-making problem. Lower values of both ξ^* and CR signify that the judgments taken by the four decision makers are almost unbiased and consistent.

TABLE 32.26 Best-to-Others and Others-to-Worst Matrixes for the Four Decision Makers

				Decision Maker 1				
	C_1	C_2	C_3	C_4	C_5	C_6	C_7	C_8
Best-to-Others	1	3	6	7	8	4	3	6
Others-to-Worst	8	6	3	6	1	5	6	4
				Decision Maker 2				
	C_1	C_2	C_3	C_4	C_5	C_6	C_7	C_8
Best-to-Others	1	3	4	4	5	8	6	5
Others-to-Worst	8	5	4	4	3	1	2	4
				Decision Maker 3				
	C_1	C_2	C_3	C_4	C_5	C_6	C_7	C_8
Best-to-Others	1	2	4	5	4	4	5	9
Others-to-Worst	9	7	5	2	4	5	3	1
	Decision Maker 4							
	C_1	C_2	C_3	C_4	C_5	C_6	C_7	C_8
Best-to-Others	1	4	3	6	4	5	7	6
Others-to-Worst	7	5	4	2	3	3	1	4

TABLE 32.27 BWM-based LP Model for Decision Maker 1

Min ξ

Subject to:

$$\left|\frac{w_1}{w_2}-8\right| \le \xi, \left|\frac{w_1}{w_3}-6\right| \le \xi, \left|\frac{w_1}{w_4}-7\right| \le \xi, \left|\frac{w_1}{w_5}-3\right| \le \xi, \left|\frac{w_1}{w_6}-4\right| \le \xi, \left|\frac{w_1}{w_7}-3\right| \le \xi, \left|\frac{w_1}{w_8}-6\right| \le \xi, \left|\frac{w_2}{w_5}-6\right| \le \xi,$$

$$\left|\frac{w_3}{w_5}-3\right| \le \xi, \left|\frac{w_4}{w_5}-6\right| \le \xi, \left|\frac{w_6}{w_5}-5\right| \le \xi, \left|\frac{w_7}{w_5}-6\right| \le \xi, \left|\frac{w_8}{w_5}-4\right| \le \xi$$

$$w_1 + w_2 + w_3 + w_4 + w_5 + w_6 + w_7 + w_8 = 1,\ w_j \ge 0, \text{ for all } j$$

32.8 FULL CONSISTENCY METHOD (FUCOM)

Like AHP and BWM, FUCOM [31] also depends on pair-wise comparison between the considered criteria, and the derived results are validated based on deviation from maximum consistency. The criteria weights estimated using this method are influenced by the subjective judgments of the participating decision makers. It employs less number of pair-wise comparisons ($n - 1$ in total) (where n is the number of criteria) as compared to AHP method. Thus, the chances of error during comparisons and involvement of the decision makers are greatly reduced in this method. Unlike other methods, it shows minor deviations in the estimated criteria weights from the optimal solutions. It is also least affected by the problem of redundant pair-wise comparison, as encountered in other subjective methods. The steps involved in this method are presented as below [32]:

TABLE 32.28 Determination of the Criteria Weights

	Decision Maker 1	Decision Maker 2	Decision Maker 3	Decision Maker 4	Aggregated Weight
C_1	0.342	0.364	0.331	0.357	0.362
C_2	0.151	0.143	0.193	0.110	0.152
C_3	0.075	0.107	0.097	0.147	0.107
C_4	0.065	0.107	0.077	0.074	0.082
C_5	0.029	0.086	0.097	0.110	0.075
C_6	0.113	0.037	0.097	0.088	0.080
C_7	0.151	0.071	0.077	0.039	0.078
C_8	0.075	0.086	0.031	0.074	0.064
ξ^*	0.1097	0.0642	0.0560	0.0845	
CR	0.0245	0.0144	0.0107	0.0226	

- **Step 1:** From a set of criteria, i.e., $C = \{C_1, C_2, \ldots, C_n\}$, they are ranked by the decision maker based on their relative significance. Thus, the criterion with the maximum significance is ranked first, having the highest weight. The following ranking of the criteria based on their expected weights is obtained:

$$C_{j(1)} > C_{j(2)} > \ldots > C_{j(k)} \tag{32.35}$$

where; k is the rank allotted to the considered criterion.

- **Step 2:** The ranked criteria are then compared and the comparative priority ($a_{k/(k+1)}$), ($k = 1, 2, \ldots, n$) is developed. It represents the advantage of criterion having $C_{j(k)}$ rank over the criterion with rank $C_{j(k+1)}$. The following vector comprising the comparative priorities of the criteria is thus attained:

$$\Phi = (a_{1/2}, a_{2/3}, \ldots, a_{k/(k+1)}) \tag{32.36}$$

- **Step 3:** The weights of the criteria $(w_1, w_2, \ldots, w_n)^T$ are calculated. These criteria weights should fulfill the following two conditions:
  a. the ratio of criteria weights should be equal to the comparative priority among the observed criteria, ($a_{k/(k+1)}$) i.e.:

$$\frac{w_k}{w_{k+1}} = a_{k/(k+1)} \tag{32.37}$$

b. the criteria weights should satisfy the condition of mathematical transitivity, i.e., $a_{k/(k+1)} \otimes a_{(k+1)/(k+2)}) = a_{k/(k+2)}$.

Full consistency, i.e., minimum deviation from full consistency (DFC) (χ) can only be attained if the transitivity condition is fulfilled. For optimal criteria weights, the condition of maximum consistency, i.e., $\chi = 0$, must be satisfied. The mathematical model for computing the optimal criteria weights can be formulated as follows:

$\min \chi$

Subject to:

$$\left| \frac{w_{j(k)}}{w_{j(k+1)}} - a_{k/(k+1)} \right| \leq \chi, \forall j$$

$$\left| \frac{w_{j(k)}}{w_{j(k+2)}} - a_{k/(k+1)} \otimes a_{(k+1)/(k+2)} \right| \leq \chi, \forall j \tag{32.38}$$

$$\sum_{j=1}^{n} w_j = 1, \forall j$$

After solving Eqn. (32.38), the corresponding criteria weights $(w_1, w_2, \ldots, w_n)^T$ and degree of DFC (χ) are determined.

To demonstrate the application of FUCOM as a criteria weight measurement technique, a decision-making problem involving four decision makers and nine evaluation criteria is considered. Depending on varied levels of knowledge and expertise, those nine criteria are ranked by the individual decision makers based on their varying significance (importance).

- **Decision Maker 1:** $C_2 > C_3 > C_1 = C_4 > C_6 > C_8 = C_9 > C_7 > C_5$
- **Decision Maker 2:** $C_2 = C_3 > C_4 > C_5 > C_1 = C_6 > C_7 > C_8 = C_9$
- **Decision Maker 3:** $C_2 = C_3 > C_4 > C_5 > C_1 = C_7 > C_6 > C_8 > C_9$
- **Decision Maker 4:** $C_1 = C_2 > C_3 = C_4 > C_5 = C_8 > C_9 > C_6 > C_7$

It is observed that during the allocation of relative importance to the considered criteria by the decision makers, some of the criteria have equal significance. In these cases, the '>' sign is replaced by the equality sign. Now, based on a 9-point scale, the ranked criteria are pairwise compared by the individual decision makers, where 1 denotes the best-ranked criterion, while 9 represents the worst-ranked criterion. Table 32.29 exhibits the priorities of the criteria assigned by the individual decision makers. Using these priorities of the criteria, the corresponding comparative priorities are computed. Table 32.30 depicts the comparative criteria priorities estimated based on the judgments of decision maker 1.

TABLE 32.29 Priorities of Criteria Assigned by the DMs

Decision Maker 1		Decision Maker 2		Decision Maker 3		Decision Maker 4	
Criteria	**Priority**	**Criteria**	**Priority**	**Criteria**	**Priority**	**Criteria**	**Priority**
C_2	1	C_2	1	C_2	1	C_1	1
C_3	1.5	C_3	1	C_3	1	C_2	1
C_1	2	C_4	1.2	C_4	1.5	C_3	1.5
C_4	2	C_5	3	C_5	2	C_4	1.5
C_6	3	C_1	4	C_1	3	C_5	3
C_8	3.5	C_6	4	C_7	3	C_8	3
C_9	3.5	C_7	4.5	C_6	4	C_9	3.5
C_7	4	C_8	5	C_8	4.5	C_6	4
C_5	5	C_9	5	C_9	5	C_7	5

TABLE 32.30 Comparative Priorities of the Criteria for Decision Maker 1

$$\frac{w_2}{w_3}=1.5, \frac{w_3}{w_1}=1.333, \frac{w_1}{w_4}=1, \frac{w_4}{w_6}=1.5, \frac{w_6}{w_8}=1.167, \frac{w_8}{w_9}=1, \frac{w_9}{w_7}=1.143, \frac{w_7}{w_5}=1.25,$$

$$\frac{w_2}{w_1}=2, \frac{w_3}{w_4}=1.333, \frac{w_1}{w_6}=1.5, \frac{w_4}{w_8}=1.75, \frac{w_6}{w_9}=1.167, \frac{w_8}{w_7}=1.143, \frac{w_9}{w_5}=1.429$$

Now, the following mathematical model for determining criteria weights using the judgments of decision maker 1 is formulated:

Min χ

Subject to:

$$\left|\frac{w_2}{w_3}-1.5\right|\le\chi, \left|\frac{w_3}{w_1}-1.333\right|\le\chi, \left|\frac{w_1}{w_4}-1\right|\le\chi, \left|\frac{w_4}{w_6}-1.5\right|\le\chi, \left|\frac{w_6}{w_8}-1.167\right|\le\chi, \left|\frac{w_8}{w_9}-1\right|\le\chi,$$

$$\left|\frac{w_9}{w_7}-1.143\right|\le\chi, \left|\frac{w_7}{w_5}-1.25\right|\le\chi, \left|\frac{w_2}{w_1}-2\right|\le\chi, \left|\frac{w_3}{w_4}-1.333\right|\le\chi, \left|\frac{w_1}{w_6}-1.5\right|\le\chi,$$

$$\left|\frac{w_4}{w_8}-1.75\right|\le\chi, \left|\frac{w_6}{w_9}-1.167\right|\le\chi, \left|\frac{w_8}{w_7}-1.143\right|\le\chi,$$

$$\left|\frac{w_9}{w_5}-1.429\right|\le\chi, w_1+w_2+w_3+w_4+w_5+w_6+w_7+w_8+w_9=1, w_j\ge 0, \forall j$$

This LP model is solved to determine the optimal criteria weights as allocated by decision maker 1. In a similar direction, the optimal criteria weights are also calculated for the three other decision makers, as exhibited in Table 32.31. Smaller values of DFC (χ^*) signify that the criteria weights determined by all the four decision makers are unbiased and consistent. Finally, using the geometric mean method, these criteria weights estimated by the participating decision makers are aggregated to derive values of the final weights for the considered criteria.

TABLE 32.31 Calculation of the Criteria Weights

	Decision Maker 1	Decision Maker 2	Decision Maker 3	Decision Maker 4	Aggregated Weight
w_1	0.124	0.058	0.074	0.211	0.106
w_2	0.249	0.233	0.222	0.211	0.235
w_3	0.166	0.233	0.222	0.141	0.192
w_4	0.124	0.194	0.148	0.141	0.154
w_5	0.05	0.078	0.111	0.07	0.076
w_6	0.083	0.058	0.056	0.053	0.063
w_7	0.062	0.052	0.074	0.042	0.058
w_8	0.071	0.047	0.049	0.07	0.060
w_9	0.071	0.047	0.044	0.061	0.056
χ^*	0.000136	0.000039	0.000130	0.000269	–

32.9 LEVEL-BASED WEIGHT ASSESSMENT (LBWA) METHOD

While applying level-based weight assessment (LBWA) method for determination of the criteria weights, it is first assumed that there exists an MCDM problem having n criteria, $S = \{C_1, C_2, \ldots, C_n\}$. It is also considered that the weights of those criteria are not known beforehand. Application of LBWA method involves the following procedural steps [33]:

- **Step 1:** Identify the most important criterion from the set of criteria $S = \{C_1, C_2, \ldots, C_n\}$. For example, for a given decision-making problem, the concerned decision maker considers criterion C_1 as the most significant criterion in the criteria set.
- **Step 2:** Group all the criteria by levels of their significance. The decision maker establishes subsets of the criteria in the following way:

- **Level S_1:** At level S_1, group the criteria from the set S whose significance is equal to the significance of the C_1 criterion or up to twice as less as the significance of the C_1 criterion.
- **Level S_2:** At level S_2, group the criteria from the set whose significance is exactly twice as less as the significance of the C_1 criterion or up to three times as less as the significance of the C_1 criterion.
- **Level S_3:** At level S_3, group the criteria from the set S whose significance is exactly three times as less as the significance of the C_1 criterion or up to four times as less as the significance of the C_1 criterion.

  ...
- **Level S_k:** At S_k level, group the criteria from the set S whose significance is exactly *k* times as less as the significance of C_1 criterion or up to $(k + 1)$ times as less as the significance of C_1 criterion.

Based on the above-developed rules, the decision maker can establish a rough classification of the considered criteria depending on their varying levels of significance. Let, the significance of C_j criterion is expressed as $s(C_j)$, where $j \in \{1, 2, \ldots, n\}$, then $S = S_1 \cup S_2 \cup \ldots \cup S_k$. Thus, for every level $i \in \{1, 2, \ldots, k\}$, the following relation becomes true:

$$S_i = \left\{C_{i_1}, C_{i_2}, \ldots, C_{i_S}\right\} = \left\{C_j \in S : i \leq s(C_j) < i+1\right\} \tag{32.39}$$

For each $p, q \in \{1, 2, \ldots, k\}$, such that $p \neq q$, the relation $S_p \cap S_q = \varnothing$ holds true. In this way, a well-defined partition of the set of criteria S is thus achieved.

➢ **Step 3:** Within the developed subsets, the performance of the considered criteria is now compared based on their relative significance. Each criterion in $C_{i_p} \in S_i$ the subset $S_i = \left\{C_{i_1}, C_{i_2}, \ldots, C_{i_S}\right\}$ is assigned with an integer $I_{i_p} \in \{0, 1, \ldots, r\}$ so that the most important criterion C_1 is assigned with $I_1 = 0$. If C_{i_p}, is observed to be more significant than C_{i_q}, then $I_p < I_q$. On the other hand, if C_{i_p}, it is equivalent to C_{i_q}, then $I_p = I_q$. The maximum value on the scale for comparing the relative significance can be defined using the following expression:

$$r = \max\left\{|S_1|, |S_2|, \ldots, |S_k|\right\} \tag{32.40}$$

- **Step 4:** Based on the maximum value of the scale for criteria comparison (r), it can be stated that the elasticity coefficient $r_0 \in N$ (where N is the set of real numbers) should also satisfy the condition as defined in Eqn. (32.40).
- **Step 5:** Compute the influence function of the criteria. For every criterion $C_{i_p} \in S_i$, the influence function can be defined using the following equation:

$$f(C_{i_p}) = \frac{r_0}{i \times r_0 + I_{i_p}} \tag{32.41}$$

where; i is the number of subsets in which the criterion is classified; r_0 is the elasticity coefficient; and is the value assigned to the criterion C_{i_p} within the observed level.
- **Step 6:** Estimation of the optimal weights of the considered criteria. Using Eqn. (32.42), the weight coefficient of the most significant criterion C_1 can be estimated:

$$w_1 = \frac{1}{1 + f(C_2) + ... + f(C_n)} \tag{32.42}$$

Finally, weights of the remaining criteria are derived while applying the following expression:

$$w_j = f(C_j).w_1 \tag{32.43}$$

where; n is the number of evaluation criteria under consideration.

In order to demonstrate the mathematical formulations of LBWA method, an arc welding process selection problem [34] is considered here. In this problem, the welding performance of three arc welding processes, e.g., shielded metal arc welding (SMAW), gas tungsten arc welding (GTAW) and gas metal arc welding (GMAW) are assessed using six criteria, i.e., weld quality (C_1), operator fatigue (C_2), skill required (C_3), cleaning required after welding (C_4), availability of consumables (C_5) and initial preparation required (C_6). Rao [34] determined the relative weights of those criteria as 0.303, 0.242, 0.182, 0.121, 0.091, and 0.061, respectively. The calculations of criteria weights using LBWA method are illustrated as below:

- **Step 1:** Identify the most important criterion from the set of evaluation criteria S = {C_1, C_2, …, C_6}. In this problem, C_1 is selected as the most important/influential criterion.
- **Step 2:** Grouping these criteria by levels of their significance. In accordance with the preferences of the decision makers, the criteria are grouped in the following subsets/levels:

- **Level S_1:** Criteria C_3 and C_5 are up to twice as less significant as criterion C_1.
- **Level S_3:** Criteria C_2 and C_6 are up to between three or four times as less important as criterion C_1.
- **Level S_4:** Criteria C_4 is between four or five times as less important as criterion C_1.

Then, based on the mentioned preferences of the decision-makers, the criteria can be grouped in the following subsets/levels:

- $S_1 = \{C_1, C_3, C_5\}$;
- $S_2 = \emptyset$;
- $S_3 = \{C_2, C_6\}$;
- $S_4 = \{C_4\}$.

➢ **Step 3:** Using Eqn. (32.40), the maximum value on the scale for comparing the relative significance is defined.

$$\left.\begin{array}{c} S_1 = \{C_1, C_3, C_5\} \\ S_2 = \emptyset \\ S_3 = \{C_2, C_6\} \\ S_4 = \{C_4\} \end{array}\right\} \Rightarrow r = \max\{|S_1|, |S_3|, |S_4|\} = 3$$

Based on the maximum value of the scale for comparison, it can be concluded that the scale for comparing the criteria ranges in the interval $I_{i_p} \in \{0, 1, 2, 3\}$. Using the scale and pre-defined set of criteria, the comparison of criteria within each individual set can now be performed:

- **Level S_1:** Based on the preferences of the decision-makers, the following relations are defined: $I_1 = 0$, $I_5 = 1$ and $I_3 = 3$.
- **Level S_3:** Within the set S_3, the following relations are defined: $I_2 = 1$ and $I_6 = 3$.
- **Level S_4:** Within the set S_4, the following relation is defined: $I_4 = 2$.

➢ **Step 4:** Based on the defined maximum value of the scale for comparison of criteria $r = 3$, the elasticity coefficient can now be stated such that $r_o > r$ (i.e., $r_o > 3$).

➢ **Step 5:** Define the influence function of the criteria. If it is known that $r_o > 3$, the value $r_o = 4$ can be determined arbitrarily. By applying Eqn. (32.41), the influence functions of the criteria are calculated as below:

$$f(C_1) = \frac{4}{1 \times 4 + 0} = 1$$

$$f(C_5) = \frac{4}{1 \times 4 + 1} = 0.8$$

$$f(C_3) = \frac{4}{1 \times 4 + 3} = 0.571$$

$$f(C_2) = \frac{4}{3 \times 4 + 1} = 0.308$$

$$f(C_6) = \frac{4}{3 \times 4 + 3} = 0.267$$

$$f(C_4) = \frac{4}{4 \times 4 + 2} = 0.222$$

- **Step 6:** Calculate the optimal values of weight coefficients for the considered evaluation criteria. Applying Eqn. (32.42), the weight coefficient of the most influential criterion (C_1) is estimated as shown below:

$$w_1 = \frac{1}{1 + 0.8 + 0.571 + 0.308 + 0.267 + 0.222} = 0.316$$

  Values of weight coefficients of the remaining criteria are now calculated employing Eqn. (32.43). Therefore, weights of the evaluation criteria for the considered arc welding process selection problem are determined as: $w_1 = 0.316$, $w_2 = f(C_2) \times w_1 = 0.308 \times 0.316 = 0.097$, $w_3 = 0.18$, $w_4 = 0.07$, $w_5 = 0.253$ and $w_6 = 0.084$. Thus, C_1 (weld quality) is the most important criterion for this problem, and C_6 (initial preparation required) and C_4 (cleaning required after welding) are the two least important criteria. Rao [34] also identified the same importance of the evaluation criteria while applying a subjective weight assessment technique.

32.10 LOGARITHM METHODOLOGY OF ADDITIVE WEIGHTS (LMAW)

The application of logarithm methodology of additive weights (LMAW) method has the following procedural steps [35]:

- **Step 1:** Formation of the initial decision matrix (X) which contains m number of alternatives and n number of criteria. On the other hand, w_j is the weight assigned to the jth criterion. This process involves the participation of k experts (E_1, E_2, ..., E_k) evaluating the criteria based on a predefined linguistic scale. For each expert, the matrix $X^e = [\vartheta_{ij}^e]_{m\times n}$ $(1 \le e \le k)$ is obtained where ϑ_{ij}^e the value is assigned from the considered linguistic scale. Now, based on the Bonferroni aggregator of Eqn. (32.44), the aggregated initial decision matrix $X = [\vartheta_{ij}]_{m\times n}$ can be derived.

$$\vartheta_{ij} = \left(\frac{1}{k(k-1)} \sum_{x=1}^{k} \left(\vartheta_{ij}^{(x)}\right)^p \sum_{\substack{y=1 \\ y\neq x}}^{k} \left(\vartheta_{ij}^{(y)}\right)^q \right)^{\frac{1}{p+q}} \tag{32.44}$$

where; ϑ_{ij} is the average value obtained using the Bonferroni aggregator; p, $q \ge 0$ are the stabilization parameters of the Bonferroni aggregator; and e is the eth expert $(1 \le e \le k)$.

- **Step 2:** Normalization of the elements of the initial decision matrix depending on the type of the criterion (beneficial or cost). The standardized matrix $Y = [\upsilon_{ij}]_{m\times n}$ can be obtained using the following two equations:

$$\upsilon_{ij} = \begin{cases} \dfrac{\vartheta_{ij} + \vartheta_j^+}{\vartheta_j^+} & \text{for beneficial criterion} \\[2ex] \dfrac{\vartheta_{ij} + \vartheta_j^-}{\vartheta_{ij}} & \text{for cost criterion} \end{cases} \tag{32.45}$$

where; $\vartheta_j^+ = \max_i(\vartheta_{ij})$, $\vartheta_j^- = \min_i(\vartheta_{ij})$; and υ_{ij} is the standardized value of the initial decision matrix.

- **Step 3:** Determine weight coefficients of the considered criteria. The participating experts from the group E = (E_1, E_2, ..., E_k) prioritize the evaluation criteria C = (C_1, C_2, ..., C_n) using the values from the predefined linguistic scale. In this process, a higher value from the linguistic scale is assigned to the criterion having higher significance, whereas, a lower linguistic value is allotted to the criterion having lower importance. Thus, a priority vector $P^e = \left(\gamma_{C_1}^e, \gamma_{C_2}^e, \ldots, \gamma_{C_n}^e\right)$ is obtained where $\gamma_{C_n}^e$ symbolizes the linguistic value assigned by expert e $(1 \le e \le k)$ to criterion C_j $(1 \le j \le n)$.

In this step, an absolute anti-ideal point (γ_{AIP}) is also defined based on the minimum values from the priority vector, and it should be lower than the smallest value from the priority vector. The value of γ_{AIP} can be defined as $\gamma_{AIP} = \left(\gamma_{\min}^e / s\right)$ and s is a number greater than the base of logarithm (A). If ln is taken as the logarithmic function, the value of s should be considered as 3.

Now, based on the following equation, the relation between elements of the priority vector and absolute anti-ideal point (γ_{AIP}) is determined.

$$\eta^e_{C_n} = \frac{\gamma^e_{C_n}}{\gamma_{AIP}} \tag{32.46}$$

Thus, a relation vector $R^e = \left(\eta^e_{C_1}, \eta^e_{C_2}, \ldots, \eta^e_{C_n}\right)$ is derived where $\eta^e_{C_n}$ denotes the value obtained from Eqn. (32.46).

Determine the vector of weight coefficients. The values of the weight coefficients of the criteria are obtained for expert e ($1 \leq e \leq k$) using the following expression:

$$w_j = \frac{\log_A\left(\eta^e_{C_n}\right)}{\log_A\left(b^e\right)}, \quad A>1 \tag{32.47}$$

where; $b^e = \prod_{j=1}^{n} \eta^e_{C_n}$.

The derived weight coefficients must satisfy the condition $\sum_{j=1}^{n} w^e_j = 1$ where; w_j^e is the weight assigned to the jth criterion by e^{th} expert.

Applying the Bonferroni aggregator, as given in Eqn. (32.48), the aggregated vector of weight coefficients $w_j = (w_1, w_2, \ldots, w_n]^T$ can be obtained.

$$w_j = \left(\frac{1}{k(k-1)}\sum_{x=1}^{k}\left(w_j^{(x)}\right)^p \sum_{\substack{y=1 \\ y\neq x}}^{k}\left(w_j^{(y)}\right)^q\right)^{\frac{1}{p+q}} \tag{32.48}$$

In order to explain the mathematical formulations of LMAW method, a piston material selection problem [36] is considered here. In this problem, based on 10 physical, mechanical, and thermal properties of the candidate materials, the most appropriate piston material is identified among a set of seven alternatives. These criteria are density (C_1) (in g/cm^3), hardness (C_2) (in BHN), elastic modulus (C_3) (in GPa), ultimate tensile strength (C_4) (in MPa), yield tensile strength (C_5) (in MPa), fatigue strength (C_6) (in MPa), elongation (C_7) (in %), thermal conductivity (C_8) (in W/mK), coefficient of thermal expansion (C_9) (in μm/m-K) and specific heat (C_{10}) (in J/g-C). While prioritizing those evaluation criteria based on the opinions of four experts, a 1–5 point rating scale is taken into account, as exhibited in Table 32.32. In these calculations, the value of the absolute anti-ideal point (γ_{AIP}) is assumed as 0.50. Now, the concerned experts evaluate the relative significance of all the criteria based on the predefined scale, as provided in Table 32.32. These evaluations of the relative significance of the 10 criteria by the four experts are provided in Table 32.33. Using the procedural steps of LMAW method, the relation between elements of the priority vector and absolute anti-ideal point (γ_{AIP}) is determined, as shown in Table 32.34. Table 32.35 provides the vector of the criteria weight coefficients. Finally, while employing Eqn. (32.48), the aggregated weights of all the evaluation criteria in this piston material selection problem are estimated, as depicted in Table 32.36. It can be noted that in this material selection problem, elongation, and thermal conductivity are the two most important properties, whereas, yield tensile strength and specific heat have the least importance in selecting the most appropriate piston material from the set of candidate alternatives.

TABLE 32.32 Scale Used in LMAW Method

Absolutely low (AL)	1
Very low (VL)	1.5
Low (L)	2
Medium-low (ML)	2.5
Equal (E)	3
Medium-high (MH)	3.5
High (H)	4
Very high (EH)	4.5
Absolutely high (AH)	5

TABLE 32.33 Expert Opinions for Criteria Importance in LMAW Method

Expert	C_1	C_2	C_3	C_4	C_5	C_6	C_7	C_8	C_9	C_{10}
Expert 1	2.5	1.5	4	2	1.5	1.5	5	3.5	2	2
Expert 2	3	1.5	4	2.5	1.5	1.5	5	4.5	1.5	1.5
Expert 3	3.5	1.5	4	2.5	1	1.5	4	5	1.5	1
Expert 4	3	1.5	4	2	1.5	1.5	5	3	2	2

TABLE 32.34 Relation between the Elements of Priority Vector and Absolute Anti-Ideal Point (γ_{AIP})

Expert	C_1	C_2	C_3	C_4	C_5	C_6	C_7	C_8	C_9	C_{10}
Expert 1	5.00	3.00	8.00	4.00	3.00	3.00	10.00	7.00	4.00	4.00
Expert 2	6.00	3.00	8.00	5.00	3.00	3.00	10.00	9.00	3.00	3.00
Expert 3	7.00	3.00	8.00	5.00	2.00	3.00	8.00	10.00	3.00	2.00
Expert 4	6.00	3.00	8.00	4.00	3.00	3.00	10.00	6.00	4.00	4.00

TABLE 32.35 Vector of Weight Coefficients in LMAW Method

Expert	C_1	C_2	C_3	C_4	C_5	C_6	C_7	C_8	C_9	C_{10}
Expert 1	0.10	0.07	0.14	0.09	0.07	0.07	0.15	0.13	0.09	0.09
Expert 2	0.12	0.07	0.13	0.10	0.07	0.07	0.15	0.14	0.07	0.07
Expert 3	0.13	0.07	0.14	0.11	0.05	0.07	0.14	0.16	0.07	0.05
Expert 4	0.12	0.07	0.13	0.09	0.07	0.07	0.15	0.12	0.09	0.09

TABLE 32.36 Aggregated Weight Coefficients in LMAW Method

Weight	C_1	C_2	C_3	C_4	C_5	C_6	C_7	C_8	C_9	C_{10}
w_j	0.117	0.072	0.136	0.098	0.065	0.072	0.150	0.140	0.081	0.070

32.11 METHOD BASED ON THE REMOVAL EFFECTS OF CRITERIA (MEREC)

This objective weighting method for estimating criteria weights in an MCDM problem is recently developed by Ghorabaee et al. [37]. It is based on studying each criterion's removal effect on the performance of alternatives to determine the corresponding criteria weights. Higher weights are assigned to those criteria having higher effects on the performance of the alternatives. In this method, at first, a suitable measure for the performance of alternatives is defined. Usually, a simple logarithmic measure is considered with equal weights to calculate the performance of the alternatives. On the other hand, absolute deviation measure is adopted to identify the effects of removing each criterion in the decision-making problem. This measure represents the difference between the overall alternative's performance and its performance in removing a criterion. The procedural steps of MEREC method are elucidated as below:

➤ **Step 1:** Develop the initial decision matrix (X).

$$X = \begin{bmatrix} x_{11} & x_{12} & \cdots & x_{1n} \\ x_{21} & x_{22} & \cdots & x_{2n} \\ \cdots & \cdots & \cdots & \cdots \\ x_{m1} & x_{m2} & \cdots & x_{mn} \end{bmatrix} \tag{32.49}$$

where; x_{ij} is the performance of ith alternative with respect to j^{th} criteria ($x_{ij} > 0$), m is the number of alternatives and n is the number of criteria.

➤ **Step 2:** Normalize the decision matrix. Using a simple linear normalization technique, all the elements of the decision matrix (X) are now normalized to form the corresponding normalized decision matrix (N). For normalization, the following equations can be utilized.

$$n_{ij} = \begin{cases} \dfrac{\underset{i}{\text{Min}}\, x_{ij}}{x_{ij}} & \text{for } j \in B \\ \dfrac{x_{ij}}{\underset{i}{\text{Max}}\, x_{ij}} & \text{for } j \in C \end{cases} \tag{32.50}$$

where; n_{ij} is the normalized value of x_{ij}, B is the set of beneficial criteria and C is the set of cost criteria. It is worthwhile to mention here that in this normalization approach, all the criteria values are transformed into minimization type just by switching between the formulas of beneficial and cost criteria.

- **Step 3:** Estimate the overall performance of the alternatives (S_i). To determine the overall performance of the alternatives (S_i), a logarithmic measure with equal criteria weights is employed. This measure is based on a non-linear function. Using the considered normalization procedure, it can be assumed that smaller n_{ij} values lead to higher performance of the alternatives (S_i). The performance of ith alternative can be calculated using the following equation:

$$S_i = \ln\left(1 + \left(\frac{1}{n}\sum\left|\ln\left(n_{ij}\right)\right|\right)\right) \tag{32.51}$$

- **Step 4:** Estimate the performance of the alternatives by removing each criterion. In this step, the same logarithmic measure is adopted to determine the performance of the alternatives by removing each criterion separately. Let, S'_{ij} denotes the overall performance of ith alternative after removal of jth criterion. The following equation is employed to determine the value of S'_{ij}.

$$S'_{ij} = \ln\left(1 + \left(\frac{1}{n}\sum_{k,k \neq j}\left|\ln\left(n_{ij}\right)\right|\right)\right) \tag{32.52}$$

- **Step 5:** Calculate the summation of absolute deviations. The removal effect of the jth criterion using the values derived in Step (3) and the removal effect of the jth criterion based on the values obtained in Steps (3) and (4) are computed. Let E_j represents the removal effect of the jth criterion. The value of E_j can be calculated applying the following equation:

$$E_j = \sum_i \left|S'_{ij} - S_j\right| \tag{32.53}$$

➢ **Step 6:** Estimation of the final criteria weights. In the last step, the priority weight (w_j) for the jth criterion is finally computed.

$$w_j = \frac{E_j}{\sum_k E_k} \tag{32.54}$$

In order to demonstrate the applications of MEREC in estimating the objective weights of a set of evaluation criteria, a material selection problem for cryogenic storage tank is considered here. In this material selection problem, the performance of seven alternative materials is appraised with respect to seven criteria. The corresponding decision matrix is exhibited in Table 32.37. Among the considered criteria, toughness index, yield strength and Young's modulus are beneficial attributes, while, density, thermal expansion, thermal conductivity, and specific heat are non-beneficial attributes. Now, based on the type of the criterion and employing Eqn. (32.50), the elements of the initial decision matrix are first linearly normalized, as shown in Table 32.38. Using Eqns. (32.51) and (32.52), the overall performance of the alternatives (S_i) and performance of the alternatives by removing each criterion (S'_{ij}) are respectively computed. The values of the overall performance of the alternatives (S_i) and performance of the alternatives by removing each criterion (S'_{ij}) are respectively exhibited in Tables 32.39 and 32.40. Using these two measures, the removal effect of the jth criterion (E_j) is computed. The values of E_j are shown in Table 32.41 which finally lead to the estimation of the objective weights of the considered evaluation criteria for this cryogenic storage tank material selection problem. It can be noticed from Table 32.41 that yield strength has the maximum importance, followed by thermal conductivity. For this material selection problem, thermal expansion has the minimum importance.

TABLE 32.37 Decision Matrix for the Cryogenic Storage Tank Material Selection Problem [3]

Material	Toughness Index	Yield Strength	Young's Modulus	Density	Thermal Expansion	Thermal Conductivity	Specific Heat
Al 2024-T6	75.5	420	74.2	2.8	21.4	0.37	0.16
Al 5052-O	95	91	70	2.68	22.1	0.33	0.16
SS 301-FH	770	1365	189	7.9	16.9	0.04	0.08
SS 310-3AH	187	1120	210	7.9	14.4	0.03	0.08
Ti-6Al-4V	179	875	112	4.43	9.4	0.016	0.09
Inconel 718	239	1190	217	8.51	11.5	0.31	0.07
70Cu-30Zn	273	200	112	8.53	19.9	0.29	0.06

Source: Reprinted with permission from Ref. [3]. © 2021 Elsevier.

TABLE 32.38 Normalized Decision Matrix for MEREC

Material	Toughness Index	Yield Strength	Young's Modulus	Density	Thermal Expansion	Thermal Conductivity	Specific Heat
Al 2024-T6	1	0.2167	0.9434	0.3283	0.9683	1	1
Al 5052-O	0.7947	1	1	0.3142	1	0.8919	1
SS 301-FH	0.0981	0.0667	0.3704	0.9261	0.7647	0.1081	0.5
SS 310-3AH	0.4037	0.0813	0.3333	0.9261	0.6516	0.0811	0.5
Ti-6Al-4V	0.4218	0.104	0.625	0.5193	0.4253	0.0432	0.5625
Inconel 718	0.3159	0.0765	0.3226	0.9977	0.5204	0.8378	0.4375
70Cu-30Zn	0.2766	0.455	0.625	1	0.9005	0.7838	0.375

TABLE 32.39 Overall Performance of the Alternatives in MEREC Approach

Material	S_i
Al 2024-T6	0.5208
Al 5052-O	0.3188
SS 301-FH	1.2004
SS 310-3AH	1.1173
Ti-6Al-4V	1.1649
Inconel 718	0.9664
70Cu-30Zn	0.677

TABLE 32.40 Performance of the Alternatives by Removing Each Criterion in MEREC

Material	Toughness Index	Yield Strength	Young's Modulus	Density	Thermal Expansion	Thermal Conductivity	Specific Heat
Al 2024-T6	0.5208	0.2632	0.5122	0.34	0.5161	0.5208	0.5208
Al 5052-O	0.2762	0.3188	0.3188	0.0825	0.3188	0.2978	0.3188
SS 301-FH	1.0083	0.9725	1.1227	1.1946	1.18	1.0172	1.1469
SS 310-3AH	1.0402	0.8875	1.0231	1.111	1.0816	0.8873	1.0589
Ti-6Al-4V	1.0952	0.9707	1.1276	1.1125	1.0959	0.884	1.119
Inconel 718	0.8503	0.686	0.8525	0.9662	0.9022	0.9494	0.8845
70Cu-30Zn	0.4988	0.5716	0.6155	0.677	0.6636	0.6456	0.544

TABLE 32.41 Performance of the Alternatives by Removing Each Criterion in MEREC

Weight	Toughness Index	Yield Strength	Young's Modulus	Density	Thermal Expansion	Thermal Conductivity	Specific Heat
E_j	0.6759	1.2954	0.3933	0.4819	0.2074	0.7637	0.3728
w_j	0.2375	0.4551	0.1382	0.1693	0.0729	0.2683	0.1310

KEYWORDS

- **criteria importance**
- **decision-making**
- **entropy method**
- **intercriteria correlation**
- **pivot pair-wise relative criteria importance assessment**
- **surface damage**
- **tolerance**

REFERENCES

1. Shannon, C. E., (2001). A mathematical theory of communication. *ACM Mobile Computing and Communications Review, 5,* 3–55.
2. Li, X., Wang, K., Liu, L., Xin, J., Yang, H., & Gao, C., (2011). Application of the entropy weight and TOPSIS method in safety evaluation of coal mines. *Procedia Engineering, 26,* 2085–2091.
3. Dehghan-Manshadi, B., Mahmudi, H., Abedian, A., & Mahmudi, R., (2007). A novel method for materials selection in mechanical design: Combination of non-linear normalization and a modified digital logic method. *Materials & Design, 28*(1), 8–15.
4. Fontela, E., & Gabus, A., (1976). The DEMATEL Observer. *DEMATEL 1976 Report.* Battelle Geneva Research Center, Geneva, Switzerland.
5. Gabus, A., & Fontela, E., (1973). *Perceptions of the World Problematique: Communication Procedure, Communicating with Those Bearing Collective Responsibility.* DEMATEL Report No. 1, Battelle Geneva Research Center, Geneva, Switzerland.
6. Tamura, H., & Akazawa, K., (2005). Structural modeling and systems analysis of uneasy factors for realizing safe, secure, and reliable society. *Journal of Telecommunication and Information Technology, 3,* 64–72.
7. Chen, Y. C., Lien, H. P., & Tzeng, G. H., (2009). Measures and evaluation for environment watershed plans using a novel hybrid MCDM model. *Expert Systems with Applications, 37*(2), 926–938.
8. Hamidi, N., Yousefi, P., Rahimi, A., & Jabari, F., (2012). A hybrid of Borda and DEMATEL for productivity improvement. *Management Science Letters, 2,* 2757–2764.
9. Kobryń, A., (2017). DEMATEL as a weighting method in multi-criteria decision analysis. *Multiple Criteria Decision Making, 12,* 153–166.
10. Gupta, R., & Naqvi, S. K., (2017). A framework for applying CSFs to ERP software selection: An extension of fuzzy TOPSIS approach. *International Journal of Intelligent Information Technologies, 13*(2), 41–62.
11. Stanujkic, D., Karabasevic, D., & Zavadskas, E. K., (2015). A framework for the selection of a packaging design based on the SWARA method. *Engineering Economics, 26*(2), 181–187.

12. Kersuliene, V., Zavadskas, E. K., & Turskis, Z., (2010). Selection of rational dispute resolution method by applying new stepwise weight assessment ratio analysis (SWARA). *Journal of Business Economics and Management, 11*(2), 243–258.
13. Durairaj, S., Satthiyadekar, K., & Ilangkumaran, M., (2016). Selection of alternative fuel for electrical power generator using hybrid multi-criteria decision-making technique. *UPB Scientific Bulletin Series C, 78*(1), 247–258.
14. Stanujkic, D., Karabasevic, D., Zavadskas, E. K., Smarandache, F., & Cavallaro, F., (2019). An approach to determining customer satisfaction in traditional Serbian restaurants. *Entrepreneurship and Sustainability Issues, 6*(3), 1127–1138.
15. Stanujkic, D., Zavadskas, E. K., Karabasevic, D., Smarandache, F., & Turskis, Z., (2017). The use of the pivot pair-wise relative criteria importance assessment method for determining the weights of criteria. *Romanian Journal of Economics, 20*(4), 116–133.
16. Kazancoglu, Y., & Burmaoglu, S., (2013). ERP software selection with MCDM: Application of TODIM method. *International Journal of Business Information Systems, 13*(4), 435–452.
17. Diakoulaki, D., Mavrotas, G., & Papayannakis, L., (1995). Determining objective weights in multiple criteria problems: The CRITIC method. *Computers and Operations Research, 22*(1), 763–770.
18. Kazan, H., & Ozdemir, O., (2014). Financial performance assessment of large-scale conglomerates via TOPSIS and CRITIC methods. *International Journal of Management and Sustainability, 3*(4), 203–224.
19. Madić, M., & Radovanović (2015). Ranking of some commonly used non-traditional machining processes using ROV and CRITIC methods. *UPB Scientific Bulletin Series D, 77*(2), 193–204.
20. Yalçin, N., & Ünlü, U., (2018). A multi-criteria performance analysis of initial public offering (IPO) firms using CRITIC and VIKOR methods. *Technological and Economic Development of Economy, 24*(2), 534–560.
21. Tuş, A., & Adalı, E. A., (2019). The new combination with CRITIC and WASPAS methods for the time and attendance software selection problem. *OPSEARCH, 56*, 528–538.
22. Khabbaz, R. S., Manshadi, B. D., Abedian, A., & Mahmudi, R., (2009). A simplified fuzzy logic approach for materials selection in mechanical engineering design. *Materials and Design, 30*, 687–697.
23. Gineviciu, R., (2011). A new determining method for the criteria weights in multicriteria evaluation. *International Journal of Information Technology & Decision Making, 10*(6), 1067–1095.
24. Yazdani, M., (2018). New approach to select materials using MADM tools. *International Journal of Business and Systems Research, 12*(1), 25–42.
25. Pitchipoo, P., Vincent, D. S., Rajini, N., & Rajakarunakaran, S., (2014). COPRAS decision model to optimize blind spot in heavy vehicles: A comparative perspective. *Procedia Engineering, 97*, 1049–1059.
26. Yurdakul, M., & Cogun, C., (2003). Development of a multi-attribute selection procedure for non-traditional machining processes. *Proceedings of the Institution of Mechanical Engineers, Journal of Engineering Manufacture, 217*(7), 993–1009.
27. Rezaei, J., (2015). Best-worst multi-criteria decision-making method. *Omega, 53*, 49–57.
28. Rezaei, J., Nispeling, T., Sarkis, J., & Tavasszy, L., (2016). A supplier selection life cycle approach integrating traditional and environmental criteria using the best worst method. *Journal of Cleaner Production, 135*, 577–588.

29. Beemsterboer, D. J. C., Hendrix, E. M. T., & Claassen, G. D. H., (2018). On solving the best-worst method in multi-criteria decision-making. *IFAC Paper, 51*(11), 1660–1665.
30. Kheybari, S., Kazemi, M., & Rezaei, J., (2019). Bioethanol facility location selection using best-worst method. *Applied Energy, 242*, 612–623.
31. Pamučar, D., Stević, Ž., & Sremac, S., (2018). A new model for determining weight coefficients of criteria in MCDM models: Full consistency method (FUCOM). *Symmetry, 10*. doi: 10.3390/sym10090393.
32. Pamučar, D., Lukovac, V., Božanić, D., & Komazec, N., (2018). Multi-criteria FUCOM-MAIRCA model for the evaluation of level crossings: Case study in the Republic of Serbia. *Operational Research in Engineering Sciences: Theory and Applications, 1*, 108–129.
33. Žižović, M., & Pamučar, D., (2019). New model for determining criteria weights: Level based weight assessment (LBWA) model. *Decision Making: Applications in Management and Engineering, 2*(2), 126–137.
34. Rao, R. V., (2007). Welding process selection for a given application. *Decision Making in the Manufacturing Environment: Using Graph Theory and Fuzzy Multiple Attribute Decision Making Methods* (pp. 215–220). Springer-Verlag.
35. Pamučar, D., Žižović, M., Biswas, S., & Božanić, D., (2021). New logarithm methodology of additive weights (LMAW) for multi-criteria decision-making: Application in logistics. *FACTA Universitatis Series: Mechanical Engineering*. https://doi.org/10.22190/FUME210214031P.
36. Lakshmi, C. K., & Srinivas, K., (2019). Sensitive analysis on selection of piston material using MADM techniques. *Journal of Mechanical Engineering, 69*(4), 45–56.
37. Ghorabaee, M. K., Amiri, M., Zavadskas, E. K., Turskis, Z., & Antucheviciene, J., (2021). Determination of objective weights using a new method based on the removal effects of criteria (MEREC). *Symmetry, 13*. https://doi.org/10.3390/sym13040525.

INDEX

E

O

P

Q

R

U

V

W

Y